Quantitative Real-time PCR in Applied Microbiology

Edited by

Martin Filion

Department of Biology
Université de Moncton
Moncton, NB
Canada

Caister Academic Press

Caister Academic Press
Norfolk, UK

www.caister.com

British Library Cataloguing-in-Publication Data
A catalogue record for this book is available from the British Library

ISBN: 978-1-908230-01-0

Cover design adapted from Figure 1.3

Printed and bound in Great Britain

Contents

Contributors

Nicholas Ashbolt
US Environmental Protection Agency
Office of Research and Development
Cincinnati, OH
USA

ashbolt.nicholas@epa.gov

Stephen A. Bustin
Barts and the London School of Medicine and Dentistry
Queen Mary University of London
London
UK

s.a.bustin@qmul.ac.uk

Luca Cocolin
Di.Va.P.R.A.
Agricultural Microbiology and Food Technology Sector
Faculty of Agriculture
University of Turin
Turin
Italy

lucasimone.cocolin@unito.it

Catherine E. Dandie
CSIRO Land and Water
Glen Osmond, SA
Australia

cathy.dandie@csiro.au

Michelle Davey
Atlantic Cancer Research Institute
Moncton, NB
Canada

michelled@canceratl.ca

Martin Filion
Department of Biology
Université de Moncton
Moncton, NB
Canada

martin.filion@umoncton.ca

Vijay J. Gadkar
Department of Biology
Université de Moncton
Moncton, NB
Canada

vijay.gadkar@umoncton.ca

Claudia Goyer
Potato Research Centre
Agriculture and Agri-Food Canada
Fredericton, NB
Canada

claudia.goyer@agr.gc.ca

Trine Hansen
National Food Institute
Technical University of Denmark
Søborg
Denmark

trih@food.dtu.dk

Jeffrey Hoorfar
National Food Institute
Technical University of Denmark
Søborg
Denmark

jhoo@food.dtu.dk

Mathilde H. Josefsen
National Food Institute
Technical University of Denmark
Søborg
Denmark

mhjo@food.dtu.dk

Mikael Kubista
TATAA Biocenter
Gothenburg
Sweden;
Department of Biotechnology
Czech Academy of Science
Prague
Czech Republic

mikael.kubista@tataa.com

Mark Laflamme
Department of Fisheries and Oceans Canada
Moncton, NB
Canada

mark.laflamme@dfo-mpo.gc.ca

Charlotta Löfström
National Food Institute
Technical University of Denmark
Søborg
Denmark

chalo@food.dtu.dk

Tania Nolan
Sigma-Aldrich
Haverhill
UK

tania.nolan@sial.com

Michael W. Pfaffl
Physiology
Centre for Life and Food Sciences Weihenstephan
Technical University of Munich
Freising-Weihenstephan
Germany

michael.pfaffl@wzw.tum.de

Kalliopi Rantsiou
Di.Va.P.R.A.
Agricultural Microbiology and Food Technology Sector
Faculty of Agriculture
University of Turin
Turin
Italy

kalliopi.rantsiou@unito.it

Eyjólfur Reynisson
Matís Ltd – Icelandic Food and Biotech R&D
Reykjavík
Iceland

eyjolfur.reynisson@matis.is

Vendula Rusnakova
Department of Biotechnology
Czech Academy of Science
Prague
Czech Republic

vendula.rusnakova@img.cas.cz

Hodon Ryu
US Environmental Protection Agency
Office of Research and Development
Cincinnati, OH
USA

ryu.hodon@epa.gov

Jorge Santo Domingo
US Environmental Protection Agency
Office of Research and Development
Cincinnati, OH
USA

santodomingo.jorge@epa.gov

Mary Schoen
US Environmental Protection Agency
Office of Research and Development
Cincinnati, OH
USA

schoen.mary@epa.gov

Björn Sjögreen
Center for Applied Scientific Computing
Lawrence Livermore National Laboratory
Livermore, CA
USA

sjogreen2@llnl.gov

David Svec
TATAA Biocenter
Gothenburg
Sweden;
Department of Biotechnology
Czech Academy of Science
Prague
Czech Republic

david.svec@tataa.com

Lia C.R.S. Teixeira
Professor Paulo de Góes Microbiology Institute
Federal University of Rio de Janeiro
Rio de Janeiro, RJ
Brazil

liarteixeira@gmail.com

Ales Tichopad
Medical Faculty Pilsen
Charles University of Prague
Prague
Czech Republic;
Physiology Weihenstephan
Technical University of Munich
Freising-Weihenstephan
Germany

tichopad@ceeor.com

Dan Tulpan
Institute for Information Technology
National Research Council Canada
Biology Department
University of Moncton
Moncton, NB
Canada

dan.tuplan@nrc-cnrc.gc.ca

Etienne Yergeau
Biotechnology Research Institute
National Research Council of Canada
Montréal, QC
Canada

Etienne.Yergeau@cnrc-nrc.gc.ca

Sara Zaccara
Clinical Biochemistry
Department of Clinical Physiopathology
Viale Morgagni
Florence
Italy

sarazaccara.87@gmail.com

Current Books of Interest

Bioremediation of Mercury	2012
Rhabdoviruses	2012
Horizontal Gene Transfer in Microorganisms	2012
Microbial Ecological Theory: Current Perspectives	2012
Two-Component Systems in Bacteria	2012
Foodborne and Waterborne Bacterial Pathogens	2012
Yersinia: Systems Biology and Control	2012
Stress Response in Microbiology	2012
Bacterial Regulatory Networks	2012
Systems Microbiology: Current Topics and Applications	2012
Quantitative Real-time PCR in Applied Microbiology	2012
Bacterial Spores: Current Research and Applications	2012
Small DNA Tumour Viruses	2012
Extremophiles: Microbiology and Biotechnology	2012
Bacillus: Cellular and Molecular Biology	2012
Microbial Biofilms: Current Research and Applications	2012
Bacterial Glycomics: Current Research, Technology and Applications	2012
Non-coding RNAs and Epigenetic Regulation of Gene Expression	2012
Brucella: Molecular Microbiology and Genomics	2012
Molecular Virology and Control of Flaviviruses	2012
Bacterial Pathogenesis: Molecular and Cellular Mechanisms	2012
Bunyaviridae: Molecular and Cellular Biology	2011
Emerging Trends in Antibacterial Discovery: Answering the Call to Arms	2011
Epigenetics: A Reference Manual	2011
Metagenomics: Current Innovations and Future Trends	2011
Nitrogen Cycling in Bacteria: Molecular Analysis	2011
The Biology of Paramyxoviruses	2011
Helicobacter pylori	2011
Microbial Bioremediation of Non-metals: Current Research	2011
Lactic Acid Bacteria and Bifidobacteria: Current Progress in Advanced Research	2011
Viruses and Interferon: Current Research	2011
Essentials of Veterinary Parasitology	2011
Hepatitis C: Antiviral Drug Discovery and Development	2011
Streptomyces: Molecular Biology and Biotechnology	2011
Alphaherpesviruses: Molecular Virology	2011
Recent Advances in Plant Virology	2011
Vaccine Design: Innovative Approaches and Novel Strategies	2011
Salmonella: From Genome to Function	2011
PCR Troubleshooting and Optimization: The Essential Guide	2011
Insect Virology	2010
Environmental Microbiology: Current Technology and Water Applications	2011

Preface

Since its introduction in the mid-1990s, quantitative real-time PCR (qPCR) has revolutionized almost all areas of microbiology. It has become a common technique used in most laboratories interested in clinical, environmental, food and industrial microbiology. This versatile technology keeps evolving, and new platforms, chemistries, approaches and software are becoming available, making qPCR more sensitive, faster and affordable than ever before.

This book aims to provide all microbiologists, both novice and experienced users, with the latest information available on qPCR technology. The initial chapters cover the main technical aspects of qPCR, from a thorough description of the technique to choosing an instrument, fluorescent chemistries, experimental approaches, controls, validation and software. In the second part of the book, the focus is put on how to design qPCR assays for quantifying microorganisms from different substrates and conditions in a culture-independent manner. Finally, the final chapters focus on microbial gene expression and describe the use of RT-qPCR for quantifying transcripts from complex environmental conditions and for validating microbial microarray data. Future trends in RT-qPCR are also discussed, as well as their implication in applied microbiology. I hope that this book will find its place on the bench in microbiology laboratories, so that users will come back to it from time to time, perhaps stimulating the development of novel innovative qPCR approaches to address their specific needs.

Martin Filion

Part I

Technical considerations

An Introduction to the Real-time Polymerase Chain Reaction

1

Stephen A. Bustin, Sara Zaccara and Tania Nolan

Abstract

The real-time, fluorescence-based, quantitative polymerase chain reaction (qPCR) has become the benchmark technology for the detection of nucleic acids in every area of microbiology, biomedical research, biotechnology and in forensic applications. Unlike conventional ('legacy') PCR, which is a qualitative end-point assay, qPCR allows accurate quantification of amplified DNA in real time, during the exponential phase of the reaction. The cost of instruments and reagents is well within reach of individual laboratories, assays are easy to perform, capable of high throughput and combine high sensitivity with reliable specificity. It is possible to achieve accurate and biologically meaningful quantification if meticulous attention is paid to the details of every step of the qPCR assay, starting with sample selection, acquisition and handling through to assay design, validation and optimization. The growing awareness of the need for standardization, quality control and the significant problems associated with inadequate reporting of the assay has resulted in the publication of guidelines for minimum information for the publication of qPCR experiments (MIQE; Bustin *et al.*, 2009).

Introduction

The capacity of the polymerase chain reaction (PCR) to amplify many million-fold any known DNA fragment from complex genetic material, in a short time (Saiki *et al.*, 1985), has revolutionized all areas of the life sciences, making it the most widely used molecular technology today. However, conventional PCR has several disadvantages; it is labour-intensive, not easily automated or adapted for high-throughput applications, and quantification of nucleic acids is arduous. This is because amplification typically reaches a yield plateau where the PCR product is no longer being doubled at each cycle owing to reagent constraints and end product inhibition. That plateau will differ for each assay owing to the different reaction kinetics for each sample, but since conventional PCR is an endpoint assay, samples are typically analysed after reaching this linear phase of the reaction. Consequently, post-PCR gel electrophoresis analysis shows a broadly similar amount of product DNA, independent of the initial amount of template. This makes quantification by legacy PCR difficult since it requires the establishment of additional empirical quantification parameters, e.g. competitive PCR, that vary with each assay and are tedious to reproduce.

The introduction of novel reagents, detection chemistries and instrumentation platforms has modified legacy PCR to create a technology, real-time, fluorescence-based quantitative PCR (qPCR) that addresses these issues. The use of qPCR allows monitoring of the progress of PCR in real time using fluorescent reporter molecules. In this way parallel and immediate amplification, detection and quantification of PCR products is possible, firmly establishing qPCR as the method of choice for the detection of nucleic acids (Lu *et al.*, 2008; Le Guillou-Guillemette and Lunel-Fabiani, 2009; Lopez *et al.*, 2009; Smith and Osborn, 2009; Zhang *et al.*, 2009). The features that make qPCR so universally popular are:

- Speed: assay reaction times are typically

measured in tens of minutes, but can be even shorter.

- Convenience: it is a homogeneous assay, hence there is no need for post-amplification processing.
- Simplicity: the assay requires target, buffer, two primers, a fluorescent reporter and a single enzyme.
- Sensitivity: single copy targets can be detected, if not quantified.
- Specificity: a well-designed assay is specific for a single target, but mismatch-tolerant assays are easily designed.
- Robustness: a wide range of reaction conditions will yield results.
- High throughput: thousands of assays can be carried out on a single run.
- Quantification: accuracy typically covers a wide dynamic range (8–11 orders of magnitude).
- Familiarity: PCR has been around since 1985 and its advantages and disadvantages are well understood.
- Cost: all assay reagents are inexpensive; together with the trend towards smaller reaction volumes, the costs per assay can be very low.

Three essential innovations underlie this technology:

1. A procedure had to be established that allowed the concurrent amplification and detection of specific DNA sequences. The original arrangement involved the addition of a DNA intercalator, ethidium bromide (EtBr) to the PCR (Higuchi *et al.*, 1992) together with the use of a thermal cycler that has an ultraviolet light source and a computer-controlled CCD camera (Higuchi *et al.*, 1993). A plot of increased fluorescence – caused by EtBr binding to newly synthesized DNA – against cycle number allowed the calculation of the quantity of starting material. The drawback of this approach was that this measurement could not easily distinguish between specific and non-specific products.
2. The second development contributed that specificity by adding a separate oligonucleotide probe, labelled with ^{32}P at the 5′ end and designed to hybridize within the target amplicon during the amplification. Successful amplification of the target sequence generated a substrate suitable for the double-strand-specific 5′–3′ exonuclease activity of Taq DNA polymerase (Holland *et al.*, 1991) and resulted in cleavage of the probe. However, measurement of the ^{32}P-labelled cleavage products was neither simple nor practical, requiring their separation by thin layer chromatography.
3. The invention of fluorogenic probes (Lee *et al.*, 1993) made it possible to combine the above innovations in a fluorescence-based hydrolysis probe assay, also known as 5′-nuclease or 'TaqMan' assay, the first practical qPCR method (Heid *et al.*, 1996).

Taken together, these features make qPCR the molecular-enabling technology *par excellence* (Bustin, 2008c) and have transformed the approach to PCR-based quantification of nucleic acids (Bustin, 2004) and, excitingly with recent modifications, for proteins (Reuter *et al.*, 2009).

Principles and basics

Assay characteristics

The output of a qPCR is the 'amplification plot', which is used to quantify DNA targets and provides key information with regards to the quality of the qPCR. The amplification plot is characterized by a series of phases; the background noise of the detection system prior to sufficient amplification signal is recorded as the baseline, this is followed by a period of exponential amplification where the detected signal is correlated to increase in DNA concentration until the plateau phase is reached (Fig. 1.1). All qPCR assays are characterized by the quantification cycle (Cq; Bustin *et al.*, 2009) (also known as the threshold cycle or C_t), the fractional PCR cycle used for quantification. The Cq defines the cycle at which amplification of a PCR product is first detected and is inversely correlated to the logarithm of the initial copy number. There are different ways of determining a Cq, but in general data are collected from the first PCR cycles, where

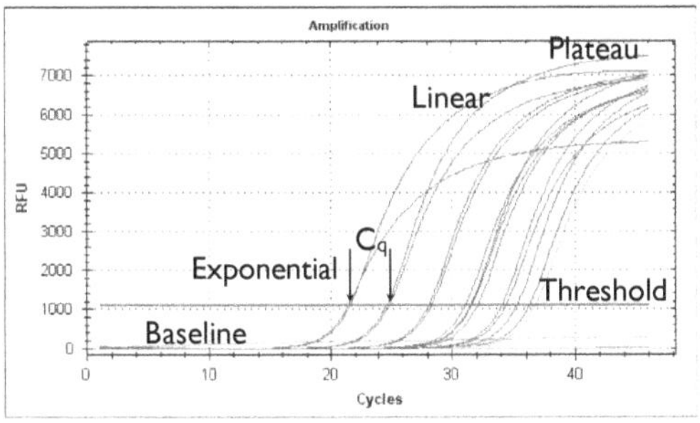

Figure 1.1 qPCR phases. *The baseline region:* the initial cycles of qPCR during which amplification proceeds but there is no change in fluorescence signal as it is below the detection threshold of the instrument. *The exponential phase:* during this phase the doubling of product at every cycle (assuming 100% reaction efficiency) creates exponential signal growth. The quantification cycle, Cq, is used for determining initial target copy numbers. *The linear phase:* reaction components are becoming limited, resulting in reduced amplification reaction efficiency and the signal no longer growing exponentially. *The plateau phase:* this phase, also known as end-point detection, is where traditional PCR products are detected. Reaction components have been exhausted and no more fluorescence is generated. Since each sample is characterized by different reaction kinetics, each reaction plateau at different points. This can be seen from the replicates that have similar Cqs but widely different endpoint fluorescence values.

there is little change in fluorescence signal, and are used to calculate the background fluorescence and the standard deviation. These are subtracted from subsequent results when setting the baseline, resulting in an normalized fluorescence emission intensity. Most systems use a threshold method to define the Cq: this involves identifying the intersection point between the amplification plot and the fractional cycle where fluorescence levels are significantly above the baseline within the exponential region of the amplification curve (Fig. 1.1). The fewer cycles it takes to reach a detectable level of fluorescence, the greater the initial copy number. However, although most analysis software packages have automated threshold settings, the choice of threshold can be made by the operator and introduces a subjective element into real-time quantification. This has led to the development of alternative methods, e.g. sigmoidal curve fitting using non-linear regression (Rutledge and Stewart, 2008), that analyse the curve shape using a variety of algorithms that predict the cycle at which a positive signal occurs. Regardless of method used, the Cq is dependent on the starting template copy number, the efficiency of the PCR amplification and the background fluorescence.

A well-optimized assay will be linear over a range of at least six logs of template concentration, with efficiency close to 100% and high reproducibility between technical replicates (Fig. 1.2A). Importantly, the linearity of the serial dilution of template defines the dynamic range of detection for the assay and only data within these limits are acceptable for experimental samples. Amplification efficiency is best determined by generating a standard curve using serial dilutions of a template and determining the slope from the linear regression of a plot of Cq (y-axis) versus log template quantity (Fig. 1.2B). The template may be any suitable material such as cDNA, genomic DNA, PCR product or a synthetic DNA oligo manufactured for the purpose. Analysis of the standard curve assumes that perfect doubling (or consistent amplification) occurs with each amplification cycle. Thus, the spacing of the fluorescence curves will be determined by the equation 2n = dilution factor, where n is the number of cycles between curves. For example, with a 10-fold serial dilution of DNA, 2n = 10. Therefore, n = 3.32 and the Cq values should be separated by approximately 3.32 cycles.

Amplification efficiency, *E*, is calculated from the slope of the standard curve using the formula $E = 10^{(-1/\text{slope})}$, which is usually converted into a percentage efficiency [% Efficiency = $(E - 1) \times 100\%$]. In a perfect scenario where doubling occurs at each cycle, $E = 10^{(-1/-3.32)} = 2$ and the % efficiency = $(2 - 1) \times 100\% = 100\%$. In practice, amplification efficiency will be around 90–105% or result in a gradient of around −3.2 to −3.5. It is important to note that the presence of inhibitors can result in an apparent increase in efficiency. This is because samples with the highest concentration of template also have the highest level of inhibitors, and therefore display a greater lag between Cq values than samples with lower template concentrations and lower levels of inhibitors. As a result, the absolute value of the slope decreases and the calculated efficiency appears to increase.

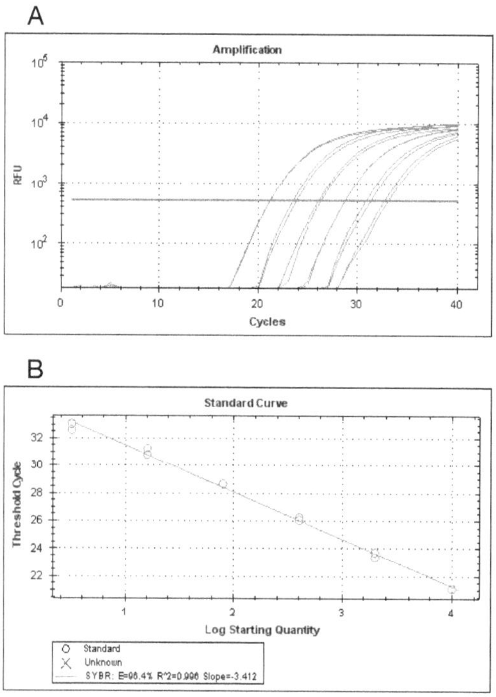

Figure 1.2 (A) Wide dynamic range over several orders of magnitude of template starting concentration. A 10-fold serial dilution of human total RNA/DNA was amplified using iQ supermix (BioRad) and incorporation of SYBR Green I dye. Duplicates of each reaction were monitored (RFU, relative fluorescence units). (B) Standard curve. The closer the slope is to –3.323, the closer the amplification efficiency is to the 100%, showing good reproducibility, with the R^2 being 0.996.

Similarly, inaccurate pipetting can lead to data suggestive of higher or lower efficiency and so it is important to assess the correlation coefficient in order to determine the linearity and reproducibility of the assay.

During assay evaluation and optimization, technical replicates of each template dilution should be processed in parallel in order to establish that the assay is reproducible. A stable assay will demonstrate a correlation coefficient of $R^2 > 0.99$ over at least six logs and with three replicates.

Detection chemistries

A wide range of chemistries utilizing fluorescent reporter molecules has been developed to permit the monitoring of PCR assays in real time. The simplest chemistries use either non-specific fluorescent dyes, e.g. SYBR Green I or EvaGreen®, that bind to any double-stranded DNA or primers labelled with a fluorophore that emit light only when incorporated into amplification products. More complex chemistries make use of template-specific fluorescent probes (e.g. Molecular beacons or hydrolysis probes) using single or dual probes with appropriately positioned fluorescent reporters and quenchers. Although less common, others use only a reporter and make use of the quenching properties of the DNA.

DNA-binding dyes

SYBR Green I dye is the most widely used double-strand DNA-specific reporter. In solution, unbound dye exhibits very little fluorescence; during amplification, fluorescence is dramatically enhanced when the dye is bound to the nascent double-stranded DNA (Morrison *et al.*, 1998) (Fig. 1.3A). When monitored in real-time, this results in an increase in the fluorescence signal that can be observed during the polymerization step, and that decreases when the DNA is denatured. Consequently, fluorescence measurements are performed at the end of the elongation step of every PCR cycle to monitor the increasing amount of amplified DNA. Specificity is dependent entirely on the primers, but the PCR product can be verified by plotting fluorescence as a function of temperature to generate a melting curve of the amplicon (Ririe *et al.*, 1997). This is done by slowly elevating the temperature above the amplicon's melting temperature (T_m) and measuring the fluorescence. Since the T_m of the amplicon depends markedly on its nucleotide composition and length, it is possible to identify the signal obtained from the correct product. A characteristic melting peak at the amplicon's melting temperature (T_m) will distinguish it from amplification artefacts that generally melt at lower temperatures in broader peaks (Fig. 1.3B). A potential drawback is that multiple dye molecules bind to a single amplified molecule and consequently the amount of signal generated, following excitation, is dependent on the mass of double-stranded DNA produced in the reaction. Assuming the same amplification efficiencies, amplification of a longer product will generate

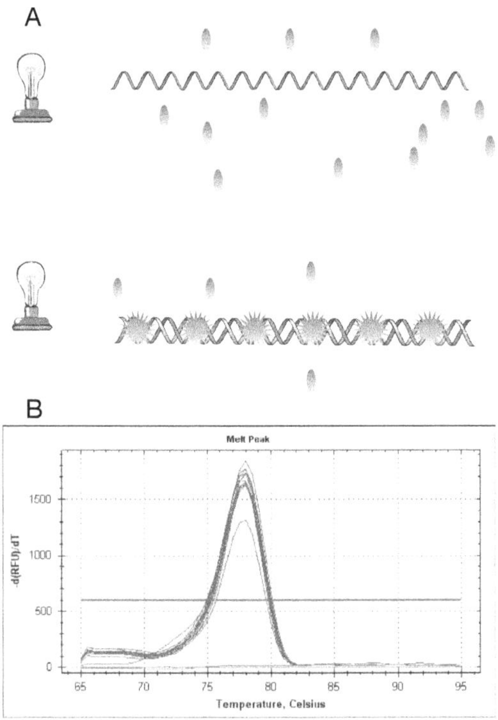

Figure 1.3 (A) The simplest and most economical detection system uses the double-stranded DNA-specific dye SYBR Green I. SYBR Green I dye binds double-stranded DNA and emits light upon excitation. Thus, as a PCR product accumulates, fluorescence increases. The main advantages are that it is simple to use and inexpensive. The disadvantage is that SYBR Green I will bind to any double-stranded DNA in the reaction, including primer-dimers and other non-specific reaction products, which results in an overestimation of the target concentration. (B) Melting curve analysis. The melting temperature of a DNA double helix depends on its base composition and its length. All PCR amplicons for a particular primer pair should have the same melting temperature as shown here; if not, there are likely to be contamination, primer dimers or other problems. If the peaks are not similar, problems with contamination, mispriming or primer-dimer artefacts are indicated.

more signal than a shorter one. If amplification efficiencies are different, quantification can be somewhat inaccurate. This is in contrast to the use of fluorogenic probes, in which the signal from a single fluorophore is observed for each amplified molecule synthesized, regardless of its length.

Probe-based assays

There are a number of chemistries available for probe-based detection of DNA targets; all make use of the same principle: a fluorescent signal is only generated if the amplicon-specific probe hybridizes to its complementary target. Depending on the chemistry used, the increase in fluorescence may be reported at the annealing step when the probe is hybridized (e.g. Molecular beacons) or at the end of the polymerization step when probe hydrolysis has resulted in the physical separation of fluorophore and quencher (e.g. hydrolysis probes).

The hydrolysis probe assay (also known as 5′-nuclease or 'TaqMan') is the most widely used probe-based assay (Gibson *et al.*, 1996) (Fig. 1.4). As its success relies on the 5′-nuclease activity of the DNA polymerase required to hydrolyse the hybridization probe bound to its target amplicon (hence its alternative name 5′-nuclease assay), the correct choice of DNA polymerase is important (Kreuzer *et al.*, 2000). Since the polymerase will cleave the probe only while it remains hybridized to its complementary strand, the temperature conditions of the polymerization phase of the PCR must be adjusted to ensure probe binding. Most probes have a T_m of around 70°C; therefore, the hydrolysis probe method uses a combined annealing and polymerization step at 60–62°C, thus ensuring that the probe remains bound to its target during the primer extension step. It also ensures maximum 5′–3′ exonuclease activity of the DNA polymerases.

Hybridization probe chemistry uses two probes to maximize specificity (Wittwer *et al.*, 1997) (Fig. 1.5). One of the probes carries at its 3′-end a fluorescein donor, which emits green fluorescent light when excited by a light source. Its emission spectrum overlaps the excitation spectrum of an acceptor fluorophore that is attached to the 5′-end of the second probe. This probe must be blocked at its 3′-end to prevent its extension during the annealing step. Excitation of the donor results in fluorescence resonance energy transfer (FRET) to the acceptor and the emission of red fluorescent light. In solution, the two dyes are apart, and since the energy transfer depends on the spacing between the two dye molecules, only background fluorescence is emitted

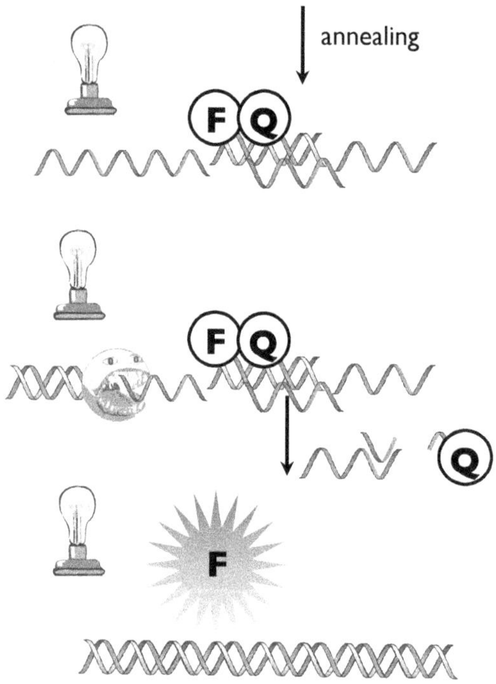

Figure 1.4 Hydrolysis probes. If the specific target sequence is present, the probe anneals downstream from one of the primer sites. During the extension phase of PCR, it is displaced and cleaved by the 5′ nuclease activity of Taq DNA polymerase. This cleavage of the probe separates the reporter dye from quencher dye, resulting in an increased reporter dye signal. Primer extension then continues to the end of the template strand and additional reporter dye molecules are cleaved from their respective probes with each cycle. This generates an increase in fluorescence intensity that is proportional to the amount of PCR amplicon produced.

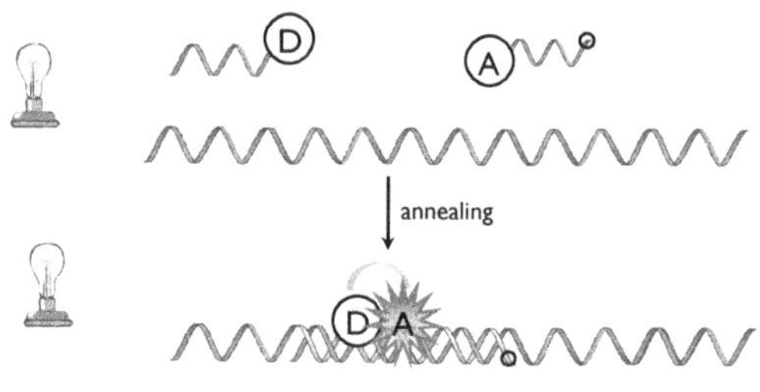

Figure 1.5 Hybridization probes. Hybridization probes are designed as a pair of oligonucleotides with one labelled at its 3′-end with a donor, and the other labelled at its 5′-end with an acceptor fluorphore. As FRET decreases with the sixth power of distance, hybridization probes have to be designed to hybridize to adjacent regions of the template DNA, separated by 1–5 nucleotides. During the denaturation step, PCR template, primers and probes are single-stranded and separate. At the annealing temperature, primers and probes bind to their specific target regions and the donor dye comes into close proximity of the acceptor dye. Energy emitted from the donor dye excites the acceptor dye, which now emits light that is detected and recorded. At the PCR melting temperature, the probes melt off and DNA is ready for the next cycle.

by the donor. Following the denaturation step, both probes hybridize to their target sequence in a head-to-tail arrangement during the annealing step. This brings the two dyes in close proximity to one another and the fluorescein can transfer its energy at high efficiency. The intensity of the light of longer wavelength emitted by the second dye is measured, with increasing amounts of measured fluorescence being proportional to the amount of DNA synthesized during the PCR. A fluorescent signal is detected only as a result of two independent probes hybridizing to their correct target sequence. This increases specificity and generates additional flexibility for probe design. Furthermore, since the probes are not hydrolysed, fluorescence is reversible and allows the generation of melting curves.

Molecular beacons are DNA hybridization probes that form a stem-and-loop structure; the loop portion of the molecule is complementary to the target nucleic acid molecule and the stem is formed by the annealing of complementary arm sequences on the ends of the probe sequence (Tyagi and Kramer, 1996) (Fig. 1.6). A fluorescent marker is attached to the end of one arm and a quencher is attached to the end of the other arm. In solution, free Molecular beacons adopt a hairpin structure and the stem keeps the arms in close proximity, resulting in efficient quenching of the fluorophore. When Molecular beacons encounter a complementary target at the annealing temperature, they undergo a conformational transition that forces the stem apart and results in the formation of a probe/target hybrid that is longer and more stable than the stem (Bonnet *et al.*, 1999). This separates the fluorophore and the quencher, leading to the emission of fluorescence, which can be detected, whereas any free Molecular

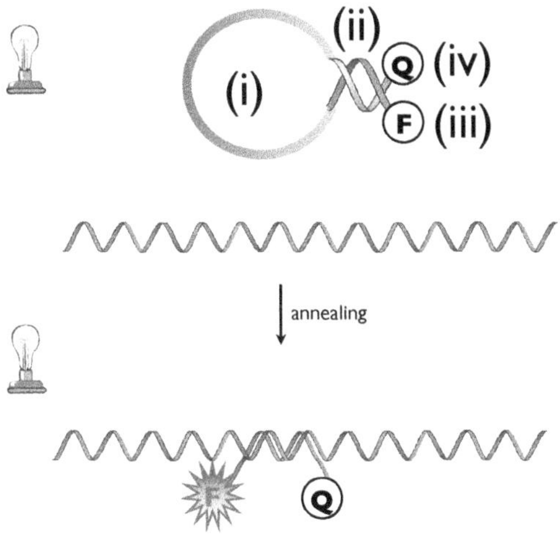

Figure 1.6 Molecular beacons. A Molecular beacon consists of four parts: (i) a loop that is complementary to the target sequence; (ii) a stem that is typically 7 bp long with sequences that are complementary to each other; (iii) a fluorophore that fluoresces in the presence of a complementary target and (iv) a quencher that prevents the fluorophore from emitting light when the Molecular beacon is in a closed loop shape. In the absence of a complementary target sequence, the Molecular beacon remains closed and there is no appreciable fluorescence. When the probe sequence in the loop hybridizes to its target, a conformational reorganization occurs that separates the quencher from the fluorophore, restoring fluorescence which is monitored and reported during each annealing step.

beacons remain closed and non-fluorescent. If the target DNA sequence does not exactly match the sequence of these conventional Molecular beacons, hybridization and fluorescence will not occur, as the thermodynamic properties of the Molecular beacons favour the formation of a hairpin form rather than continued hybridization to a less than perfectly matched target sequence. Conversely, more recently introduced 'sloppy' Molecular beacons possess relatively long probe sequences, enabling them to form hybrids with amplicons despite the presence of mismatched base pairs (El-Hajj *et al.*, 2009). This is useful for screening of families of microorganisms that have some sequence divergence. Furthermore, by using multiple sloppy Molecular beacons with different probe sequences and labelled with different fluorophores, different T_m values can be determined simultaneously, allowing potential screening of numerous microorganisms in a single assay (Chakravorty *et al.*, 2010).

Of the other probe-based systems, Scorpions (Whitcombe *et al.*, 1999) are the most interesting, although their use is fairly specific. Scorpions convert sequence-specific priming and probing into a unimolecular event, making the signalling reaction extremely fast by having the fluorophore attached to the 5′ end of an oligonucleotide that is made up of two separate sequences: (i) a probe sequence complementary to the extension product of the primer, (ii) a target specific primer. Both are linked via a non-amplifiable monomer to prevent copying of the probe sequences during the polymerization step of the PCR. The quencher is linked to the 3′-end of a second oligonucleotide that is fully complementary to the probe sequence. Following extension of the Scorpion primer, and after denaturation and re-annealing, the specific probe sequence is able to hybridize to its complement within the same strand of DNA. Fluorescence is no longer quenched and a fluorescent signal is observed. This unimolecular hybridization is kinetically more favourable than the quencher oligonucleotide reannealing with the Scorpion and allows the introduction of rapid cycling conditions together with a significantly stronger signal strength when compared with both hydrolysis probes and Molecular beacons (Thelwell *et al.*, 2000).

There are several other probe chemistries, all with their own advantages and disadvantages. These include Hybeacons, which require only a single fluorophore and make use of the quenching properties of DNA. This makes them easy to design and synthesize (French *et al.*, 2002). Light-up probes are composed of thiazole orange conjugated to peptide nucleic acid (PNA) and combine the excellent hybridization properties of PNA, which allows the use of shorter probes, with the extraordinary fluorescence enhancement of asymmetrical cyanine dyes upon binding to nucleic acids (Timcheva *et al.*, 1997). Eclipse probes are linear probes that have a minor groove binder (MGB) and quencher on the 5′-end and the fluorophore on the 3′-end (Afonina *et al.*, 2002).

Real-time PCR assays generally use symmetric

primers. However, this results in the reactions typically slowing down and entering the plateau phase in a stochastic manner, because reannealing of the template strands gradually outcompetes primer and probe binding to the template strands. This can be a particular problem when the aim is to detect very low copy number targets, for example in single cell analysis. Asymmetric PCR potentially circumvents the problem of amplicon strand reannealing by using unequal primer concentrations. However, asymmetric amplification is much less efficient and requires extensive optimization to identify the optimal primer ratios, the amounts of starting material, and the number of amplification cycles that can generate reasonable amounts of product for individual template/target combinations. A qPCR variant termed linear-after-the-exponential PCR (LATE-PCR) uses unequal primer concentrations but takes into account the effect of the actual primer concentrations on primer T_m (Sanchez *et al.*, 2004). LATE-PCR features a correction for the fact that the T_m of the limiting primer is often several degrees below the T_m of excess primer and allows the asymmetric PCR to proceed as efficiently as symmetric PCR. Furthermore, single strand amplicons are generated with predictable kinetics for many cycles beyond the exponential phase of the PCR. This permits uncoupling of primer annealing from product detection. As a result, the T_m of the probe no longer needs to be higher than the T_m of either primer. This permits the use of low-T_m probes, which are inherently more allele-discriminating, generate lower background, and can be used at saturating concentrations without interfering with the efficiency of amplification, making them particularly suited for allelic discrimination and the detection of nucleic acids from single cells (Pierce and Wangh, 2011). At the same time, the generation of single-stranded amplification products allows rapid and accurate sequencing of the PCR amplicons (Jia *et al.*, 2010).

Finally, the introduction of DNA analogues such as locked nucleic acids (LNA) (Kumar *et al.*, 1998) has simplified probe design and improved assay quality. Structural studies have shown LNA to be an RNA mimic, fitting seamlessly into an A-type duplex geometry. In LNA, the furanose ring conformation is restricted by a methylene bridge that connects the 2′-oxygen position of ribose to the 4′-carbon. This bridge reduces the conformational flexibility of the ribose, which provides superior affinity and specificity in binding complementary sequences of DNA or RNA. At the same time, this binding affinity is decreased substantially when the hybrids contain even a single mismatched base pair; the change in T_m caused by a mismatch is significantly greater with a LNA/DNA duplex than with a DNA/DNA duplex, resulting in enhanced specificity for SNP/mutation analyses (Johnson *et al.*, 2004). Using appropriate dyes, it is possible to use probes as short as seven nucleotides long for mismatch discrimination, making it possible to generate universal genotyping reagents. It is also possible to substitute DNA with LNA in primers, with the potential to improve primer specificity and performance (Latorra *et al.*, 2003).

qPCR evolution

qPCR technology continues to develop in response to the need for ever-more specific, convenient and cost-effective applications, with several recent improvements refining qPCR assay performance and accuracy as well as simplifying data analysis:

1 The introduction of more reliable and labour-saving, yet less expensive reagents, coupled with the launch of affordable, but more robust qPCR thermal cyclers make reaction assembly simpler and improve flexibility, performance and consistency of qPCR assays. GMP-compliant manufacturing and formulation facilities permit the production of oligonucleotides with rigorous quality control and the documentation required for clinical trials and diagnostic expression analysis.
2 Automated workflows (Grisold and Kessler, 2006; Marshall *et al.*, 2007), more reliable multiplexing assays (Westh *et al.*, 2009; Beck *et al.*, 2010), the launch of instruments using 384- and 1536-well formats and of miniaturized qPCR systems (Morrison *et al.*, 2006), including microfluidic digital PCR (Zhang and Xing, 2007), has resulted in the

expansion of qPCR technology into high throughput applications. Miniaturization offers several potential advantages such as short assay time, low reagent usage and rapid heating/cooling rates, as well as integration of multiple processing modules to further reduce size and power consumption (Neuzil *et al.*, 2006; Marcus *et al.*, 2006). High-density array-based formats and support for faster assay conditions enables shorter assay times, allowing the parallel screening of hundreds of targets at vastly reduced reagent cost where qPCR can provide an alternative to DNA microarrays for accurate expression analysis with high sensitivity (Dixon *et al.*, 2009; Wagner and Arany, 2009) or even be used for routine diagnostic services (Gunson *et al.*, 2008). They also allow powerful multiway profiling, where quantification of reporter genes permits the characterization of samples as functions of multiple factors such as time and genetic make-up (Stahlberg *et al.*, 2008).

3 The increased emphasis on high throughput profiling ties in with the increased understanding for the need for constant and consistent quality qPCR data control. This has resulted in the development of software programs that aim to minimize subjective human input and maximize objective multilevel analysis. The publication of numerous advanced algorithms (Burns *et al.*, 2005; Larionov *et al.*, 2005; Jin *et al.*, 2006; Nordgard *et al.*, 2006; Smith *et al.*, 2007; Batsch *et al.*, 2008; O'Connor and Runquist, 2008; Platts *et al.*, 2008; Rutledge and Stewart, 2008; Spiess *et al.*, 2008; Verderio *et al.*, 2008) and their emergent inclusion in the development of more intuitive analysis software helps with assay set-up (Gallup and Ackermann, 2008), project management and data analysis (Hellemans *et al.*, 2007) to allow robust, accurate and biologically relevant quantification (see below).

Assay considerations

The relative simplicity of a qPCR assay means that it is not difficult to obtain numerous data points from many samples in a short amount of time. The difficulty is deciding whether a Cq reported by the qPCR instrument is real, or whether it is the result of a problem with the assay (Fig. 1.7).

Sample handling and processing

Sample acquisition, handling and processing routines constitute obvious sources of experimental variability and have a significant effect on the power of the assay to deliver biologically relevant qPCR results. Unsurprisingly, the literature is full of different protocols describing methods for sample collection, storage and nucleic acid extraction, with little attempt to standardize any of the procedures. Since such samples are used for ever-increasingly sophisticated analysis, it is important to verify the variability introduced during sample handling. This is particularly important for experiments targeting RNA. While DNA is remarkably stable, RNA molecules are highly unstable and so mRNA profiles may be altered during sample collection, storage or processing. As a general guide, the faster RNA is stabilized the better; therefore sample preparation and speed of extraction are key to successful RNA extraction (Bustin, 2000). Conventionally, successful RNA extraction requires that samples are processed immediately after collection from the biological source (e.g. fresh tissue) or are snap frozen in liquid nitrogen (fresh-frozen). However, there are practical issues surrounding either of these approaches: for example, in a clinical setting, biological material may be left at room temperature for several minutes or hours prior to stabilization and sample storage and transport to the laboratory, which are major concerns.

Assay design

Another critical limitation of qPCR assays is associated with assay design in general and primer design in particular. One obvious question addresses the specificity of the target to be amplified, since a search of any sequence repository for a specific nucleic acid target often reveals a number of variants, making it challenging to deduce which particular microorganism, DNA or RNA has been targeted. Providing an accession number is the most basic, yet frequently omitted piece of information in publications. Common starting points for qPCR target searches are the National

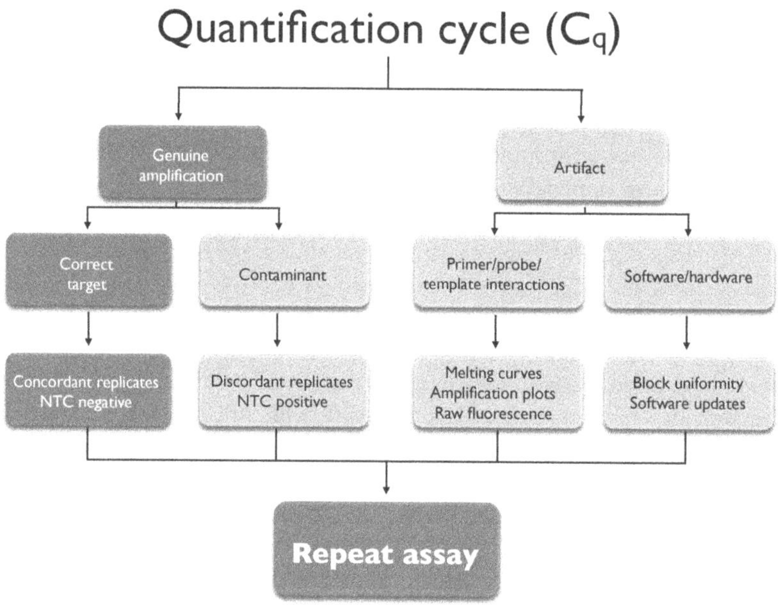

Figure 1.7 Considerations following the reporting of a Cq. The Cq may be a result of genuine amplification or due to some artefact. If genuine, it may still be caused by contamination, which is an ever-present danger for any nucleic acid amplification procedure. It is essential to take note of discordant replicates and positive NTCs, since these frequently allow an immediate assessment of the validity of a result. Alternatively, positive Cq values can be the result of primer dimer amplification or other oligonucleotide/template interactions or some software or hardware problem. Analysis of amplification plots, melt curves and raw fluorescence data are often sufficient to allow an informed decision and there are testing procedures available that check instrument function. In any case, it is essential to repeat an assay, ideally using biological replicates, especially when very small changes in target copy numbers are being quantified.

Center for Biotechnology Information (NCBI) nucleotide search site (http://www.ncbi.nlm.nih.gov/sites/nuccore) or gene search (http://www.ncbi.nlm.nih.gov/gene) websites.

It is now well established that amplification efficiency, and hence assay sensitivity, is directly associated with good primer design (Nolan *et al.*, 2006a; Bustin and Nolan, 2009). In theory, it should be easy to utilize the tens of thousands of published qPCR assays to obtain suitable primer and probe sequences. The qPrimerDepot (http://primerdepot.nci.nih.gov) provides RT-qPCR primers for 99.96% of human RefSeq sequences, most of which cross an exon-exon border which overlaps one of the largest introns. An alternative choice is RTPrimerDB (http://medgen.ugent.be/rtprimerdb/), which contains more than 8000 qPCR assays for over 5500 genes. When an assay is submitted there is a requirement for appropriate validation and optimization data, with the option for further users to leave feedback (Pattyn *et al.*, 2003, 2006; Lefever *et al.*, 2009b). The database provides a freely accessible data retrieval system and an in silico assay evaluation pipeline for all listed qPCR assays. Specifically, assay reports contain gene information, oligonucleotide sequences, reaction conditions, publication information, users' experimental evaluation, feedback and submitter's contact details. Additional information is provided on the alignment of primer and probe sequences on known transcript variants of a gene, along with SNP positions and peptide domain information as well as the secondary structure of the amplicon sequence.

The use of such properly validated assays has numerous advantages, obviating the requirement for time-consuming primer design and empirical optimization and critically, introducing more uniformity and standardization among different laboratories. RTPrimer DB is particularly flexible and can be queried using the official gene name or symbol, Entrez or Ensembl Gene identifier, SNP identifier, or oligonucleotide sequence. Queries can be restricted to a particular application, e.g. RNA quantification, genomic DNA copy number quantification, SNP detection, organism or detection chemistry.

However, no database can accommodate every gene or variant; new variants may have been identified, or suitable assays have never been designed for the organism of interest. Consequently, there are a large number of conditions where good assay design know-how is essential. In practice this means designing several primer combinations for each target and testing the various permutations to obtain the optimal combination. Although this has been well documented and is easily and cheaply implemented (Nolan *et al.*, 2006a), it is still an aspect of assay design that is all too frequently neglected and often results in primers being used that have not been optimized, both in terms of design as well as reaction conditions.

In many ways, the requirements for design of a PCR or qPCR primer are similar, in that it is undesirable to include repeats of the same base or palindromic motifs within the primer sequence and GC content should be around 50%. The degree of clamping at the 3′ should be sufficient to secure hybridization but not to promote non-specific priming. It is a good general guide to aim for two of five bases at the 3′ to be G or C and the remaining three to be A or T. Primer dimers are avoided by rejecting primers with three or more bases matches at the 3′ end and six bases or more matching within the rest of the primer. A measure of the Gibbs free energy (ΔG) can be helpful to estimate primer secondary structure stability. A highly negative value indicates a stable structure; hence primers with negative self dimerization scores are likely to form primer dimers. Similarly, individual primers adopting a hairpin structure with negative ΔG values should be rejected. At the very least, the stem must not remain closed at the annealing temperature of the primer. Fig. 1.8 shows examples of two primers targeting two different mRNAs designed by the authors using Beacon Designer (Premier Biosoft; see below). Both generate very robust assays and both assays result in single melting curve peaks characteristic of unique amplification products.

In contrast to conventional PCRs, the qPCR amplicon length is ideally kept to within 60–150 bp. Amplification of a shorter fragment is much more efficient than of a larger fragment and so keeping the amplicon short drives the reaction towards doubling of the target at each cycle and therefore ensures a greater accuracy in quantification. In addition, the target sequence should be free of secondary structure; this can be predicted by entering the sequence into a folding prediction program such as mfold (Zuker, 2003). Amplification of folded regions is less efficient that of open structures because folding inhibits priming and elongation of the sequence by DNA polymerase.

Probe design and location differs slightly depending on the application. Hydrolysis probes are designed to be located close to the 3′ of the primer on the same strand when being used for quantification; for SNP detection they are positioned centrally in the amplicon and such that the SNP is in the centre of the probe. This is also the case when designing Molecular beacons or LNA-modified probes. All probes have an annealing temperature of 7–10°C higher than the primers; as with primers it is important to ensure that the probe does not contain repetitive regions, and will not adopt secondary structure. In addition, runs of four repeats of the same base should be avoided and the fluorophore should not be attached to a G because this naturally suppresses fluorescence.

Inhibition of the PCR assay by factors co-purified during nucleic acid extraction can be an important factor contributing to poor assay performance (Nolan *et al.*, 2006b). It has recently been shown that the general assumption that inhibitors affect all PCRs to the same extent, i.e. that the reaction of interest and the control reaction are equally susceptible to inhibition, is wrong (Huggett *et al.*, 2008). Instead it appears that susceptibility to inhibition is highly variable between reactions, inhibitors and targets. These findings have serious implications for all PCR-based gene expression studies, including the relatively new PCR array method, and for both qualitative and quantitative PCR-based molecular diagnostic assays, suggesting that careful consideration

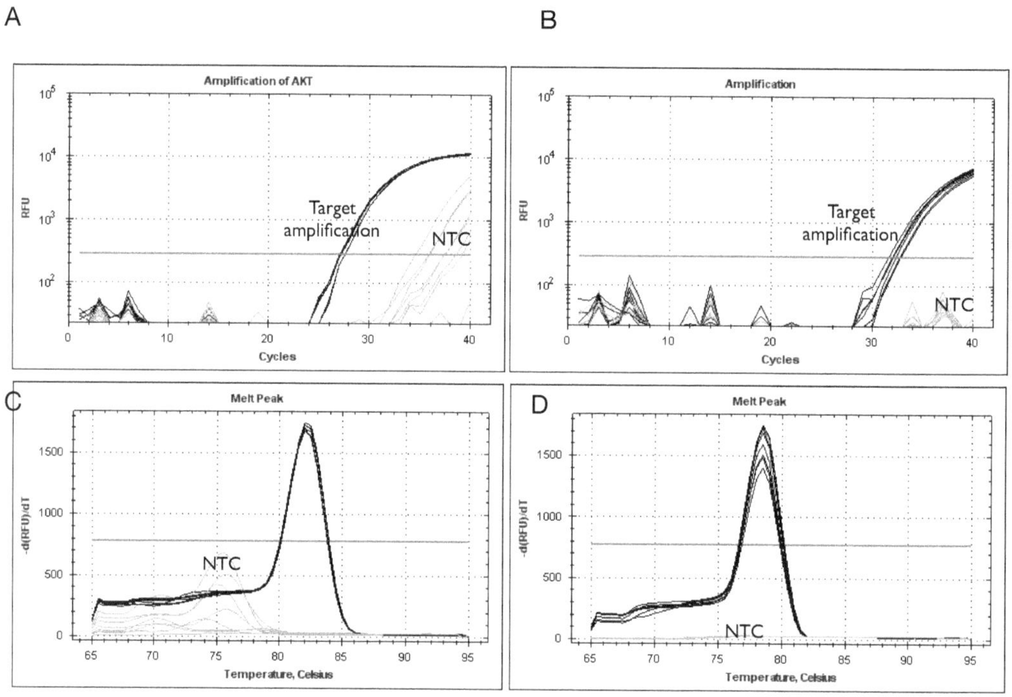

Figure 1.8 Robust qPCR assay design. (A) The primers initiate similarly efficient amplification reactions with SYBR Green I dye as the reporter over a wide temperature annealing range (55–65°C). Positive signals are seen in the no template controls (NTC), but can be minimized by choosing the appropriate annealing temperature. (B) The melting profile clearly distinguishes correctly amplified amplicon from the NTC. (C) Again the assay is characterized by its tolerance to a wide temperature gradient, although clearly there is a little more variability than seen in A. However, at none of the temperatures do the NTC samples generate positive fluorescence signals. (D) A single melt peak demonstrates that the assay generates unique PCR amplicons. The lower G/C content and shorter length of the amplicon is reflected in the lower temperature of the melt peak.

should be given to inhibition compatibility when conducting PCR analyses.

It is becoming routine practice to use specific software packages for qPCR assay design, especially when these automatically screen for structure within the amplicon, oligos and repetitive sequences. Beacon Designer (www.premierbiosoft.com), which is used to design all qPCR assays, not just Molecular beacons, can be used for RNA expression or genotyping assays and the associated package Allele ID also incorporates a function to select sequences from homology line-up plots. Alternatively, there are free design services available from oligonucleotide synthesis companies, such as Sigma who will provide a custom design to the specification of the experiment. There are also commercially available assays, but it is important to ensure that the assay sequences are available in accordance with new guidelines on the requirements for reporting qPCR assays (MIQE guidelines; Bustin *et al.*, 2009).

RT-qPCR considerations

qPCR is the ideal application for the detection and discrimination of microorganisms, mutations or SNP analyses. This is because sample handling and DNA template preparation, assay protocols and data interpretation are all relatively straightforward. In contrast, although RT-qPCR is widely used for the detection of cellular mRNA (Pignot *et al.*, 2008), micro RNAs (Schmittgen *et al.*, 2008), as well as microorganisms (Lu *et al.*, 2008), there are significant biological as well as technical limitations that make its use much more problematic (Bustin, 2008b). A significant contribution to variability in gene expression studies, and one that constitutes 'baseline' variability, arises from the fact that *in vivo* RNA is subject to constant degradation by complex interactions of deadenylation and decapping enzyme complexes as well as exonucleases and endonucleases (Coller and Parker, 2004). There are numerous reasons for technical variability although, unlike biological variability, these can be addressed by the use of appropriate standing operating procedures. The most obvious causes are inconsistent sample selection, handling, and RNA isolation. RNA extraction is a laborious process, often involving solid samples that must be ground to a fine powder while ensuring the sample remains frozen to minimize the activity of nucleases after cell lysis. In many cases, tissue disruption is accompanied by simultaneous addition of a strong protein denaturation agent containing a chaotropic compound such as guanidinium salt. This serves to inactivate nucleases and solubilize cell membranes. Particularly challenging sample types include formalin-fixed and paraffin-fixed archival sections. These are typically treated with proteinase K to degrade the tissue structure prior to guanadinium-based extractions. The nucleic acid purified from these samples is usually very fragmented, around 200 bases, and RNA is difficult to work with due to cross-linking between bases and also to proteins (Lewis *et al.*, 2001; Lewis and Maughan, 2004).

Crucially, the accuracy of gene expression profiling is highly dependent on RNA quality, both in terms of its integrity as well as in terms of the lack of inhibitors co-purified during the extraction procedures (Gallup and Ackermann, 2006; Nolan *et al.*, 2006a,b). Unfortunately, insufficient attention is paid to the analysis of RNA quality, and a survey of the working practices of 100 experienced qPCR users revealed that around half did not adequately quality assess their RNA (Bustin, 2005). A recent survey of papers published in 2007/08 revealed that >60% of papers did not even mention mRNA quality (Bustin, 2008a) and an analysis of papers published in 2010 showed that the higher the impact factor of the journal, the less likely it is that any relevant technical information is included (Huggett and Bustin, 2011).

Technical problems are clearly illustrated by the incorrect RT-qPCR data analysis that resulted in the retraction (Bohlenius *et al.*, 2007) of a paper with Science magazine 'Breakthrough of the Year' status (Huang *et al.*, 2005). One of the most egregious examples of the enormous implications for the health and lives of individuals that result from inappropriate use of this technology is provided by the use of RT-qPCR data that purported to demonstrate the presence of measles virus (MV) RNA in children with developmental disorders (Uhlmann *et al.*, 2002). It provided sustenance to the controversy surrounding the triple measles, mumps and rubella (MMR) virus vaccine, as the data were interpreted as providing hard evidence for a link between MMR, gut pathology and autism. However, a detailed analysis of the raw data underlying that report carried out by one of the authors, acting as an expert witness to the UK High Court and the US Vaccine Court, revealed that these data were obtained amongst a catalogue of mistakes, inaccuracies and inappropriate analysis methods as well as contamination and poor assay performance (Bustin, 2008d). A reanalysis of the data concluded that the assay had been detecting DNA and since MV is an RNA-only virus, the RT-qPCR data had been erroneously interpreted. A recent paper that included two of the authors from the initial report was unable to reproduce the original findings and concluded that there was no link between autism and enteropathy (Hornig *et al.*, 2008). Sadly, despite this and other evidence (Afzal *et al.*, 2006; D'Souza *et al.*, 2007), the authors of the 2002 report have not retracted their original findings, in spite of the immense repercussions their original report has had upon public health.

Clearly, it is essential that the quality of RNA should be determined as a matter of routine whenever possible. Since it is advisable to use approximately the same amount of RNA for cDNA synthesis when comparing different samples, quantification of RNA in extracted samples is important. There are several procedures for the measure of nucleic acid concentration including spectrophotometric or Nanodrop analysis, the Lab901/Agilent/BioRad/Qiagen microfluidics-based systems, or Ribogreen fluorescence detection. Unfortunately, each system produces a different result and so it is imperative that a single system is selected and data obtained using the different methods are not directly compared (Doma and Parker, 2007), especially when performing RT reactions. The purity of the sample includes a consideration of the extent of genomic DNA contamination that can interfere with RT reactions and also cause an increased estimate of copy number. As discussed above, it is important to establish that the RNA sample does not contain factors that may inhibit downstream RT reactions or the qPCR assay.

Since all currently available thermostable polymerases are DNA dependent, RNA must be converted ('reverse transcribed') into cDNA prior to its amplification reaction. Unfortunately, some of the assumptions carried through from legacy RT-PCR are now known to be unsound: primarily that the process of RT is reproducibly quantitative, linear and maintains proportionality between genes when the cDNA for each is produced. Furthermore, cDNA synthesis protocols are highly diverse and utilize different types of RNA-dependent DNA polymerase enzymes that are characterized by low fidelity. Commercial suppliers often provide blends of multiple enzymes to maximize the benefit from the best features of each. Nonetheless, different enzyme, buffer and RT temperature combinations have dramatic effects on cDNA synthesis, resulting in huge differences in cDNA yield of different targets (Stahlberg *et al.*, 2004a,b). There are several choices for cDNA priming strategies that involve reverse transcription from target-specific primers, oligo-dT and anchored variants thereof, various random priming procedures as well as combinations of the above. A specific primer targeting the RNA sequence can be used to direct RT towards the specific region of interest. In many cases, this will be one of the PCR primers, and conventional 'one-step' protocols include a RT step with the reaction containing both primers. Priming with a target specific primer is particularly vulnerable to template secondary structure and so the assay design must include steps to ensure that the primer targets a region that is predicted to be open under the RT conditions determined for the primer. Alternatively, it may desirable to produce cDNA that is representative of the entire RNA population. This is achieved by priming all transcripts using either a primer consisting of a string of thymidine residues (oligo-dT) (for eukaryotic organisms only) or collection of primers of random sequence (random primers). Reliable cDNA priming using oligo-dT is critically dependent on RNA integrity and targets both the 3′ polyA tract of mRNA as well as A-rich regions of mRNA and other RNA molecules. In order to increase specificity, many oligo-dT primers have two additional bases at the 3′ end ('anchored' primers) and RT proceeds from these towards the 5′-end of the RNA. Random priming will reverse transcribe mostly rRNA and may lose sensitivity, but is the most efficient way of cDNA priming an RNA sample for the analysis of multiple targets. Unfortunately, the choice of priming strategy results in large differences in cDNA yield for given targets (Stahlberg *et al.*, 2004a). Specifically, targets present at lower copy number (e.g. those typically of great interest) appear to be more vulnerable to cDNA yield concentration fluctuations, while those at higher copy number (e.g. often adopted as reference genes) are less variable. Similar phenomena have also been observed when using oligo-dT priming. For this reason it is important to validate the reproducibility of RT protocols, ensuring that protocols are designed such that all samples can be processed together or that an experimental calibrator is included so that batch-to-batch variations in the relative yield of each target can be corrected (Lacey *et al.*, 2005). Fig. 1.9 demonstrates that it is possible to obtain highly reproducible RT-qPCR results, if consistent concentrations of a high quality RNA sample and a robust assay are used by a careful operator.

Normalization

From the above, it is clear that the problems associated with inherent variability of RNA, variability of extraction protocols that may co-purify RT or PCR inhibitors, as well as different reverse transcription and PCR efficiencies, mean that it is important that an accurate method of normalization is chosen to control for experimental variation introduced during the multistage process of carrying out a qPCR assay. Unfortunately, normalization remains one of qPCRs most difficult problems to address when designing a qPCR

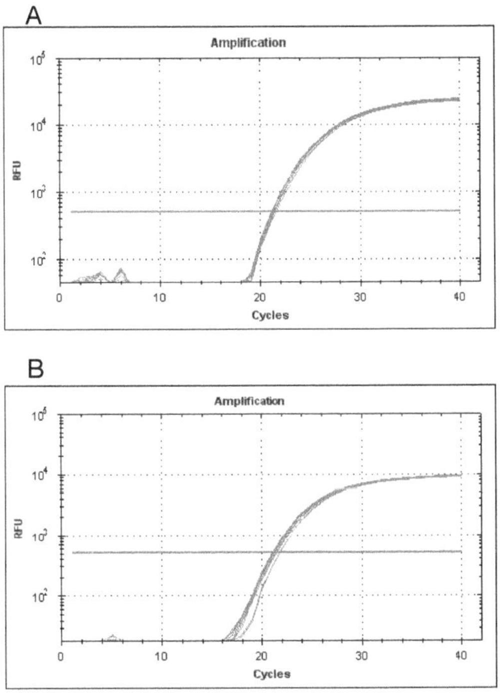

Figure 1.9 Reliability and repeatability of RT-qPCR. (A) Five separate RT reactions were set up simultaneously and subjected to separate PCR assays. The resulting C_qs are virtually identical (mean Cq=21.24±0.07). (B) Single RT reactions were set up on five consecutive days, before being subjected to separate PCR assays. Although there is a little more variation, the Cq value are essentially the same (21.27±0.08).

experiment. Several strategies have been proposed that range from ensuring that a similar sample size is chosen to using multiple internal reference genes (Huggett *et al.*, 2005). These approaches are not mutually exclusive and can be incorporated into a protocol at many stages.

Normalization by relating the quantity of a target gene to total sample mass or volume is a legacy approach remaining from northern blotting techniques. In this case gel loading and sample transfer to filter paper were validated by probing for rRNA or a housekeeping gene such as GAPDH. It was assumed that expression of this housekeeping gene was stable between individuals, experimental conditions or physiological states. While still an integral part of RT-qPCR measurements, there are several disadvantages to measuring mRNA levels relative to sample mass or volume. For example, in a comparison of samples of different origin, it is incorrect to assume that the same tissue mass contains similar cell numbers, or that the relative distribution of cells is equal. Normalization of the specific target quantity to total RNA may underestimate the expression of target genes in some samples. This approach may be more suitable when a precise number and similar cells are targeted; even then this approach is not ideal. A related technique, again reminiscent of northern blotting, is to normalize target quantity to DNA or RNA concentration. While measuring gene copy number relative to input DNA concentration is a perfectly valid approach, the situation is more complicated for transcript quantification. When RNA concentration is determined, the vast majority of the RNA component is ribosomal RNA (rRNA). Transcription of rRNA is independent of the transcription of mRNA since it is transcribed by different enzymes. Furthermore, as rRNA makes up ~80% of the RNA fraction, quantification of rRNA would mask subtle changes in the mRNA component, which typically constitutes 2–5%. In addition, this approach does not take into account variations in the template RNA quality. Nonetheless, while not ideal, there are situations where there are no other alternatives but to measure relative to total RNA and in many cases the data are reliable. Some analysis packages such as GenEx software from MultiD (www.multid.se) allow the total RNA concentration to be evaluated as a normalization technique alongside other approaches such as those below. One theoretical solution would be to purify mRNA and quantify this. Unfortunately, the purification process introduces inaccuracies and an extra processing step that is undesirable and in many cases there is insufficient RNA sample to purify mRNA.

A very common approach to correct for sample differences is to express the target concentration relative to that of a single internal reference gene usually with little or no validation. Although still used, this approach was largely discredited a long time ago (Bustin, 2000, 2002). In order for single reference gene normalization to be valid, the expression of the reference gene must remain constant between all experimental samples. It is rare to find such a convenient target and all too common and misguided is to select this gene at random, without validation; GAPDH, β-actin and 18S or 16S being particular favourites. When the reference gene is not stably expressed between all samples, a ratio of the target gene of interest to the reference gene will reflect the expression changes of both targets. This is unhelpful when the expression behaviour of neither target has been defined and can lead to inaccuracies and false results. An alternative approach, which is gaining momentum, is to express the data relative to the behaviour of multiple reference genes and use geometric averaging to measure and control for error induced trends in the relative reference gene expression. There are various approaches such as those offered by geNorm (Vandesompele *et al.*, 2002) that allow the quantity of potential reference genes to be compared in selected groups of experimental and control samples. These use the combined variation of several genes to produce a stable reference factor for normalization and can be very accurate, potentially providing 0.5-fold resolution (Hellemans *et al.*, 2004). In an analogous manner, gene expression analysis of miRNA molecules in eukaryotic cells requires accurate and reliable normalization to relevant targets and the mean expression value of all expressed microRNAs in a given sample has been shown to provide optimal normalization for miRNA qPCR data (Mestdagh *et al.*, 2009). Clearly, such extensive reference gene validation is taxing on time, sample and resource. In future, alternative approaches such as normalization to expressed ALU repeats (EARs) or other repeats may provide less labour-intensive normalization strategies (Marullo *et al.*, 2010).

Data analysis

The successful generation of high-quality qPCR experimental runs and acquisition of Cq values is followed by the critical data analysis step. The increasing penetration of 384-well and higher throughput qPCR instruments, together with PCR array methods such as the Biotrove and Fluidigm systems has resulted in data analysis becoming one of the biggest bottlenecks in qPCR experiments. Coupled with this is the fact that many biologists may not be able to analyse and interpret appropriately the quantitative data from their qPCR assays. Hence it has become increasingly obvious that, even for research applications, data analysis, as an essential component of the complete assay, should be carried out using dedicated software. A wide range of analysis packages can be accessed from http://www.gene-quantification.de/download.html.

One of the first tools was REST (Relative Expression Software Tool) (Pfaffl *et al.*, 1998), which was developed to address the problem of comparing target gene expression levels relative to reference genes indifferent of samples. Importantly, these calculations also accommodated differences in the efficiency of each of the PCR assays. The use of ratios for gene expression measurements makes it complex to perform traditional statistical analysis, as ratio distributions do not have a standard deviation. The latest REST (http://www.gene-quantification.de/rest-2009.html) is a standalone software package and has been extended with advanced algorithms using randomization and bootstrapping techniques to take into account issues of multiple reference gene normalization and quantification of the uncertainty inherent in any measurement of expression ratios. The program also allows measurement of not only the statistical significance of deviations but also their likely magnitude, even in the presence of outliers. A visually attractive graphing ability provides a visual representation of variation for each gene, highlighting potential issues such as distribution skew.

In terms of power and sophistication, two commercial software tools stand out: GenEx and qBase plus.

GenEx (http://www.multid.se/) is a powerful package for processing and analysing qPCR data. It offers a vast range of tools, ranging from basic data editing and management to advanced, cutting-edge data analysis including selection and validation of reference genes, classification of samples, gene grouping, monitoring of time dependent processes and basic and sophisticated statistical and graphic capacity. Statistical features include parametric and non-parametric tests, clustering methods, principal component analysis and artificial neural networks. Most qPCR experiment- processes have a nested design with technical replicates at each step, where each step introduces more variance to the data. The noise contribution from each step can be analysed and used to optimize the experimental design for future experiments. One interesting feature is the experimental design optimization module, which calculates how to minimize the total variation for the experimental design with consideration of the experimental budget. GenEx can be used for the analysis of both relative quantification (ΔΔCq) and 'absolute' quantification, i.e. relative to a dilution curve, with powerful presentation tools generating visually attractive illustrations. For accurate and robust normalization of qPCR data, GenEx has the advantage of incorporating both NormFinder (Andersen *et al.*, 2004) and geNorm (Vandesompele *et al.*, 2002) in the software. geNorm (http://medgen.ugent.be/~jvdesomp/genorm/) is used to select an optimal set of reference genes from a larger set of candidate genes. It is critical that the tested candidate genes are not co-regulated. The geNorm process performs a reiterative calculation of the M-value, which describes the variation of a potential reference gene compared with all other candidate genes. The gene with highest M-value is eliminated until there are only two reference genes left. Usually the last two or three remaining genes are recommended as the optimum candidate reference genes. NormFinder (http://www.mdl.dk/publicationsnormfinder.htm) is also used to find the optimum reference genes but, in contrast to geNorm, takes information of expression in groupings of samples into account. The result is an optimum pair of reference genes that might have compensating expression, so that one gene is slightly overexpressed in one group, but the other gene is correspondingly underexpressed in the same group.

qBase plus (http://www.biogazelle.com/products/qBase plus) is a relatively new software tool developed by members of the group that invented the geNorm algorithms (Vandesompele *et al.*, 2002; Hellemans *et al.*, 2007). It has the capacity to determine PCR replicate variation, assessment of positive and negative control samples, determination of reference gene expression stability and evaluation of deviating normalization factors.

Technical replicates (repeated measurements of the same sample in the same run) are useful as they allow for quality control of the technical reproducibility of the qPCR data, provide better accuracy, and allow the generation of results when individual qPCR reactions failed. Low reproducibility of the qPCR reaction is often the first indication of an unstable assay that requires further optimization. qBase plus and GenEx automatically deal with technical replicates or repeated measurements, outliers can be excluded so they do not contribute to a significant bias and large errors. However, it is important to note that biological variability is often much larger than RT technical variability, which is also much larger than PCR technical variability.

The second type of quality control allows an evaluation of the positive and negative sample controls. An amplification signal in the NTC sample indicates a potential contamination issue or formation of primer dimers; hence both packages flag suspicious no-template control (NTC) samples based on a user defined or default thresholds.

Reference gene or normalization factor stability is the third type of quality control. The user is able to choose a minimal acceptable normalization factor stability by defining a threshold value for two indicators of expression stability of the used reference genes; the geNorm expression stability value of the reference gene (M) (Vandesompele *et al.*, 2002) and the coefficient of variation (CV) of the relative quantities (Hellemans *et al.*, 2007). qBase plus has extended and improved the functionality of geNorm by allowing ranking of candidate reference genes up to the single most stable gene, by combining the calculation of relative quantities and geNorm analysis and by allowing the processing of experiments with missing data in a way that has the lowest impact on the overall analysis through intelligent retention of as many data points as possible.

Inspection of normalization factors is a very useful feature and constitutes a fourth type of quality control. qBase plus displays the geometric mean of selected reference genes (the normalization factor) for each sample. These factors should be similar for all samples if approximately equal amounts of equal quality input material were used. High variability of the normalization factors indicates large differences in starting material quantity or quality, or a problem with one of the reference genes, which may be either not stably expressed or not adequately measured.

There are alternatives, usually web-based analysis tools available: QPCR (https://esus.genome.tugraz.at/rtpcr/) allows storage, management, and analysis of qPCR data (Pabinger *et al.*, 2009). It comprises a parser to import data from qPCR instruments and includes a variety of analysis methods to calculate cycle-threshold and amplification efficiency values. The analysis pipeline includes technical and biological replicate handling, incorporation of sample or gene specific efficiency, normalization using single or multiple reference genes, inter-run calibration, and fold change calculation. Moreover, the application supports an assessment of error propagation throughout all analysis steps and allows conducting of statistical tests on biological replicates. Results can be visualized in customizable charts and exported for further investigation. Calculation of amplification efficiencies for RT-PCR experiments (CAmpER (http://camper.cebitec.uni-bielefeld.de) is a more limited web-based tool for the basic analyses of qPCR assays run on the Roche Lightcycler, the Biorad Opticon and the Qiagen Rotorgene.

MIQE

As discussed above, technological advances mean that there is an ever-increasing choice of platforms, chemistries, protocols as well as applications and targets for qPCR analysis. Whilst this serves to underline the pre-eminence of qPCR as a technique, is exciting and generating a vast amount of data, it has also become clear that a high percentage of publications utilizing qPCR technology, and especially those aiming to profile cellular RNA levels, report poorly designed, executed and interpreted experiments and results (Bustin, 2010). Considerations of mRNA transcription, *in vivo* stability, regulation by miRNAs, tissue-specificity of splice variants, allele-specific differences in expression, the lack of concordance between most mRNAs and their specified proteins, the critical importance of post-translational modifications and questions of tissue heterogeneity all describe serious issues that are not being addressed in an adequate manner.

Trust in the accuracy and integrity of the scientific literature is an essential prerequisite for maintaining scientific excellence and advancing knowledge. This requires urgent consideration by researchers, reviewers and editors who need to agree a basic set of quality criteria and adhere to elementary procedures that are prerequisites for the publication of reliable and reproducible data. Such a list must include delineating minimum quality standards for template preparation, validation and consistent use of cDNA priming methods, enzymes, protocols and, equally critically, appropriate analysis of data. Furthermore, it is entirely unacceptable that most publications do not address the critical issue of RNA quality assessment. It is equally intolerable that data are not normalized in a proper way. In addition, it is vital that the conventions underlying data acquisition and analysis become more transparent. Consequently, it is essential that the editors of scientific publications consider issuing agreed checklists specifying the key information to be included when reporting experimental results. There are significant efforts under way to organize such 'minimum information' checklists, with the 'Minimum information for biological and biomedical investigations' (MIBBI) project (www.mibbi.org) offering a common portal aimed at promoting gradual data integration (http://mibbi.sourceforge.net).

The problem of inadequate information being provided with publications is being addressed by the minimum information for publication of quantitative real-time PCR experiments (MIQE) guidelines (Bustin *et al.*, 2009). MIQE constitutes a set of publication guidelines for researchers, journal reviewers and editors that list the minimum information required to allow potential reproduction, as well as unambiguous quality assessment, of a qPCR-based experiment. MIQE is modelled on analogous rules drawn up for DNA microarray analysis (Brazma *et al.*, 2001), proteomics experiments (Taylor *et al.*, 2007), genome sequence specification (Field *et al.*, 2008) and those under discussion for RNAi work (Haney, 2007) and metabolomics (Sansone *et al.*, 2007), initiatives coordinated under the auspices of MIBBI (Taylor *et al.*, 2008).

The four key areas of standardization that

define any qPCR experiment are study design, technical detail, analysis methods and statistics. MIQE addresses these under a set of captions that describe a large number of individual elements: 'Experimental design, sample, nucleic acids, reverse transcription, target, primers and probes, assay details, PCR cycling and data analysis'. At first sight, these look daunting, arduous and overexacting. In practice, it is clear that most, if not all of these parameters, describe information that would be obtained as a matter of course during the experimental design, optimization and validation stages. Importantly, there is a clear hierarchy with some parameters, labelled 'E' (essential) in the published guidelines, indispensable for an adequate description of the qPCR assay, whereas other components, labelled 'D' (desirable) more peripheral, yet constituting an effective foundation for the realization of best practice protocols. Of course, these parameters are not set in stone and are open for discussion (Bustin *et al.*, 2010), but the guidelines encompass a core of indispensable information that should be accessible to reviewers and readers.

Finally, another development complementary to MIQE concerns the problems associated with attempting to share qPCR data between different laboratories and users. Currently, easy data interchange between instrument software and third-party data analysis packages, colleagues and collaborators, and between authors, peer reviewers, journals and readers, is difficult. One solution to this problem is the development of the Real-time PCR Data Markup Language (RDML), a structured and universal data standard for exchanging quantitative PCR (qPCR) data (Lefever *et al.*, 2009a). The aim is to enable transparent exchange of annotated qPCR data by providing sufficient information for a third party to be able to understand the experimental set-up, re-analyse the data and interpret the results. RDML files contain all measured data as well as information about the samples and targets being analysed; thus this data standard provides sufficient information to allow the reader to understand the experimental set-up, re-analyse the data and interpret the results. A useful tool for converting instrument-specific data to the RDML format for convenient data interchange is found on the RDML website (http://medgen.ugent.be/rdml/chooseTool.php).

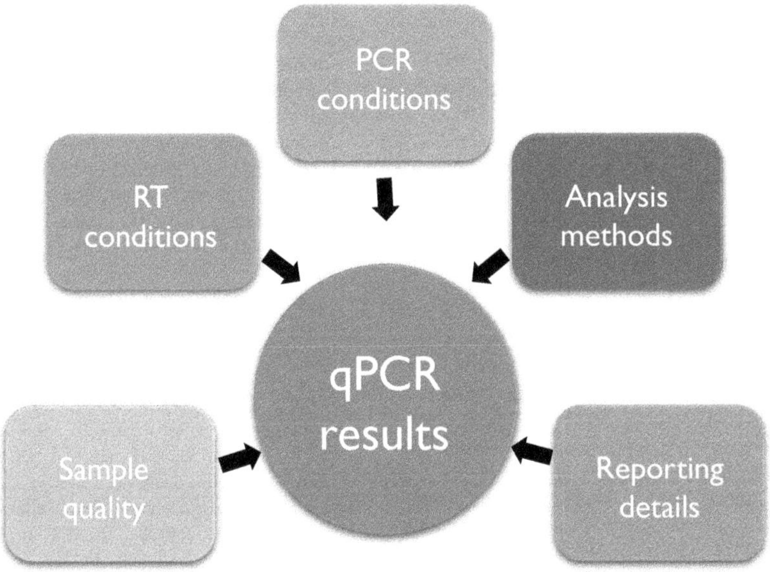

Figure 1.10 Successful qPCR results depend on samples being acquired, handled and processed appropriately, on suitable RT and PCR conditions being established, fit-for-purpose analysis methods being applied and, finally, experimental protocols being reported in a transparent and complete manner.

Conclusions

Some technologies arrive at a time that is 'just right' and drive the development of new and unexpected applications. Advances in enzymes, buffers, chemistries and instrumentation have allowed qPCR to progress from a cutting edge technology used as a research tool in specialist laboratories to becoming a routine technique across the life sciences, making it a technology whose time has come (Walker, 2002). Currently, it provides the best compromise between ease of use, sensitivity, reliability and specificity of any molecular diagnostic assay.

Biologically relevant qPCR results depend on several parameters that must all come together in an appropriate, controlled and transparent manner (Fig. 1.10). The main issues remaining are those concerning the potential for observing false positive or negative results, nucleic acid quality control issues, the requirement for good assay design and optimization and transparency of protocols and analysis methodologies. Therefore, it cannot be over emphasized that there remains a constant need for assay validation to ascertain the reliability and reproducibility of established, modified or newly developed technologies for routine analysis. Finally, the development of lab-on-a-chip devices combining miniaturized reactors, advanced microfluidics and sophisticated detection technologies will open up a whole new field for qPCR applications, bringing new opportunities and new challenges.

References

Afonina, I.A., Reed, M.W., Lusby, E., Shishkina, I.G., and Belousov, Y.S. (2002). Minor groove binder-conjugated DNA probes for quantitative DNA detection by hybridization-triggered fluorescence. Biotechniques 32, 940–949.

Afzal, M.A., Ozoemena, L.C., O'Hare, A., Kidger, K.A., Bentley, M.L., and Minor, P.D. (2006). Absence of detectable measles virus genome sequence in blood of autistic children who have had their MMR vaccination during the routine childhood immunization schedule of UK. J. Med. Virol. *78*, 623–630.

Andersen, C.L., Jensen, J.L., and Orntoft, T.F. (2004). Normalization of real-time quantitative reverse transcription-PCR data: a model-based variance estimation approach to identify genes suited for normalization, applied to bladder and colon cancer data sets. Cancer Res. *64*, 5245–5250.

Batsch, A., Noetel, A., Fork, C., Urban, A., Lazic, D., Lucas, T., Pietsch, J., Lazar, A., Schomig, E., and Grundemann, D. (2008). Simultaneous fitting of real-time PCR data with efficiency of amplification modeled as Gaussian function of target fluorescence. BMC Bioinformatics *9*, 95.

Beck, E.T., Jurgens, L.A., Kehl, S.C., Bose, M.E., Patitucci, T., LaGue, E., Darga, P., Wilkinson, K., Witt, L.M., Fan, J., *et al.* (2010). Development of a rapid automated influenza A, influenza B, and respiratory syncytial virus A/B multiplex real-time RT-PCR assay and its use during the 2009 H1N1 swine-origin influenza virus epidemic in Milwaukee, Wisconsin. J. Mol. Diagn. *12*, 74–81.

Bohlenius, H., Eriksson, S., Parcy, F., and Nilsson, O. (2007). Retraction. Science *316*, 367.

Bonnet, G., Tyagi, S., Libchaber, A., and Kramer, F.R. (1999). Thermodynamic basis of the enhanced specificity of structured DNA probes. Proc. Natl. Acad. Sci. U.S.A. *96*, 6171–6176.

Brazma, A., Hingamp, P., Quackenbush, J., Sherlock, G., Spellman, P., Stoeckert, C., Aach, J., Ansorge, W., Ball, C.A., Causton, H.C., *et al.* (2001). Minimum information about a microarray experiment (MIAME)-toward standards for microarray data. Nat. Genet. *29*, 365–371.

Burns, M.J., Nixon, G.J., Foy, C.A., and Harris, N. (2005). Standardisation of data from real-time quantitative PCR methods – evaluation of outliers and comparison of calibration curves. BMC Biotechnol. *5*, 31.

Bustin, S.A. (2000). Absolute quantification of mRNA using real-time reverse transcription polymerase chain reaction assays. J. Mol. Endocrinol. *25*, 169–193.

Bustin, S.A. (2002). Quantification of mRNA using real-time reverse transcription PCR (RT-PCR): trends and problems. J. Mol. Endocrinol. *29*, 23–39.

Bustin, S.A. (2004) A-Z of Quantitative PCR (IUL Press, La Jolla, CA).

Bustin, S.A. (2005). Real-time, fluorescence-based quantitative PCR: a snapshot of current procedures and preferences. Expert Rev. Mol. Diagn. *5*, 493–498.

Bustin, S.A. (2008a). Molecular medicine, gene-expression profiling and molecular diagnostics: putting the cart before the horse. Biomarkers Med. *2*, 201–207.

Bustin, S.A. (2008b). Real-time polymerase chain reaction – towards a more reliable, accurate and relevant assay. Eur. Pharm. Rev. *6*, 19–27.

Bustin, S.A. (2008c). Real-time quantitative PCR-opportunities and pitfalls. Eur. Pharm. Rev. *4*, 18–23.

Bustin, S.A. (2008d). RT-qPCR and molecular diagnostics: no evidence for measles virus in the GI tract of autistic children. Eur. Pharm. Rev. Dig. *1*, 11–16.

Bustin, S.A. (2010). Why the need for qPCR publication guidelines? The case for MIQE. Methods *50*, 217–226.

Bustin, S.A., and Nolan, T. (2009). qPCR Assay design. Eur. Pharm. Rev. *3*, 26–32.

Bustin, S.A., Benes, V., Garson, J.A., Hellemans, J., Huggett, J., Kubista, M., Mueller, R., Nolan, T., Pfaffl, M.W., Shipley, G.L., *et al.* (2009). The MIQE guidelines: minimum information for publication of quantitative real-time PCR experiments. Clin. Chem. *55*, 611–622.

Bustin, S.A., Beaulieu, J.F., Huggett, J., Jaggi, R., Kibenge, F.S., Olsvik, P.A., Penning, L.C., and Toegel, S. (2010). MIQE precis: Practical implementation of minimum standard guidelines for fluorescence-based quantitative real-time PCR experiments. BMC Mol. Biol. *11*, 74.

Chakravorty, S., Aladegbami, B., Burday, M., Levi, M., Marras, S.A., Shah, D., El-Hajj, H.H., Kramer, F.R., and Alland, D. (2010). Rapid universal identification of bacterial pathogens from clinical cultures by using a novel sloppy Molecular beacon melting temperature signature technique. J. Clin. Microbiol. *48*, 258–267.

Coller, J., and Parker, R. (2004). Eukaryotic mRNA decapping. Annu. Rev. Biochem. 73, 861–890.

D'Souza, Y., Dionne, S., Seidman, E.G., Bitton, A., and Ward, B.J. (2007). No evidence of persisting measles virus in the intestinal tissues of patients with inflammatory bowel disease. Gut *56*, 886–888.

Dixon, J.M., Lubomirski, M., Amaratunga, D., Morrison, T.B., Brenan, C.J., and Ilyin, S.E. (2009). Nanoliter high-throughput RT-qPCR: a statistical analysis and assessment. Biotechniques 46, ii–viii.

Doma, M.K., and Parker, R. (2007). RNA quality control in eukaryotes. Cell *131*, 660–668.

El-Hajj, H.H., Marras, S.A., Tyagi, S., Shashkina, E., Kamboj, M., Kiehn, T.E., Glickman, M.S., Kramer, F.R., and Alland, D. (2009). Use of sloppy Molecular beacon probes for identification of mycobacterial species. J. Clin. Microbiol. *47*, 1190–1198.

Field, D., Garrity, G., Gray, T., Morrison, N., Selengut, J., Sterk, P., Tatusova, T., Thomson, N., Allen, M.J., Angiuoli, S.V., *et al.* (2008). The minimum information about a genome sequence (MIGS) specification. Nat. Biotechnol. *26*, 541–547.

French, D.J., Archard, C.L., Andersen, M.T., and McDowell, D.G. (2002). Ultra-rapid DNA analysis using HyBeacon probes and direct PCR amplification from saliva. Mol. Cell Probes *16*, 319–326.

Gallup, J.M., and Ackermann, M.R. (2006). Addressing fluorogenic real-time qPCR inhibition using the novel custom Excel file system 'FocusField2-6GallupqPCRSet-upTool-001' to attain consistently high fidelity qPCR reactions. Biol. Proced. Online *8*, 87–152.

Gallup, J.M., and Ackermann, M.R. (2008). The 'PREXCEL-Q Method' for qPCR. Int. J. Biomed. Sci. *4*, 100–120.

Gibson, U.E., Heid, C.A., and Williams, P.M. (1996). A novel method for real time quantitative RT-PCR. Genome Res. *6*, 995–1001.

Grisold, A.J., and Kessler, H.H. (2006). Use of hybridization probes in a real-time PCR assay on the LightCycler for the detection of methicillin-resistant *Staphylococcus aureus*. Methods Mol. Biol. *345*, 79–89.

Gunson, R.N., Bennett, S., Maclean, A., and Carman, W.F. (2008). Using multiplex real time PCR in order to streamline a routine diagnostic service. J. Clin. Virol. *43*, 372–375.

Haney, S.A. (2007). Increasing the robustness and validity of RNAi screens. Pharmacogenomics *8*, 1037–1049.

Heid, C.A., Stevens, J., Livak, K.J., and Williams, P.M. (1996). Real time quantitative PCR. Genome Res. *6*, 986–994.

Hellemans, J., Preobrazhenska, O., Willaert, A., Debeer, P., Verdonk, P.C., Costa, T., Janssens, K., Menten, B., Van Roy, N., Vermeulen, S.J., *et al.* (2004). Loss-of-function mutations in LEMD3 result in osteopoikilosis, Buschke-Ollendorff syndrome and melorheostosis. Nat. Genet. *36*, 1213–1218.

Hellemans, J., Mortier, G., De Paepe, A., Speleman, F., and Vandesompele, J. (2007). qBase relative quantification framework and software for management and automated analysis of real-time quantitative PCR data. Genome Biol. *8*, R19.

Higuchi, R., Dollinger, G., Walsh, P.S., and Griffith, R. (1992). Simultaneous amplification and detection of specific DNA sequences. Biotechnology *10*, 413–417.

Higuchi, R., Fockler, C., Dollinger, G., and Watson, R. (1993). Kinetic PCR analysis: real-time monitoring of DNA amplification reactions. Biotechnology *11*, 1026–1030.

Holland, P.M., Abramson, R.D., Watson, R., and Gelfand, D.H. (1991). Detection of specific polymerase chain reaction product by utilizing the 5′–3′ exonuclease activity of Thermus aquaticus DNA polymerase. Proc. Natl. Acad. Sci. U.S.A. *88*, 7276–7280.

Hornig, M., Briese, T., Buie, T., Bauman, M.L., Lauwers, G., Siemetzki, U., Hummel, K., Rota, P.A., Bellini, W.J., O'Leary, J.J., *et al.* (2008). Lack of association between measles virus vaccine and autism with enteropathy: a case–control study. PLoS ONE 3, e3140.

Huang, T., Bohlenius, H., Eriksson, S., Parcy, F., and Nilsson, O. (2005). The mRNA of the *Arabidopsis* gene FT moves from leaf to shoot apex and induces flowering. Science *309*, 1694–1696.

Huggett, J., and Bustin, S.A. (2011). Standardisation and reporting for nucleic acid quantification. Accred. Qual. Assur. *16*, 399–405.

Huggett, J., Dheda, K., Bustin, S., and Zumla, A. (2005). Real-time RT-PCR normalisation; strategies and considerations. Genes Immun. *6*, 279–284.

Huggett, J.F., Novak, T., Garson, J.A., Green, C., Morris-Jones, S.D., Miller, R.F., and Zumla, A. (2008). Differential susceptibility of PCR reactions to inhibitors: an important and unrecognised phenomenon. BMC Res. Notes *1*, 70.

Jia, Y., Osborne, A., Rice, J.E., and Wangh, L.J. (2010). Dilute-'N'-Go dideoxy sequencing of all DNA strands generated in multiplex LATE-PCR assays. Nucleic Acids Res. *38*, e119.

Jin, N., He, K., and Liu, L. (2006). qPCR-DAMS: a database tool to analyze, manage, and store both relative and absolute quantitative real-time PCR data. Physiol. Genomics *25*, 525–527.

Johnson, M.P., Haupt, L.M., and Griffiths, L.R. (2004). Locked nucleic acid (LNA) single nucleotide polymorphism (SNP) genotype analysis and validation using real-time PCR. Nucleic Acids Res. *32*, e55.

Kreuzer, K.A., Bohn, A., Lass, U., Peters, U.R., and Schmidt, C.A. (2000). Influence of DNA polymerases on quantitative PCR results using TaqMan(TM)probe format in the LightCycler(TM) instrument. Mol. Cell Probes *14*, 57–60.

Kumar, R., Singh, S.K., Koshkin, A.A., Rajwanshi, V.K., Meldgaard, M., and Wengel, J. (1998). The first analogues of LNA (locked nucleic acids): phosphorothioate-LNA and 2′-thio-LNA. Bioorg. Med. Chem. Lett. *8*, 2219–2222.

Lacey, H.A., Nolan, T., Greenwood, S.L., Glazier, J.D., and Sibley, C.P. (2005). Gestational profile of Na^+/H^+ exchanger and Cl–/HCO3– anion exchanger mRNA expression in placenta using real-time QPCR. Placenta *26*, 93–98.

Larionov, A., Krause, A., and Miller, W. (2005). A standard curve based method for relative real time PCR data processing. BMC Bioinformatics *6*, 62.

Latorra, D., Arar, K., and Hurley, J.M. (2003). Design considerations and effects of LNA in PCR primers. Mol. Cell Probes *17*, 253–259.

Le Guillou-Guillemette, H., and Lunel-Fabiani, F. (2009). Detection and quantification of serum or plasma HCV RNA: mini review of commercially available assays. Meth. Mol. Biol. *510*, 3–14.

Lee, L.G., Connell, C.R., and Bloch, W. (1993). Allelic discrimination by nick-translation PCR with fluorogenic probes. Nucleic Acids Res. *21*, 3761–3766.

Lefever, S., Hellemans, J., Pattyn, F., Przybylski, D.R., Taylor, C., Geurts, R., Untergasser, A., and Vandesompele, J. (2009a). RDML: structured language and reporting guidelines for real-time quantitative PCR data. Nucleic Acids Res. *37*, 2066–2069.

Lefever, S., Vandesompele, J., Speleman, F., and Pattyn, F. (2009b). RTPrimerDB: the portal for real-time PCR primers and probes. Nucleic Acids Res. *37*, D942–5.

Lewis, F., and Maughan, N.J. (2004). Extraction of Total RNA from Formalin-Fixed Paraffin-Embedded Tissue. In A-Z of Quantitative PCR, Bustin, S.A., ed (IUL, La Jolla, CA), pp. 591–603.

Lewis, F., Maughan, N.J., Smith, V., Hillan, K., and Quirke, P. (2001). Unlocking the archive – gene expression in paraffin-embedded tissue. J. Pathol. *195*, 66–71.

Lopez, M.M., Llop, P., Olmos, A., Marco-Noales, E., Cambra, M., and Bertolini, E. (2009). Are molecular tools solving the challenges posed by detection of plant pathogenic bacteria and viruses? Curr. Issues Mol. Biol. *11*, 13–46.

Lu, X., Holloway, B., Dare, R.K., Kuypers, J., Yagi, S., Williams, J.V., Hall, C.B., and Erdman, D.D. (2008). Real-time reverse transcription-PCR assay for comprehensive detection of human rhinoviruses. J. Clin. Microbiol. *46*, 533–539.

Marcus, J.S., Anderson, W.F., and Quake, S.R. (2006). Parallel picoliter rt-PCR assays using microfluidics. Anal. Chem. *78*, 956–958.

Marshall, R., Chernesky, M., Jang, D., Hook, E.W., Cartwright, C.P., Howell-Adams, B., Ho, S., Welk, J., Lai-Zhang, J., Brashear, J., *et al.* (2007). Characteristics of the m2000 automated sample preparation and multiplex real-time PCR system for detection of Chlamydia trachomatis and Neisseria gonorrhoeae. J. Clin. Microbiol. *45*, 747–751.

Marullo, M., Zuccato, C., Mariotti, C., Lahiri, N., Tabrizi, S.J., Di Donato, S., and Cattaneo, E. (2010). Expressed Alu repeats as a novel, reliable tool for normalization of real-time quantitative RT-PCR data. Genome Biol. *11*, R9.

Mestdagh, P., Van Vlierberghe, P., De Weer, A., Muth, D., Westermann, F., Speleman, F., and Vandesompele, J. (2009). A novel and universal method for microRNA RT-qPCR data normalization. Genome Biol. *10*, R64.

Morrison, T., Hurley, J., Garcia, J., Yoder, K., Katz, A., Roberts, D., Cho, J., Kanigan, T., Ilyin, S.E., Horowitz, D., *et al.* (2006). Nanoliter high throughput quantitative PCR. Nucleic Acids Res. *34*, e123.

Morrison, T.B., Weis, J.J., and Wittwer, C.T. (1998). Quantification of low-copy transcripts by continuous SYBR Green I monitoring during amplification. Biotechniques *24*, 954–962.

Neuzil, P., Zhang, C., Pipper, J., Oh, S., and Zhuo, L. (2006). Ultra fast miniaturized real-time PCR: 40 cycles in less than six minutes. Nucleic Acids Res. *34*, e77.

Nolan, T., Hands, R.E., and Bustin, S.A. (2006a). Quantification of mRNA using real-time RT-PCR. Nature Protocols *1*, 1559–1582.

Nolan, T., Hands, R.E., Ogunkolade, B.W., and Bustin, S.A. (2006b). SPUD: a qPCR assay for the detection of inhibitors in nucleic acid preparations. Anal. Biochem. *351*, 308–310.

Nordgard, O., Kvaloy, J.T., Farmen, R.K., and Heikkila, R. (2006). Error propagation in relative real-time reverse transcription polymerase chain reaction quantification models: the balance between accuracy and precision. Anal. Biochem. *356*, 182–193.

O'Connor, W., and Runquist, E.A. (2008). Error minimization algorithm for comparative quantitative PCR analysis: Q-Anal. Anal. Biochem. *378*, 96–98.

Pabinger, S., Thallinger, G.G., Snajder, R., Eichhorn, H., Rader, R., and Trajanoski, Z. (2009). QPCR: Application for real-time PCR data management and analysis. BMC Bioinformatics *10*, 268.

Pfaffl, M., Schwarz, F., and Sauerwein, H. (1998). Quantification of insulin-like growth factor-1 (IGF-1) mRNA: modulation of growth intensity by feeding results in inter- and intra-tissue-specific differences of IGF-1 mRNA expression in steers. *106*, 514–521.

Pierce, K.E., and Wangh, L.J. (2011). LATE-PCR and allied technologies: real-time detection strategies for rapid, reliable diagnosis from single cells. Meth. Mol. Biol. *688*, 47–66.

Pignot, G., Bieche, I., Vacher, S., Guet, C., Vieillefond, A., Debre, B., Lidereau, R., and Amsellem-Ouazana, D. (2008). Large-scale Real-time Reverse Transcription-PCR approach of angiogenic pathways in human transitional cell carcinoma of the bladder: identification of VEGFA as a major independent prognostic marker. Eur. Urol. *56*, 678–688.

Platts, A.E., Johnson, G.D., Linnemann, A.K., and Krawetz, S.A. (2008). Real-time PCR quantification using a variable reaction efficiency model. Anal. Biochem. *380*, 315–322.

Reuter, T., Gilroyed, B.H., Alexander, T.W., Mitchell, G., Balachandran, A., Czub, S., and McAllister, T.A. (2009). Prion protein detection via direct immuno-quantitative real-time PCR. J. Microbiol. Methods *78*, 307–311.

Ririe, K.M., Rasmussen, R.P., and Wittwer, C.T. (1997). Product differentiation by analysis of DNA melting curves during the polymerase chain reaction. Anal. Biochem. *245*, 154–160.

Rutledge, R.G., and Stewart, D. (2008). A kinetic-based sigmoidal model for the polymerase chain reaction and its application to high-capacity absolute quantitative real-time PCR. BMC Biotechnol. *8*, 47.

Saiki, R.K., Scharf, S., Faloona, F., Mullis, K.B., Horn, G.T., Erlich, H.A., and Arnheim, N. (1985). Enzymatic amplification of beta-globin genomic sequences and restriction site analysis for diagnosis of sickle cell anemia. Science *230*, 1350–1354.

Sanchez, J.A., Pierce, K.E., Rice, J.E., and Wangh, L.J. (2004). Linear-after-the-exponential (LATE)-PCR: an advanced method of asymmetric PCR and its uses in quantitative real-time analysis. Proc. Natl. Acad. Sci. U.S.A. *101*, 1933–1938.

Sansone, S.A., Fan, T., Goodacre, R., Griffin, J.L., Hardy, N.W., Kaddurah-Daouk, R., Kristal, B.S., Lindon, J., Mendes, P., Morrison, N., *et al.* (2007). The metabolomics standards initiative. Nature Biotechnol. *25*, 846–848.

Schmittgen, T.D., Lee, E.J., Jiang, J., Sarkar, A., Yang, L., Elton, T.S., and Chen, C. (2008). Real-time PCR quantification of precursor and mature microRNA. Methods *44*, 31–38.

Smith, C.J., and Osborn, A.M. (2009). Advantages and limitations of quantitative PCR (Q-PCR)-based approaches in microbial ecology. FEMS Microbiol. Ecol. *67*, 6–20.

Smith, M.V., Miller, C.R., Kohn, M., Walker, N.J., and Portier, C.J. (2007). Absolute estimation of initial concentrations of amplicon in a real-time RT-PCR process. BMC Bioinformatics *8*, 409.

Spiess, A.N., Feig, C., and Ritz, C. (2008). Highly accurate sigmoidal fitting of real-time PCR data by introducing a parameter for asymmetry. BMC Bioinformatics *9*, 221.

Stahlberg, A., Elbing, K., Andrade-Garda, J.M., Sjogreen, B., Forootan, A., and Kubista, M. (2008). Multiway real-time PCR gene expression profiling in yeast *Saccharomyces cerevisiae* reveals altered transcriptional response of ADH-genes to glucose stimuli. BMC Genomics *9*, 170.

Stahlberg, A., Hakansson, J., Xian, X., Semb, H., and Kubista, M. (2004a). Properties of the reverse transcription reaction in mRNA quantification. Clin. Chem. *50*, 509–515.

Stahlberg, A., Kubista, M., and Pfaffl, M. (2004b). Comparison of reverse transcriptases in gene expression analysis. Clin. Chem. *50*, 1678–1680.

Taylor, C.F., Field, D., Sansone, S.A., Aerts, J., Apweiler, R., Ashburner, M., Ball, C.A., Binz, P.A., Bogue, M., Booth, T., *et al.* (2008). Promoting coherent minimum reporting guidelines for biological and biomedical investigations: the MIBBI project. Nat. Biotechnol. *26*, 889–896.

Taylor, C.F., Paton, N.W., Lilley, K.S., Binz, P.A., Julian, R.K., Jr., Jones, A.R., Zhu, W., Apweiler, R., Aebersold, R., Deutsch, E.W., *et al.* (2007). The minimum information about a proteomics experiment (MIAPE). Nat. Biotechnol. *25*, 887–893.

Thelwell, N., Millington, S., Solinas, A., Booth, J., and Brown, T. (2000). Mode of action and application of Scorpion primers to mutation detection. Nucleic Acids Res. *28*, 3752–3761.

Timcheva, I.I., Maximova, V.A., Deligeorgiev, T.G., Gadjev, N.I., Sabnis, R.W., and Ivanov, I.G. (1997). Fluorescence spectral characteristics of novel asymmetric monomethine cyanine dyes in nucleic acid solutions. FEBS Lett. *405*, 141–144.

Tyagi, S., and Kramer, F.R. (1996). Molecular beacons: probes that fluoresce upon hybridization. Nat. Biotechnol. *14*, 303–308.

Uhlmann, V., Martin, C.M., Sheils, O., Pilkington, L., Silva, I., Killalea, A., Murch, S.B., Walker-Smith, J., Thomson, M., Wakefield, A.J., *et al.* (2002). Potential viral pathogenic mechanism for new variant inflammatory bowel disease. Mol. Pathol. *55*, 84–90.

Vandesompele, J., De Preter, K., Pattyn, F., Poppe, B., Van Roy, N., De Paepe, A., and Speleman, F. (2002). Accurate normalization of real-time quantitative RT-PCR data by geometric averaging of multiple internal control genes. Genome Biol. *3*, 0034.1–0034.11.

Verderio, P., Pizzamiglio, S., Gallo, F., and Ramsden, S.C. (2008). FCI: an R-based algorithm for evaluating uncertainty of absolute real-time PCR quantification. BMC Bioinformatics 9, 13.

Wagner, B.K., and Arany, Z. (2009). High-throughput real-time PCR for detection of gene-expression levels. Meth. Mol. Biol. *486*, 167–175.

Walker, N.J. (2002). A technique whose time has come. Science *296*, 557–559.

Westh, H., Lisby, G., Breysse, F., Boddinghaus, B., Chomarat, M., Gant, V., Goglio, A., Raglio, A., Schuster, H., Stuber, F., *et al.* (2009). Multiplex real-time PCR and blood culture for identification of bloodstream pathogens in patients with suspected sepsis. Clin. Microbiol. Infect. *15*, 544–551.

Whitcombe, D., Theaker, J., Guy, S.P., Brown, T., and Little, S. (1999). Detection of PCR products using self-probing amplicons and fluorescence. Nat. Biotechnol. *17*, 804–807.

Wittwer, C.T., Herrmann, M.G., Moss, A.A., and Rasmussen, R.P. (1997). Continuous fluorescence monitoring of rapid cycle DNA amplification. Biotechniques 22, 130–138.

Zhang, C., and Xing, D. (2007). Miniaturized PCR chips for nucleic acid amplification and analysis: latest advances and future trends. Nucleic Acids Res. 35, 4223–4237.

Zhang, X.X., Zhang, T., and Fang, H.H. (2009). Antibiotic resistance genes in water environment. Appl. Microbiol. Biotechnol. *82*, 397–414.

Zuker, M. (2003). Mfold web server for nucleic acid folding and hybridization prediction. Nucleic Acids Res. *31*, 3406–3415.

Instrumentation and Fluorescent Chemistries Used in Quantitative Polymerase Chain Reaction

2

Mathilde H. Josefsen, Charlotta Löfström, Trine Hansen, Eyjólfur Reynisson and Jeffrey Hoorfar

Abstract

The polymerase chain reaction has revolutionized the world of scientific research and its broad application has caused a tremendous development of versatile PCR instruments and chemistries to fit its purpose. This chapter provides the reader with a general introduction to the basics of real-time PCR instrumentation, including the thermal and optical systems and the software. Performance parameters such as temperature uniformity, accuracy and ramp speed as well as reaction format, optical systems, calibration of dyes, software and comparison between different real-time PCR platforms will be discussed from a user perspective leading to an instrument selection guide. Differences between fluorescent DNA binding dyes and target-specific fluorescently labelled primers or probes for detection of amplicon accumulation will be discussed, along with the properties and applications of the most frequently applied chemistries. The fluorophores and quenchers used for primer and probe labelling and their compatibility will be presented, and finally the future challenges and trends within the field of qPCR instrumentation will be discussed.

Introduction

In 1983, Dr Kary Mullis conceptualized PCR, a technique that would come to revolutionize the scientific research arena of the twentieth century (Lo and Chan, 2006). Also, in the field of applied microbiology, the sensitivity, specificity and rapidness of PCR has prompted its worldwide implementation and routine application. Before the discovery of the thermo-stable DNA polymerase and the development of thermal cycling instruments, running a PCR entailed moving the sample between water baths of different temperatures and supplying fresh DNA polymerase after each denaturation step. Since then, much effort has been devoted to developing the field of PCR and its application as well as developing and improving thermal cycling instruments. The first generation of thermal cyclers was only capable of DNA amplification, requiring subsequent detection of amplicons by gel electrophoresis. The major limitations of this end-point detection approach included lack of quantification, poor sensitivity and post PCR processing, which brought about the risk of carryover contamination. The development of the real-time PCR (Higuchi *et al.*, 1992) and the accompanying instrumentation (1996) presented the ability to monitor the accumulation of amplicons in real-time. Detection in real-time has the advantage of being more sensitive and specific, offering the possibility of quantification, and since the amplicons are not processed post PCR, the risk of carryover contamination is reduced. This chapter will provide the reader with a thorough description of the real-time PCR instrumentation, focusing on suitable applications and including state-of-the-art microchip-based PCR systems.

Independent of the level of prior experience with real-time PCR, selecting a thermal cycling instrument can be a difficult challenge. In the abundance of manufacturers and instrument models with different technical specifications, it can be difficult to identify a thermal cycler that meets the specific requirements of the laboratory. In the instrumentation section of this chapter the capacity, technical specifications and suitable application of the different thermal cyclers will be discussed, providing the reader with a guide to instrument selection.

The accumulation of amplicons can be monitored in real time either by incorporating fluorescent DNA binding dyes or target-specific fluorescently labelled primers or probes. The different approaches have their justification depending on the PCR assay, and advantages and drawbacks will be discussed in the following sections. A variety of different DNA binding dyes, primer chemistries and probe chemistries have been developed for real-time PCR, sharing the common property of producing a change in emission of fluorescence upon amplicon accumulation. The properties and applications of the alternative primer and probe chemistries will be discussed in this chapter.

Real-time PCR instrumentation

The number of commercially available real-time PCR instruments has increased remarkably since the introduction of the first instruments (ABI Prism 7700 from Applied Biosystems and the LightCycler from Roche) (Wittwer *et al.*, 1997a). The machines now cover the whole range from very simple one-channel instruments to sophisticated high-throughput systems capable of multiplexing up to six channels, as well as portable devices enabling point-of-care diagnostics (for details on available instruments see Tables 2.1–2.3). In the following sections, the most commonly used platforms for real-time PCR available on the market today will be introduced followed by a guide on how to select an instrument suitable for the needs of a particular laboratory. Finally, the future trends of the instrument development will be discussed, including some novel lab-on-a-chip technologies and devices for digital PCR.

Basics about real-time PCR instruments

Real-time PCR instrument systems consist of three basic parts (Javorschi-Miller and Delgado Orlic, 2011):

1 a thermal system to perform temperature cycling;
2 an optical system to emit light necessary for activation of the fluorophore(s) combined with a system to capture the generated fluorescence;
3 software to control the instrument operation, and collect and analyse the data generated.

For each of these points, there are several technical solutions available.

Thermal system

An important part of the thermocycler is the thermal system that automates the temperature cycling required for PCR. The majority of real-time PCR platforms employ blocks that rely on the Peltier effect for heat transfer. Alternative ways to transfer heat include the use of heated air (Roche's LightCycler 2.0), a resistive heater with air cooling and centrifugation (Qiagen's Rotorgene Q), a block filled with thermal conductive fluid agitated by the use of electromagnetic paddles (Illumina's Eco) or the use of direct heating in combination with an electrically conducting polymer (InSyte and SynChron from Biogene). See Table 2.1 for more platforms. Each of these techniques has its own properties and is suitable for different purposes. When evaluating thermal systems, factors such as uniformity, speed and reaction format can be important to take into consideration.

Temperature uniformity and accuracy

The uniformity of heat transfer and temperature accuracy has been identified as a crucial factor to obtain reliable PCR results, and differences in temperature performance between brands and types of cyclers have been demonstrated to influence PCR performance (Saunders *et al.*, 2001; Yang *et al.*, 2005; Kim *et al.*, 2008). For the highest accuracy, the temperature must remain uniform across the entire heat block, ensuring that all samples are processed equally. For demanding applications such as high-resolution melting (HRM) analysis, the instrument must be able to distinguish a temperature difference of 0.2°C (Wittwer *et al.*, 2003). Most instruments have thermal uniformity between ±0.2°C and ±0.5°C (Table 2.1). However, there are some exceptions;

the Qiagen Rotorgene Q demonstrates a uniformity of ±0.01°C (using air cooling) and the Illumina Eco which applies a block-based system with uniformity of ±0.1°C.

Ramp speed

Another aspect of the thermal system performance is speed, which is dependent on the ramping rate of the system. The ramping rates vary significantly depending on the type of heat block or system used. Conventional Peltier-based block cyclers are relatively slow, due to the thermal inertia in the solid block itself (Table 2.1). Furthermore, the plastic coating of the reaction vessels forms a layer of insulation that further slows down the heat transfer. Typically, these kinds of blocks have ramping rates of 1.5–2.5°C/s for heating, generally with cooling rates being slightly slower. Improvements leading to sample ramp rates of 2.5–8°C/s have been made using several different approaches: (i) silver and gold plated blocks (e.g. Illumina Eco, with a thermal conductive fluid in its silver hollow block), (ii) use of fast DNA polymerases and other reagents (e.g. ABI 7500 FAST using FAST reagents) and (iii) use of an heat-equalizing layer between the heating and cooling units (e.g. ThermaBase in Roche's LightCycler 480 and 1536). Alternative technologies to blocks are used by a number of platforms, generating ramp speeds as high as 40°C/s. These techniques include, for example: (i) a stationary turbulent air-heated glass capillary format (LightCycler 2.0), (ii) a centrifugal air-heated plastic tube format (Rotorgene), (iii) a high-thermal-conductivity ceramic heating plate plastic tube format (SmartCycler), and (iv) silicon heating elements (In-Check from STMicroelectronocs) (Table 2.1). To increase the uniformity between different parts of the block, a few platforms use multiple independent temperature blocks (e.g. StepOne Plus from Applied Biosystems and BioRad's CFX384 and CFX96) or even independent temperature control of each well (InSyte and SynChron from biogene).

Reaction format

There are a number of different reaction formats used for the different platforms. Several of the available instruments use a standard 96-well block format with 0.2-ml wells. Some of them also offer the possibility to use 384- and/or 1536-well blocks within the same system. In the 96-well blocks, it is also possible to use individual 0.2-ml tubes or racks of tubes if a lower throughput is needed. These blocks have the possibility to use tubes and racks available from many suppliers, which makes it cheaper compared with cyclers where instrument-specific vessels are used (e.g. LightCycler glass capillaries, SmartCycler and Rotorgene specific plastics). Furthermore, platforms with up to 48-wells (StepOne, Eco and MiniOpticon) have been introduced, offering a convenient solution between handling single tubes and using 96-well plates. Some of these allows for rapid cycling, thus making it possible to perform multiple runs in a short period of time. For laboratories that need a high-throughput, there are a number of platforms based on 384- and 1536-well blocks, as well as array-based formats available. At the other end of the scale, robust portable systems exist with a limited sample capacity (1–32 samples), which can perform in-field analysis within the hour (Table 2.3).

Optical systems

The optical system of the real-time PCR machines consists of two main parts: a light source for excitation and a detector to monitor the emitted fluorescence. For both of these, there are a range of options available.

Light source for excitation

The excitation light source can be classified into broad or narrow spectrum. In broad spectrum, light source filters can be used to narrow the light down to a suitable wavelength range for a particular fluorophore. This gives a wider choice of available fluorophores, but the sensitivity of the system might be reduced because the filters limit the light that can pass through the system. Another disadvantage is that the bulbs decay over time and need to be replaced. Examples of platforms that use this approach are the 7500 from Applied Biosystems and ExiCycler from Bioneer. Some platforms employ narrow spectrum light sources including laser (e.g. 7900HT from Applied Biosystems and InSyte from Biogene) and light-emitting diodes (LEDs) (e.g. StepOne from Applied Biosystems and MiniOpticon from Bio-Rad) (Table 2.2).

Table 2.1 Thermal system specifications of selected commercially available real-time PCR systems

Model	Company	Thermal control	Average heating ramp temperature (°C/s)	Accuracy (°C)	Uniformity (°C)	Temperature range (°C)
Mx3000p	Agilent	Peltier	2.5a	±0.25	±0.25	25–99
Mx3005p	Agilent	Peltier	2.5a	NA	±0.25	25–99
qTOWER	Analytik Jena	NA	12a	NA	NA	NA
7300	Applied Biosystems	Peltier	1.1	±0.25	±0.5	4–100
7500	Applied Biosystems	Peltier	2.5a	±0.25	±0.5	4–100
7500 Fast	Applied Biosystems	Peltier	3.5 (fast mode), 1.6 (standard mode)	±0.25	±0.5	4–100
7900HT Fast	Applied Biosystems	Peltier	3.0 (fast mode), 1.6 (standard mode)	±0.25	±0.5	4–100
OpenArray	Applied Biosystems	Flat Block	4	NA	≤0.5	35- 99.9
StepOne	Applied Biosystems	Peltier	2.2 (fast mode), 1.6 (standard mode)	±0.25	±0.5	4–100
StepOnePlus	Applied Biosystems	Peltier	2.2 (fast mode), 1.6 (standard mode)	±0.25	±0.5	4–100
ViiA 7	Applied Biosystems	Peltier	NA	NA	±0.5	NA
InSyte	Biogene	Integrated thermal control	10–20a	NA	NA	Ambient-100
SynChron	Biogene	Integrated thermal control	10–20a	±0.3	NA	Ambient-100
TOPtical	Biometra	Silver block	5	±0.1	±0.15 – 0.50	3–99

Exicycler 96	Bioneer	Peltier	2.5a	±0.3	±0.3	4.0–99.9
CFX384	Bio-Rad	Peltier	2	±0.2	±0.4	0–100
CFX96	Bio-Rad	Peltier	3.3	±0.2	±0.4	0–100
MiniOpticon	Bio-Rad	Peltier	2.5a	±0.2	±0.4	35–99
GeneXpert	Cepheid	Solid state heater	10a	±1.0	NA	50–95
SmartCycler II	Cepheid	Solid state heater, forced air cooling	10a	±0.5	NA	NA
BioMark	Fluidigm	Peltier	>2	NA	NA	4–99
LightScanner32	Idaho Technology	Heated air	NA	NA	NA	40–99
R.A.P.I.D	Idaho Technology	Heated air	20*a*	NA	NA	ambient-95
RAZOR EX	Idaho Technology	Heat plate	NA	±1.0	0.5–1.0	56–94
Eco	Illumina	Hollow silver block	5.5	±0.1	±0.1	35–100
Liat Analyser	IQuum	Heated segments	NA	NA	NA	NA
Rotorgene Q	QIAGEN	Centrifugal rotor with heated air	>15	±0.25	±0.01	ambient-99
LightCycler 1536	Roche	Peltier with Therma-Base	4.8	NA	NA	20–95
LightCycler 2.0	Roche	Heated air	~2.5	±0.3	±0.4	40–98
LightCycler 480	Roche	Peltier with Therma-Base	96-well: 4.4; 384-well: 4.8	NA	NA	20–95
In-Check	STMicroelectronics	Silicon heating elements	13	±0.2	NA	NA
SmartChip	Wafergen	NA	1.5	NA	±0.5	NA

NA, information not available.
[a]Maximum heating rate.

Table 2.2 Specifications of optical systems of selected commercially available real-time PCR systems

Model	Light source	Excitation spectrum	Detector	Channels (emission filters, nm)	Dye calibration
Mx3000p	QTH lamp	350–750 nm	Single PMT	4 (350–700)	Any dye compatible with filters
Mx3005p	QTH lamp	350–750 nm	Single PMT	4 (350–700)	Any dye compatible with filters
qTOWER	NA	5	NA	5 (520, 545, 580, 605, 670, +4 FRET)	Any dye compatible with filters
7300	TH lamp	1	CCD camera	4	FAM, SYBR Green I, VIC, JOE, NED, TAMRA, ROX
7500	TH lamp	5	CCD camera	5	FAM, SYBR Green I, VIC, JOE, NED, TAMRA, ROX, Texas Red, Cy3, Cy6
7500 Fast	TH lamp	5	CCD camera	5	FAM, SYBR Green I, VIC, JOE, NED, TAMRA, ROX, Texas Red, Cy3, Cy5
7900HT Fast	Argon-iron laser	488 nm	CCD camera	Continuous 500–660	FAM, SYBR Green I, VIC, JOE, NED, TAMRA, ROX
OpenArray	3 LEDs	3	CCD camera	4 (510, 535, 560, 620)	FAM, VIC, SYBR Green I, ROX
StepOne	LED	3	Photodiodes	3	FAM, SYBR Green I, VIC, JOE, ROX
StepOnePlus	LED	4	Photodiodes	4	FAM, SYBR Green I, VIC, JOE, NED, TAMRA, ROX
ViiA 7	NA	6	NA	6 (500–720)	Any dye compatible with filters
InSyte	Solid state laser	473 nm	PMT	32 (510–710, no filters)	Any dye compatible with wavelength range
SynChron	NA	NA	NA	NA	NA
TOPtical	3 LEDs	3	CPM	6 (520, 545, 580, 605, 670, 705, +4 FRET)	Any dye compatible with filters
Exicycler 96	Short arc lamp	5	CCD camera	5 (520, 550, 580, 610, 670)	Any dye compatible with filters
CFX384	5 LEDs	4	Photodiodes	4 (515–530, 560–580, 610–650, 675–690)	Any dye compatible with filters

CFX96	6 LEDs	5	Photodiodes	5 (510–530, 560–580, 610–650, 675–690, 705–730)	Any dye compatible with filters
MiniOpticon	LED	470–505 nm	Photodiodes	2 (523–543, 540–700)	FAM, SYBR Green I, TET, VIC
GeneXpert	LED	6	Photodiodes	6 (420–480, 510–535, 565–590, 606–650, 665–685, >700)	FAM, Alexa 532, Texas Red, Alexa 647
SmartCycler	LEDs	4	NA	4 (510–527, 565–590, 606–650, 670–750)	FAM, Cy3, TET, Texas Red, Cy5, Alexa 532, Alexa 647
BioMark	Xenon arc lamp	5	CCD camera	5 (525, 570, 645, 2 empty)	Any dye compatible with filters
LightScanner32	LED	470 nm	Fluorimeter	3 (530, 640, 705)	SYBR Green, hydrolysis probes, HybProbes
R.A.P.I.D	LED	470 nm	NA	3 (530, 640, 705)	NA
RAZOR EX	LED	470 ± 20 nm	NA	2 (530 ± 10, 690 ± 60)	TaqMan, Hybprobes
Eco	Dual LED	2	CCD camera	4 (505–545, 562–596, 604–644, 665–705)	Any dye compatible with filters
Liat Analyser	NA	4	Photometer	4	TaqMan, Molecular beacons
Rotorgene Q	LEDs	2–6	PMT	2–6 (460, 510, 557, 610, 660, 712)	Any dye compatible with filters
LightCycler 1536	Xenon lamp	2	CCD camera	2 (510 or 580)	Any dye compatible with filters
LightCycler 2.0	LED	470 ± 10 nm	Photodiodes	6 (530, 560, 610, 640, 670, 710)	Hybridization probes, hydrolysis probes, Molecular beacons, SYBR Green I, Simple probes, UPL
LightCycler 480	Xenon lamp	5	CCD camera	6 (488, 510, 580, 610, 640, 660)	Any dye compatible with filters
In-Check	NA	628 ± 40 nm	NA	692 ± 40	Any dye compatible with filters
SmartChip	QTH lamp	3	CCD camera	3 (550, 605, 625, 2 empty)	Any dye compatible with filters

QTH, quartz tungsten halogen; TH, tungsten halogen; CCD, charge-coupled device; PMT, photomultiplier tube; CPM, channel photomultiplier; LED, light-emitting diode; UPL, universal probelibrary; NA, information not available.

Table 2.3 Additional information about selected commercially available real-time PCR systems

Model	Format	Reaction volumes (μl)	Dimensions (cm)	Weight (kg)
Mx3000p	96-well plate/0.2 ml tubes	25 (optimal)	33 × 46 × 43	20
Mx3005p	96-well plate/0.2 ml tubes	25 (optimal)	33 × 46 × 43	20
qTOWER	96-well plate	5–20	NA	NA
7300	96-well plate/0.2 ml tubes	20–100	34 × 49 × 45	29
7500	96-well plate/0.2 ml tubes	20–100	34 × 49 × 45	34
7500 Fast	96-well (fast)	5–30	34 × 49 × 45	34
7900HT Fast	96-well (fast or standard), 384-wells, low density array	96-well: standard 25–100, fast 10–30, low-density array: 5–20	72 × 64 × 84	114
OpenArray	3 OpenArray Plates	33 nl	63 × 77 × 73	80
StepOne	48-well	10–30	25 × 51 × 49	24
StepOnePlus	96-well	10–30	25 × 51 × 49	24
ViiA 7	96, 384-well plates, TaqMan array microfluidic cards	96-well: standard 10–100, fast 5–30, 384-well: 5–20, cards: ~1	53 × 64 × 65	61
InSyte	96-well, independent temp. control of wells/8- strips	20–50	44 × 46 × 60	45
SynChron	96-well, independent temp. control of wells/special plate	10–50	37 × 53 × 43	25
TOPtical	96-well plate	5–80	28 × 38 × 43	16
Exicycler 96	96-well plate/0.2 ml tubes	20–50	36 × 54 × 47	30
CFX384	384-well	1–30 (5–20 recommended)	33 × 46 × 36	21
CFX96	96-well	1–50 (10–25 recommended)	33 × 46 × 36	21
MiniOpticon	48-well, 48 tubes, 6 × 8-tube strip	10–100 (20 recommended)	18 × 32 × 33	7
GeneXpert	1 cartridge	Depending on the sample type/ assay	11 × 34 × 34	NA
SmartCycler	16, 32, 48, 64, 80 or 96 tubes (special design)	25	31 × 31 × 25	10
BioMark	Integrated fluidic chip; up to 48 samples	>4 μl input volume, 0.85 nl reaction volume	49 × 24 × 33	NA
LightScanner32	32 capillaries	10	30 × 41 × 41	11
RAPID	32 capillaries	5–20	27 × 36 × 49	23
RAZOR EX	1 pouch (12 samples)	100	25 × 11 × 19	5
Eco	48-well	5–20	35 × 31 × 32	14
Liat Analyser	1 tube	10–200	8 × 17 × 15	2
Rotorgene Q	72 or 96 well rotor discs, tubes, strip tubes	10–30	37 × 28 × 42	14
LightCycler 1536	LC 1536 multiwell plates	0.5–2	57 × 59 × 50	56
LightCycler 2.0	32 capillaries	20, 100	45 × 30 × 40	22
LightCycler 480	LC 480 multiwell plate (96 or 384 wells)	5–20 (384-well), 10–100 (96-well)	57 × 59 × 50	56
In-Check	1 chip	25	50 × 30 × 15	10
SmartChip	SmartChip Nanowell Chip (up to 384 samples)	0.1	70 × 50 × 60	57

NA, information not available.

This approach might limit the number of available fluorophores that can be excited because of the narrow wavelength range of the emitted light. For example, a standard blue-green LED has a peak emission of 470 nm, corresponding to the excitation wavelength of FAM and SYBR Green dyes. Many of the other commonly used dyes cannot be excited at this wavelength (Table 2.4). However, one strategy to increase the number of excitable fluorophores is to use multiple LEDs with different wavelengths (e.g. TOPical from Biometra and CFX384 from Bio-Rad). Today, instruments with up to six different LEDs are available (e.g. CFX96 from Bio-Rad). The main advantage of the LEDs is that replacement is unnecessary and they keep the same intensity throughout the life-time of the diodes.

Detector

The second part of the optical system is the detector where the light emitted from the reporter fluorophores is recorded. Most systems use photo-multiplier tubes (PMTs) (e.g. Rotorgene Q from QIAGEN), photodiodes (e.g. StepOne from Applied Biosystems) or charge-coupled device (CCD) cameras (e.g. most models from Applied Biosystems; see Table 2.2). CCD cameras can record all wells simultaneously, leading to a more robust mechanical design, whereas both photodiodes and PMTs record one point at a time. Some platforms utilize multiple photodiodes or PMTs which reduces the number of moving parts (e.g. CFX384 and CFX96). The number of channels detected is between one and six. It is important to assure that crosstalk, i.e. the fluorescence from one channel being detected in another channel, is minimized. There are several recent studies describing the use of pentaplex (Molenkamp *et al.*, 2007; Kirchner *et al.*, 2010; Basuni *et al.*, 2011) and even hexaplex real-time PCR assays (Bahrdt *et al.*, 2010; Hopkins *et al.*, 2011).

Calibration of specific dyes

Also important to consider is if the platform is optimized for specific dyes. For example, some platforms are delivered pre-calibrated for certain dyes, and thus show optimal performance with these particular dyes, although dyes with similar excitation and emission wavelengths can be used (see Table 2.2). For some platforms, it is necessary to buy and run calibration kits before using the instruments.

Software

Another important part of a real-time PCR platform is the software used to control the instrument and analyse the data. This software should be as user-friendly as possible, and should be able to analyse data generated by the probe chemistry applied (see section on primer and probe chemistries). A critical part of the software is to specify from which channels that data should be recorded from. If the wrong channels and/or wells are selected, data might be lost. Some platforms therefore have the option to record data from all channels and wells simultaneously, making it impossible to lose data. For some specific applications, e.g. HRM, additional software is required that sometimes needs to be purchased at an additional cost.

Comparison of performances between different platforms

Performance factors, such as ramp speed and uniform heat distribution, are important to achieve reproducible results. Therefore, it is recommended that thermocyclers are calibrated on a regular basis as part of the quality assurance programme of the laboratory (Khan, 2011). Calibration has previously been noted as an important factor when assessing the performance of thermocyclers (Saunders *et al.*, 2001).

Differences in temperature performance, between brands and types of cyclers, have been demonstrated to influence PCR performance (Saunders *et al.*, 2001; Yang *et al.*, 2005; Kim *et al.*, 2008). Validation is important to ensure proper and reproducible results from PCR analysis, and also when transferring an assay between different thermocycler models (Schoder *et al.*, 2005). It can also reveal problems caused by poor thermocycler performance (Schoder *et al.*, 2005; Kim *et al.*, 2008).

There are a number of recent studies in the literature comparing the performances of different real-time PCR platforms analysing identical

Table 2.4 Compatibility between common fluorophores and quenchers commercially available (indicated by ✓). Adapted from Osterman (2009)

Fluorophore/ Quencher[a]	Emax	Dabcyl	BHQ-1	BHQ-2	BHQ-3	Cy5Q	Cy7Q	Iowa BlackFQ	Iowa BlackRQ	IRDye QC-1	QSY35	QSY7	QSY21	QXL520	QXL570	QXL610	QXL680	DDQ II
Coumarin	472	✓									✓							
Alexa Fluor® 488-X	517	✓	✓					✓			✓			✓				
Oregon Green448-X	518	✓	✓					✓			✓			✓				
Fluorescein (FAM)	520	✓	✓					✓		✓	✓	✓		✓				
Rhodamine Green-X	531	✓	✓					✓		✓		✓						
TET	536	✓	✓					✓		✓		✓						
Yakima yellow	549	✓	✓			✓		✓	✓	✓		✓						✓
AlexaFluor-532	553		✓			✓		✓	✓	✓		✓						✓
JOE/VIC/HEX	555		✓			✓		✓	✓	✓		✓						✓
Cy3	564		✓	✓		✓		✓	✓	✓		✓			✓			✓
BODIPY 558/568	568		✓	✓		✓		✓	✓	✓		✓	✓		✓			✓
Alexa Fluor-546	571		✓	✓		✓		✓	✓	✓		✓	✓		✓			✓
TAMRA	583		✓	✓				✓	✓	✓		✓	✓		✓			✓
ROX	608			✓				✓	✓	✓		✓	✓		✓	✓		✓
Alexa Fluor-594	616			✓				✓	✓	✓			✓			✓		✓
Texas Red	617			✓	✓			✓	✓	✓			✓			✓		✓
BODIPY 630/650	653			✓	✓		✓		✓	✓			✓					✓
Cy5	668			✓	✓		✓		✓	✓			✓				✓	✓
Alexa Fluor-647	670			✓	✓		✓		✓	✓			✓				✓	✓
IRDye 680	702				✓					✓								✓
IRDye 700DX	700				✓					✓								✓
Cy5.5	706				✓					✓								✓
Alexa Fluor 750	775									✓								
IRDye 800CW	789									✓								
IRDye 800	809									✓								

[a]Dyes and quencher are presented with commercial names.

samples. Although some studies report differences in results between cyclers (Reynisson *et al.*, 2006), most studies show a good agreement in results (Donald *et al.*, 2005; Wolff *et al.*, 2009; Allnutt *et al.*, 2010; Schlesinger *et al.*, 2010; Stephens *et al.*, 2010; Matero *et al.*, 2011). However, an important aspect to bear in mind is that although performances are similar, it is not always a straightforward procedure to transfer a real-time PCR assay from one platform to another. Cycling times and steps need to be adjusted when changing from a system with different ramping speeds, and the performance of the real-time PCR assay might be different using different systems (Bentley *et al.*, 2005; Lu *et al.*, 2010). For example, when using glass capillaries (e.g. LightCycler), additives such as bovine serum albumin are recommended to lower the DNA-binding effect on the glass walls (Teo *et al.*, 2002). The BIOMARK platform has been found incompatible with the commonly used dye SYBRGreen I, which absorbs to the material in the channels in the microfluidics system (Kubista, 2010). The use of fast PCR protocols has been reported to cause problems with lower sensitivity and increased variability (Hilscher *et al.*, 2005) when transferring assays from a 'traditional' PCR protocol.

Recent trends in the real-time PCR platform development

The recent development of novel techniques, including HRM and digital PCR, is also reflected in the instrument development and introduction of commercially available platforms. For example, the platforms available on the market that are able to perform HRM (i.e. with a thermal resolution <0.02°C/s and suitable software), have been steadily increasing during the last couple of years. However, the accuracy of HRM genotyping has been reported to vary between different instruments depending on the data acquisition, data density and thermal control systems (Herrmann *et al.*, 2006, 2007). Also noteworthy is that the number of instruments able to multiplex five or even six channels is increasing, and so do the number of PCR assays taking advantage of this possibility (Bahrdt *et al.*, 2010; Hopkins *et al.*, 2011).

Digital PCR

Digital PCR is a technique to quantify gene target numbers by PCR using limiting dilution (Sykes *et al.*, 1992; Vogelstein and Kinzler, 1999). The sample is diluted so that it only contains a very small number of target molecules and is then distributed into reaction containers that contain either zero or a single template molecule each. When amplified by PCR, the number of target molecules will correspond to the number of positive reactions. The implementation of this technique has until recently been limited due to the need for manual dispensation into 96- or 384-well plates. However, with the introduction of high-throughput instruments such as the OpenArray from Life Technologies, the LC1536 from Roche, the SmartChip from Wafergen and the BIOMARK from Fluidigm, it is now possible to perform digital PCR in an automated manner which has increased the applicability of this method.

In-field and point-of-care analysis

There is a growing demand for robust, portable instruments suitable for in-field or point-of-care analysis, for example analysing bioterror agents or hospital 'point-of-care' diagnostic assays. A number of platforms have been available for some years, including the RAZOR from Idaho Technology and the Liat Analyser from IQuum. The pre-PCR sample preparation has been a potential source of error and risk of contamination with this type of instrument, and therefore this step has been integrated in a number of novel platforms, such as the Enigma ML and XL instruments from Enigma Diagnostics, BD Max from BD Diagnostics, and GeneXpert from Cepheid. The number of samples that it is possible to analyse simultaneously varies between 1 and 48 for the different instruments, many of them delivering results within only 30–60 min.

Lab-on-a-chip (LOC) devices to perform real-time PCR have gone from being promising research devices into a commercially available product (Zhang and Ozdemir, 2009; Lee *et al.*, 2010). These devices have great potential also for point-of-care diagnostics where the main advantage is the speed; cycling times for 40 cycles have been reported as low as 6 min (Neuzil *et al.*, 2006; Lee *et al.*, 2010). New devices are currently being developed and will most probably be introduced on the market within a few years (Lee *et al.*, 2010; Markey *et al.*, 2010; Shen *et al.*, 2010; Sundberg *et al.*, 2010).

Instrument selection

It is not trivial to choose a PCR platform that is suitable for the specific needs of a laboratory. As discussed in the last section, there are many factors that need to be taken into consideration, including (Logan and Edwards, 2009; Javorschi-Miller and Delgado Orlic, 2011):

- throughput (small laboratory use vs. core facility/shared instrument/large project);
- multiplexing features/supported chemistries;
- HRM capabilities;
- sample format (plate vs. tubes/capillaries, sample volume available for analysis);
- instrument size;
- ease of use and robustness of software; and
- cost (including instrument and software, servicing, reagents, consumables).

In connection with the novel development of LOC technologies and miniaturized systems (Zhang and Ozdemir, 2009), an additional factor might be added regarding the need for portability and use in the field and/or at point-of-care. Table 2.5 summarizes the features of the different instruments. From the information given here, the user can make a decision on a suitable platform.

Fluorescent DNA-binding dyes

Real-time PCR was established about 20 years ago when Higuchi and co-workers reported the addition of ethidium bromide directly to a PCR tube and monitored the DNA accumulation externally in real-time by the fluorescent signal generated in the reaction (Higuchi *et al.*, 1992). Since then, real-time PCR technology has developed and other double stranded (ds) DNA dyes with lower toxicity and improved fluorescent properties have been presented.

SYBR Green

Among the fluorescent DNA binding dyes, SYBR Green I is used most widespread for dsDNA quantification in real-time PCR applications (Wittwer *et al.*, 1997b). When the dye binds to the minor groove of the DNA molecule, it acquires morphological alterations with energy transfer that produces fluorescence. Upon accumulation of dsDNA in the PCR process, increased fluorescence signal is generated and detected by the real-time PCR instrument. Post PCR, a dissociation curve is usually generated by a stepwise increase in temperature and continuous fluorescent reading. When the melting temperature of the dsDNA product is reached, the two dissociated strands will produce a decrease in fluorescent signal due to lower affinity of the dye to single stranded DNA (Ririe *et al.*, 1997). The reported drawbacks of SYBR Green I are its inhibitory effect on the PCR reaction at elevated concentrations (Monis *et al.*, 2005), limited dye stability (Monis *et al.*, 2005) and suggested preferential binding with high G + C areas (Giglio *et al.*, 2003).

Alternative DNA binding dyes

New dyes have entered the market in recent years with reported improvements compared with SYBR Green I. These include YO-PRO (Ishiguro *et al.*, 1995), BEBO (Bengtsson *et al.*, 2003), BOXTO (Lind *et al.*, 2006), LCGreen (Wittwer *et al.*, 2003), SYTO9 (Monis *et al.*, 2005) and EvaGreen (Mao *et al.*, 2007), to name the most recognized. Some of these dyes belong to a class called 'new generation saturation dyes' which are frequently used in HRM analysis for genotyping and mutation scanning based on the real-time PCR principle and precise melting temperature analysis. For these applications, HRM instruments are needed along with dyes that fulfil the 'saturation model'. They have the potential of being added in such high concentrations to the reagent mixture that all binding positions of the DNA molecule are occupied. Several studies have been made on comparison of these alternative dyes in real-time PCR. BOXTO was shown to have less PCR inhibitory effect than SYBR Green I, but otherwise performed similarly (Ahmad, 2007), and EvaGreen was stable and

Table 2.5 Instrument selection guide

Model	Automation	Gradient	Digital PCR	In-field robust	Cycling time[a]	HRM	Multiplex
Mx3000p	No	No	No	No	NA	No	4
Mx3005p	No	No	No	No	NA	No	5
qTOWER	No	NA	NA	NA	<25 min	NA	4
7300	No	No	No	No	<2 h	No	4
7500	No	No	No	No	<2 h	No	5
7500 Fast	No	No	No	No	30–37 min (fast), <2 h (standard)	Optional	5
7900HT Fast	Yes	No	No	No	33 min (96-well fast), 52 min (384-well fast), <2 h (others)	No	2
OpenArray	No	NA	Yes	No	<3 h	NA	Yes
StepOne	No	No	No	No	<40 min (fast), <2 h (standard)	Possible	3
StepOnePlus	No	Yes	No	No	<40 min (fast), <2 h (standard)	Possible	4
ViiA 7	Optional	No	No	No	~ 30 min (96/384-well)	Yes	Yes
InSyte	No	Yes	No	NA	NA	NA	Yes
SynChron	No	Yes	No	NA	20 min/45 cycles	NA	NA
TOPtical	No	Optional	No	NA	NA	NA	6
Exicycler 96	No	Yes	No	Optional	2 h	NA	5
CFX384	No	Yes	No	No	NA	Yes	4
CFX96	No	Yes	No	No	<30 min	Yes	5
MiniOpticon	No	Yes	No	NA	NA	No	2
GeneXpert	Yes	Yes	No	Yes	<1 h	NA	5
SmartCycler	No	Yes	No	Yes	NA	Yes	4
BioMark	NA	No	Yes	No	NA	NA	5
LightScanner32	No	NA	No	NA	<30 min	Yes	Yes
RAPID	No	No	No	Yes	<30 min	No	Yes
RAZOR EX	No	No	No	Yes	<30 min	No	No
Eco	No	No	No	NA	<40 min/40 cycles	Yes	4
Liat Analyser	Yes	No	No	Yes	12 min/30 cycles	No	Yes
Rotorgene Q	No	No	No	No	40 min/45 cycles	Yes	2, 5 or 6
LightCycler 1536	No	No	Yes	No	<50 min	Yes	1
LightCycler 2.0	No	NA	No	NA	30 min/35 cycles (20 μl), 60 min/35 cycles (100μl)	Yes	4
LightCycler 480	Yes	No	No	No	<40 min (40 cycles, 384-well)	Yes	6
In-Check	NA	NA	No	NA	<1 h	No	Yes
SmartChip	NA	No	Yes	No	~2 h (40 cycles)	No	NA

NA, information not available.
[a]Cycling time for a standard PCR program, as stated by the manufacturers.

compatible with different PCR instruments (Mao *et al.*, 2007). SYT09 was shown to provide a broader range of dye concentrations without causing PCR inhibition compared with SYBR Green I. More robust and consistently shaped DNA melting curves that were less affected by dye concentration were produced by SYT09. This dye was also more readily used for the detection of multiple amplicons by DNA melting curve analysis without any apparent bias (Monis *et al.*, 2005). Despite general encouragement for using the new dyes for HRM applications (Reed *et al.*, 2007), other scientists have reported that SYBR Green I is just as good for HRM analysis without the need for any modifications to analysis protocols (Price *et al.*, 2007; Pornprasert *et al.*, 2008).

DNA binding dyes versus probe-based detection

In recent years, probe-based real-time PCR applications have received more attention, and numerous reports on new detection systems have been published (see section on primer and probe chemistries). Although the probe-based systems have been shown to be more suitable for DNA quantification than the dsDNA dye systems (Malinen *et al.*, 2003; Ciglenecki *et al.*, 2008), the latter have the advantages of being more cost-effective and the possibility of post-PCR melting curve analysis enables verification of product specificity. Its use is further advised in the first steps of assay development when performance of primers and efficiency is being tested (Cirino *et al.*, 2007). Theoretically, dsDNA dye could provide more sensitivity to an assay compared with a probe system, since the number of fluorescent molecules per amplicon is higher and thus should produce an earlier detectable amplification signal than a corresponding probe system. In our experience, this is also the case in a relatively broad target concentration range, where dsDNA dye assays usually produce lower Ct values than corresponding probe assays. However, at low target concentrations, the risk of unspecific signals by SYBR Green I can hinder a proper linearity close to the detection limit. Combining a DNA binding dye to a TaqMan reaction has been reported where the non-specific detection by the DNA dye was used as a quality measure of the reaction, for estimation on primer-dimer formation or unspecific amplification (Lind *et al.*, 2006). For a multiplex diagnostic assay, however, a mix of probe based and dsDNA dye systems is not recommended (Cirino *et al.*, 2007).

Primer and probe chemistries

Measurement of amplicon accumulation in real-time is, regardless of the choice of chemistry, based on monitoring a change in emitted fluorescence. Besides the fluorescent DNA binding dyes described in the previous section, real-time PCR is commonly performed by applying fluorescently labelled oligonucleotides. Several different chemistries are available, and although they are adaptable to most real-time PCR thermal cyclers, the choice of chemistry should be considered before purchasing a real-time thermal cycler. Two basic categories of fluorescently labelled oligonucleotides exist; primers and probes, and examples of both will be described in the following section.

Primer-based chemistries

Monitoring the accumulation of PCR amplicons by means of a single primer can be performed by labelling the primer with a fluorophore that changes fluorescence properties upon annealing to nucleic acids and subsequent extension, as shown in Fig. 2.1, exemplified by the Light Upon eXtension LUX™ primer (Nazarenko *et al.*, 2002). Other primer chemistries, such as the AmpliFluor primers (Fig. 2.2) are labelled with both a reporter and a quencher dye, relying on a collisional quenching mechanism brought about by the hairpin conformation of the un-annealed primer (Nazarenko *et al.*, 1997).

A major advantage of the primer-mediated detection chemistries is the reduced cost compared with corresponding probes. In contrast to the unspecific DNA binding dyes, the primer chemistries also offer the possibility of multiplexing. However, since fluorescence is changed simply upon annealing and extension, unspecific amplicons and primer-dimers will also give rise to a PCR signal, rendering this approach less specific than probe-based chemistries (Mackay and Landt, 2007).

Light Upon eXtension (LUX™)

The principle of Light Upon eXtension (LUX™) primers is based on internal quenching of the fluorophore in the un-annealed primer by a nearby guanine base. The labelled primer has a hairpin conformation (Fig. 2.1), and only upon annealing

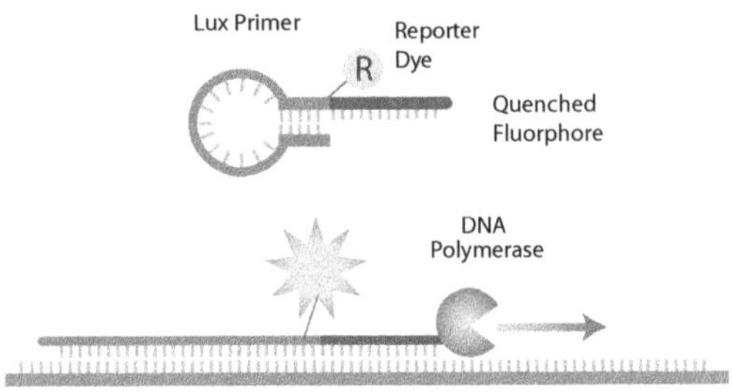

Figure 2.1 Fluorescence mechanism of the Light Upon eXtension (LUX™) primer (modified from Cepheid).

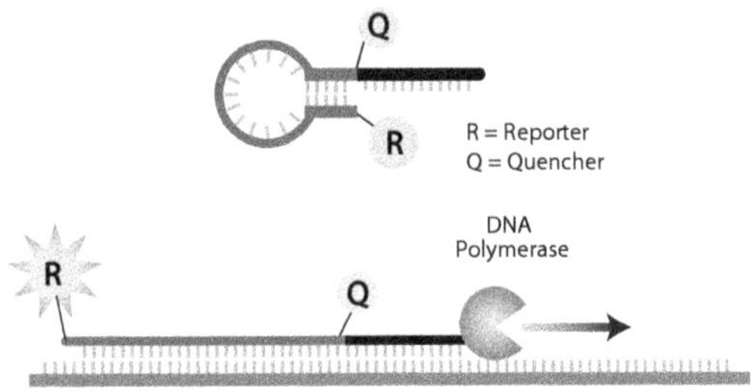

Figure 2.2 Fluorescence mechanism of the AmpliFluor® universal hairpin primer (modified from Cepheid).

and extension of this primer, can an up to 8-fold increase in fluorescence be detected (Nazarenko *et al.*, 2002). Comparison studies that include the LUX™ chemistry have been performed, and this primer-based system is reported to be a cost-effective detection alternative (McCrea *et al.*, 2007; Nordgren *et al.*, 2008; Buh Gasparic *et al.*, 2010; Chui *et al.*, 2010).

Plexor™

The Plexor™ technology distinguishes itself from other real-time PCR detection chemistries by reduced fluorescent signal upon product accumulation. One of the primers is synthesized with a modified nucleic acid coupled to a fluorophore in the 5′ end, and the reaction mixture contains complementary nucleotides that are modified to include a quencher. In subsequent PCR cycles, base pairing between the quencher-modified nucleotide and the fluorescently labelled one, on the newly synthesized DNA strand, will extinguish the fluorescent signal (Frackman *et al.*, 2005). Literature describing the application of the Plexor™ technology is limited (Atallah *et al.*, 2007; Buh Gasparic *et al.*, 2008; Lasalle *et al.*, 2011). For both the LUX™ and Plexor™ primer systems, analysis of dissociation curves can be performed to confirm the specificity.

AmpliFluor®

In the AmpliFluor® chemistry (formerly marketed as Sunrise™), three primers are applied, of which two are target specific and the third is a universal dual-labelled hairpin primer (Fig. 2.2). One of the target-specific primers has a recognition sequence for the hairpin primer incorporated, and becomes a template for this primer during the second cycle of PCR. Upon annealing of the hairpin primer and subsequent elongation, the hairpin stem will be pulled open eliminating the quenching and the fluorescence detection markedly enhanced (Nazarenko *et al.*, 1997). The application of AmpliFluor® primers for genotyping in single nucleotide polymorphism (SNP) and mutation detection studies has been described extensively, as differently labelled forward primers combined with one common reverse primer can identify sequences differing by a single nucleotide (Giancola *et al.*, 2006; Khripin, 2006; Fuhrman *et al.*, 2008).

Self-reporting DNA/PNA

Finally, a primer-based chemistry referred to as self-reporting DNA/peptide nucleic acid (PNA) primers (Fiandaca *et al.*, 2001) employ a tagged fluorophore-labelled target-specific primer, whose signal is silenced by a quencher-bearing PNA oligonucleotide. The PNA oligo is displaced upon primer annealing, and the subsequent increase in fluorescence detected. This system could be cost-effective and suitable for multiplexing, because the PNA oligo could quench the signal of any primer adapted for self-reporting by including the tag sequence and fluorophore labelling.

Probe-based chemistries

Compared with the primer-based detection chemistries, probe-based detection adds an additional level of specificity to the assay. Obtaining a fluorescent signal from the specifically designed probe ensures that the PCR amplicons accumulating are the correct ones, and that the increase in fluorescence is not caused by spurious binding of primers or primer-dimers.

Hydrolysis probes

The most widely used probe-based detection chemistry for real-time PCR remains to be the original hydrolysis probes, also known as 5′ nuclease or simply TaqMan® probes. In this chemistry, a linear, target-specific probe is designed to complement a sequence in between the forward and reverse primer, preferentially in close proximity to one of the primer binding sites. These probes are generally labelled at the 5′ end with a fluorescent reporter dye and at the 3′ end with a quencher. Hydrolysis probes are designed to have melting temperatures approximately 10°C higher than their corresponding primers, causing the probe to anneal to the DNA template prior to the primers. When the primers are subsequently extended and the probe is encountered, part of the probe will be hydrolysed by the 5′–3′ exonuclease activity of the DNA polymerase, separating the reporting dye from the quencher and allowing emission of fluorescence (Fig. 2.3).

Substituting part of the regular nucleotides of the TaqMan® probes with synthetic analogues such as locked nucleic acids (LNA) or peptide nucleic acids (PNA) allows for shorter probe designs. These analogues have a higher affinity for DNA and hence the thermal stability of the formed duplex is increased significantly (Egholm *et al.*, 1992; Kumar *et al.*, 1998). This is also the case for the next generation of TaqMan® probes, the minor groove binding (MGB) probes. In this probe type, a hyperstabilizing molecule attached to one end of the probe folds into the minor groove of the formed duplex, giving rise to increased thermal stability. Advantages of these short probe types are improved sensitivity towards single-base mismatches (specificity), ease of design and an improved signal–to-noise ratio, partly due to lower fluorescence from spurious binding and partly due to the reduced distance between the reporter and the quencher (Kutyavin *et al.*, 2000; Braasch and Corey, 2001). These advantages must be weighed against the additional cost of these more complex probe types; however, they have been reported to perform superior to other probe types and chemistries depending on the specific PCR assay (Costa *et al.*, 2004; Zhao *et al.*, 2005; Reynisson *et al.*, 2006; Josefsen *et al.*, 2009). These short probe types have been successfully applied for genotyping and detection of SNPs (Saito *et al.*, 2003; Petersen *et al.*, 2004; Li *et al.*, 2010). The advantages of the shorter probe types are not limited to the hydrolysis probe chemistries.

AllGlo™ probes are labelled with two identical reporter dyes capable of quenching each other and becoming de-quenched upon probe cleavage. This type of probe contains two signal-generating reporter dye molecules per oligo, therefore a stronger fluorescent signal can be expected. Furthermore, the cost of AllGlo™ probes is less due to the simplified manufacturing process (AlleLogic Biosciences). Use of this more recently developed probe type has not yet been reported widely.

The cycling probe technology (CPT) also belongs to the hydrolysis category, but here the probe nucleotides are substituted with modified RNA molecules. Following annealing of this DNA-RNA-DNA probe, the formed duplex is recognized and cut by RNaseH, thus separating the reporter and the quencher and releasing the fluorescence emission. This probe system does not, in contrast to the other hydrolysis probes, rely on the 5′–3′ exonuclease activity of the DNA polymerase (Duck *et al.*, 1990). The CPT was compared with eight other probe types for qualitative and quantitative applications in GMO detection and shown to have a moderate performance and a high cost, making them inappropriate for routine use (Buh Gasparic *et al.*, 2010).

The LionProbe™, in contrast, exploits the 3′–5′ exonuclease activity of a proofreading polymerase. The probe is labelled at both ends with a reporter and a quencher, respectively, and the 3′-end is designed to have the last few nucleotides mismatching the target sequence. The activity of the proofreading polymerase will cleave the reporter from the probe to correct the mismatch, and the probe now functions as a primer for the subsequent extension (Biotools).

Linear hybridization probes

In contrast to the fluorescence accumulation observed with the hydrolysis probe chemistries,

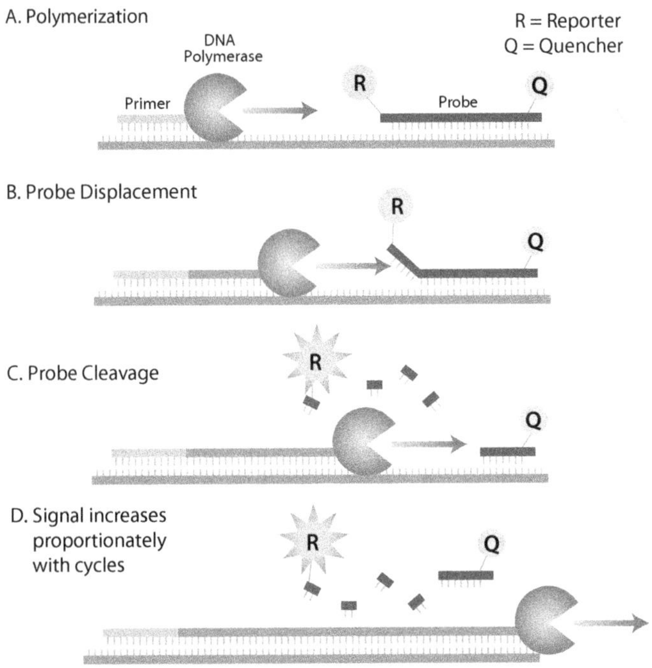

Figure 2.3 Hydrolysis of a TaqMan® probe during amplification (Cepheid).

the signal from hybridization probes is generated in each PCR cycle as the hybridization event occurs. While several probe types fall into the category of linear hybridization probes, the HybProbes® or dual hybridization probes, originally designed for application in the LightCycler instrument, are the most widely used today. In this approach, two single-dye labelled linear probes are designed to anneal adjacently on the DNA template in between the two primer binding sites. In most applications, one probe is labelled at the 3′-end with a donor dye and the other at the 5′-end with an acceptor dye. Upon hybridization of the two probes to the DNA template, the dyes are brought into close proximity and the acceptor dye will be excited by the donor dye, emitting fluorescence of a characteristic wavelength. The 5′-end labelled probe is modified and contains a PCR blocker in the 3′-end to prevent extension by the DNA polymerase. It can be challenging designing not only one, but two probes, if there is a lack of known conserved region. Likewise, the melting temperature of the probes should be identical, and in the range of 65–75°C in order to generate sufficient signal without clamping and thereby inhibiting the amplification process. An advantage of this chemistry is that it offers the possibility of performing subsequent melting curve analysis, allowing for multiplexing with only one dye, and revealing potential nucleotide polymorphisms in the DNA strand sequence under the hybridization probes (Lee *et al.*, 2009).

The displacement probes, or Yin-Yang probes as they are also referred to as, also belong to the class of linear hybridization probes (Li *et al.*, 2002). This is the probe-based counterpart of the self-reporting DNA/PNA primer described by Fiandaca *et al.* (2001). A reporter-labelled probe is silenced by a complementary, but shorter, quencher-bearing DNA strand. Upon PCR product amplification, a preferential sequence for the reporter-labelled probe will be made available for hybridization, eliminating the quenching of fluorescence (Li *et al.*, 2002).

LightUp™ Probes (Svanvik *et al.*, 2000) consists of PNAs linked to the fluorophore thiazole orange. Upon probe hybridization to a complementary sequence, the fluorescence emission will increase substantially. The SimpleProbe™ or HyBeacon™ probes for detection of mutations and genotyping, work in a similar manner to the LightUp™ probes, exhibiting enhanced fluorescence upon hybridization to a specific target (French *et al.*, 2001), allowing for identification of closely related sequences by melting point technology.

Conformational probes

Probably the most well-known types of conformational probes are the Molecular beacons (MB). The MBs are principally designed to form a secondary hairpin structure in which the fluorescence emission of the reporter is quenched due to the proximity of the quencher as shown in Fig. 2.4. The reporter and quencher-bearing stems of this conformational probe consist of G/C-rich complementary DNA strands of four to six nucleotides in length, while the probe sequence is contained in the loop region. Presence of the probe-specific target favours hybridization to this over returning to the hairpin conformation, causing the probe to unfold, and separating the reporter and the quencher, emitting detectable fluorescence (Tyagi and Kramer, 1996).

The MBs offer an exceptional specificity because the stability of the stem structure ensures that unfolding and hybridization only occurs if a perfect matching complementary annealing site is present, making them ideal for SNP analysis. In recent years, many structural variants of the original MBs have arisen such as the wavelength-shifting MB producing brighter fluorescence emission and simplifying and improving multiplex analysis, and the tripartite MB which has additional application possibilities and reduced cost of synthesis. Application of MBs is widely described and the possibilities are numerous (for a review, see Broude, 2002). A limitation of this chemistry is the high cost of synthesis, and it should be noted that MBs are unstable in the presence of DNA polymerases exhibiting exonuclease activity.

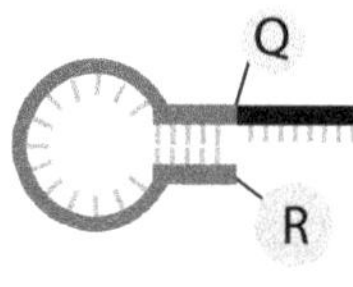

Figure 2.4. The structure of a Molecular beacon probe (modified from Cepheid).

The Scorpion™ probe system comes in two formats; uni- and bimolecular. The unimolecular Scorpion™ has a hairpin loop tail linked to the 5′-end of a primer via a PCR-blocker as shown in Fig. 2.5. This primer anneals to the template DNA, and the DNA polymerase extends it, creating a complementary probe binding site. Following the next denaturation, the probe element will fold up and anneal to this site, rather than entering the hairpin loop conformation again, because it will be more kinetically favourable. This event opens the hairpin loop, eliminating the quenching of the fluorophore and increasing the real-time PCR signal. The bi-molecular Scorpion consists of a fluorescently labelled probe coupled to a primer by a PCR-blocker. It is annealed to a complementary oligo bearing a quencher at the 3′-end. The mechanism of action is similar to that of the unimolecular in that the probe will anneal to the newly synthesized DNA template and will emit fluorescence since it is separated from the quencher. The intra molecular probing mechanism of the Scorpion™ system results in a rapid signal generation contributing to a reduced time of analysis (Whitcombe *et al.*, 1999; DxS Diagnostics). This probe system has also been used successfully for mutation detection and SNP analysis (Burucoa *et al.*, 2008; Sahoo *et al.*, 2011).

The IntraTaq probe presents a hydrolysis probe with the conformation of the Scorpion system relying on the exonuclease activity of the DNA polymerase to generate a fluorescent signal. This intra molecular probing mechanism has been shown to perform superior to equivalent linear TaqMan® probes (Solinas *et al.*, 2002).

As seen in Fig. 2.6, the MGB Eclipse™ probe also belongs to the grouping of conformational probes. An MGB moiety is attached to the 5′-end of the three-dimensional probe together with the quencher molecule and the reporter dye is linked to the 3′-end. As for the linear probes, the MGB Eclipse™ probes can be designed shorter due to the stabilizing action of the MGB molecule with similar advantages as mentioned earlier. In its non-annealed state, the short distance between the reporter and

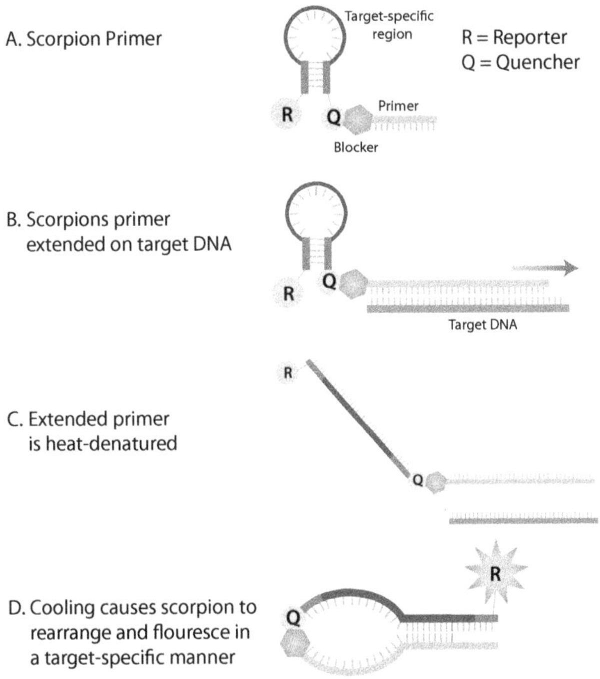

Figure 2.5 The Scorpion™ probe-based amplification process (Cepheid).

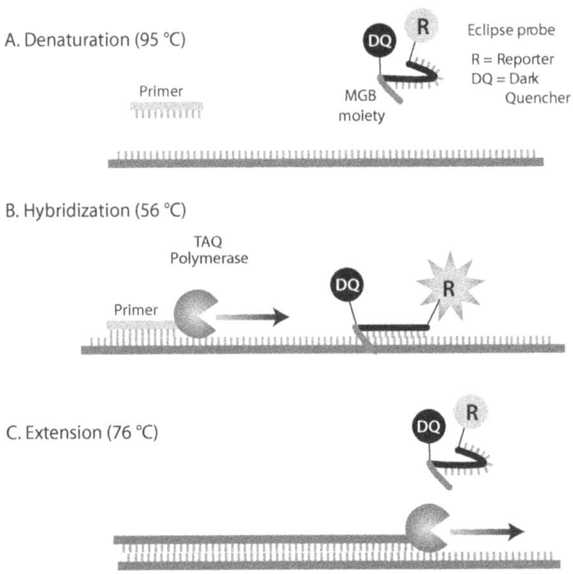

Figure 2.6 The Eclipse™ probe during the amplification process (Cepheid).

quencher will prevent emission of fluorescence, and only upon hybridization to target DNA and unfolding of the probe element will the quenching of the reporter be eliminated (Afonina *et al.*, 2002).

Fluorescent dyes and quenchers

Every probe-based real-time PCR application is dependent on molecular illumination through various types of fluorophores and quenchers

(Heid *et al.*, 1996; Cirino *et al.*, 2007). The principle of fluorescence is based on the ability of a substance to absorb excitation energy, which is usually a light source of a certain wavelength, and re-emit the energy as light at a another wavelength (Morrison, 2010). In real-time PCR, the fluorophore is in proximity to a quencher that absorbs the energy and eliminates illumination by the fluorophore upon excitation. The quenching mode can be accomplished in different manners, but in real-time PCR, it is usually either contact quenching (non-fluorescent ground state complex), which is the case with the Scorpion probe system and MBs, or energy transfer to another molecule (fluorescence resonance energy transfer, or FRET), which is the case in TaqMan probes (Morrison, 2010).

Fluorescent quenchers re-emit the energy as light (at another wave-length than the fluorophore) while other quenchers, so called dark quenchers, re-emit the energy as heat (Fig. 2.7) without affecting the temperature of the solution (May *et al.*, 2005).

Early on in real-time PCR development, the most frequent dyes were FAM quenched with TAMRA but today the dark quenchers are more commonly used (Heid *et al.*, 1996; Yang *et al.*, 2009). The major advantage of non-fluorescent quenchers is a lower fluorescent background, which improves the spectral discrimination and the ease of multiplexing. The possibility for multiplexing is further increased as all detectors available in the instrument can be used for targets of interest, rather than being occupied by light-emitting quenchers (Cirino *et al.*, 2007). Wide varieties of trademark fluorophores are available for real-time PCR but some of the most frequently applied are FAM, HEX, ROX, Cy5, Cy3, Yakima Yellow and Texas Red. Similarly, examples of dark quenchers are BHQ1–3 and Qxl, which are able to quench the entire visible spectrum, Dabcyl (green spectrum), Iowa quenchers (green-yellow-orange-red), IRDye QC-1 (visible to near infrared spectrum) and Deep Dark Quencher. Fluorophore–quencher combinations have to be chosen with respect to their absorbance and emission spectra, meaning that the absorbance spectra of the quencher must overlap with the emission wavelength of the fluorophore to ensure

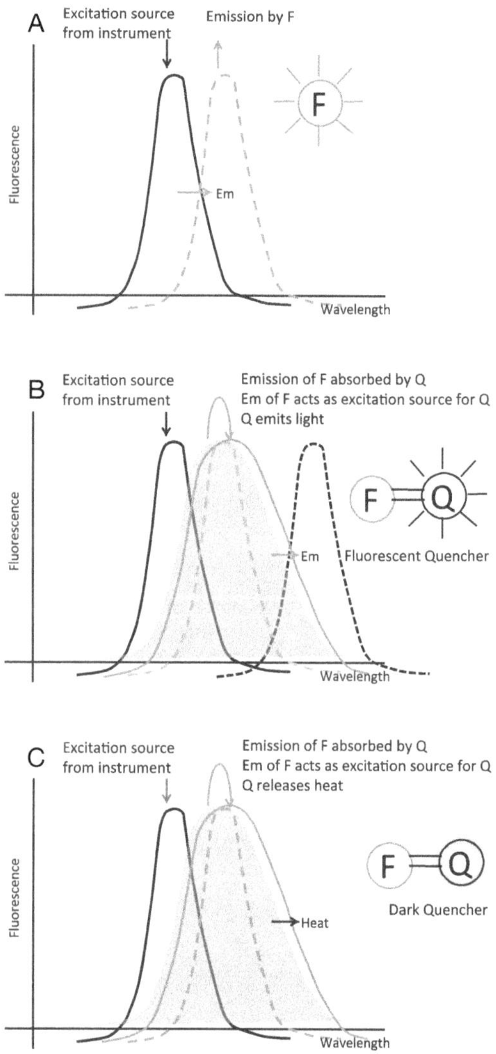

Figure 2.7 Principles of the fluorescent dyes used in a real-time PCR probe system. (A) When a fluorophore (F) is unquenched, the excitation source from the instrument results in a direct emission from the fluorophore. (B) When F is linked to a fluorescent quencher (Q) such as TAMRA, Q absorbs the emitted energy from F and results in a direct emission of Q at another wavelength than F. (C) When F is attached to a dark quencher, Q absorbs the emitted energy from F and releases it as heat.

efficient quenching (Peng *et al.*, 2009) (Table 2.4 and Fig. 2.7). The compatibility of fluorophores and instruments intended to be used must also be taken into consideration since the real-time PCR thermal cyclers are equipped with different detector channels (Table 2.4). The FAM reporter is still usually regarded as the first choice (Marras, 2006).

Multiplex real-time PCR

Multiplex real-time PCR systems offer an attractive opportunity to increase throughput and reduce reagent cost. In that manner, it is essential to carefully design the dye combinations based on the available detectors in the real-time PCR instrument to be used. The need for multiplex systems is also evident in routine diagnostics where it has been proposed to make internal amplification controls mandatory, to prevent false-negative results caused by PCR inhibitory substances (Hoorfar *et al.*, 2003). In other cases, the detection of multiple microorganisms such as different foodborne pathogens or food spoilage bacteria is advantageous (Haakensen *et al.*, 2008). Recently, an experimental strategy named 'multicolour combinatorial probe coding' (MCPC) was introduced which includes systematic labelling of large number of different probes with a limited number of differently coloured fluorophores. The identification of the 15 existing human papillomavirus types in one reaction was demonstrated, which is the maximum target number theoretically detectable in one reaction by MCPC in a four-colour channel instrument (Huang *et al.*, 2011).

Future challenges and trends

Despite the increasing number of primer and probe chemistries and real-time PCR instruments, the current applications are mostly limited to duplex detection and, in rare cases, triplex. This is far from the promised potential of real-time PCR for parallel testing of many targets in the same tube. Most scientists have alleviated this limitation by distributing the same sample in several parallel tubes, each containing different dyes. However, this is not what real-time was supposed to deliver. Instrument producers must develop true multiplex applications to support customers to take the full advantage of real-time PCR.

Scientists and end-users should develop new applications to exploit the enumeration capacity of real-time PCR. Most publications report new tests that merely detect presence or absence of certain genes. More intelligent sample preparation designs are needed to obtain quantitative data that provide more meaningful indication of the extent of the initial contamination or the extent of interventions that are necessary to deal with a situation. For instance, it has been shown to be possible to combine pre-enrichment with real-time PCR and still obtain quantitative data for foodborne pathogens (Josefsen *et al.*, 2004; Krämer *et al.*, 2011).

Interpretation of PCR results, in particular fluorescence readings close to baseline values, is very dependent on the software algorithms that come with the different instruments. The algorithms used to normalize output readings vary, in turn, among software producers. Some producers use sophisticated calculations to normalize PCR results for within sample controls (spontaneous fluorescence release, internal amplification control, etc.), sample-to-sample variations, coefficient of variations, baseline readings, and so on, while other instrument producers simply subtract a fixed value from all of the fluorescence readings to produce threshold cycle values for test samples. Thus, there is an urgent need to agree upon a standard algorithm in order to facilitate data comparison among current and future PCR platforms.

To add to the lack of standardization, the DNA or RNA standard dilutions used among different research groups to produce quantification curves could vary considerably in genome and gene copy numbers, depending on the purity, nucleic acid integrity and method of measurement. International standard organizations should agree upon a uniform guideline for the preparation and measurement of reference DNA dilutions.

There are certainly other issues that can be added to the abovementioned challenges. Nevertheless, new real-time PCR platforms and chemistries will continue to merge, in particular with connection of low-density microarray platforms that can result in true multiplex detection of hundreds of targets. To achieve this goal, scientists involved in real-life diagnostics must collaborate more closely with instrument producers, and instrument producers must welcome new application needs that are put forward by scientists.

Conclusion

The development of applications and instrumentation for real-time PCR presents the routine or research laboratory with a wealth of possibilities. Thorough considerations regarding fluorescent detection method and potential probe chemistry has to be made prior to implementing a real-time PCR assay. Selecting a real-time thermal cycler is not a straight forward task, and factors to consider include; cost (not only related to purchasing the instrument itself, but also service contracts, consumables and reagents), supported chemistries, multiplexing capacities, HRM capabilities, cycling speed, technical support, customer support, ease of use, throughput, format and footprint. In this respect it can be wise to make use of the possibility to 'try before you buy' that many companies provide today. Finally, advances in the area of fluorescent dyes and quenchers, adding to the sensitivity and spectral performance of real-time PCR assays, should be taken into consideration.

Web resources

http://www.gene-quantification.de

Manufacturers' homepages:

Idaho Technology: www.idahotech.com Illumina: www.illumina.com Roche: www.roche.com Bio-Rad: www.biorad.com BD: www.bd.com Cepheid: www.cepheid.com IQuum: www.iquum.com STMicroelectronics: www.st.com Applied Biosystems: www.appliedbiosystems.com Agilent: www.agilent.com Analytik Jena: www.analytik-jena.de Biogene: www.biogene.com Biometra: www.biometra.de Fluidigm: www.fluidigm.com Qiagen: www.qiagen.com TaKaRa: www.takara-bio.com Wafergen: www.wafergen.com

Acknowledgements

This work was supported in part by the EU-project AniBioThreat (Grant Agreement: Home/2009/ISEC/AG/191) with the financial support from the Prevention of and Fight against Crime Programme of the European Union, European Commission – Directorate General Home Affairs, in part by the European Union-funded Integrated Project BIOTRACER (Contract 036272) under the 6th RTD Framework, and in part by the Icelandic Technological fund with the era-SME programme through the FreshFish project (Contract RAN090914-1736).

References

Afonina, I.A., Reed, M.W., Lusby, E., Shishkina, I.G., and Belousov, Y.S. (2002). Minor groove binder-conjugated DNA probes for quantitative DNA detection by hybridization-triggered fluorescence. Biotechniques *32*, 940–949.

Ahmad, A.I. (2007). BOXTO as a real-time thermal cycling reporter dye. J. Biosci. *32*, 229–239.

AlleLogic Biosciences. AllGlo technology for real time PCR. Available at: http://www.allelogic.com/.

Allnutt, T.R., Ayadi, M., Berben, G., Brodmann, P., and Lee, D. (2010). Evaluation of different machines used to quantify genetic modification by real-time PCR. J. AOAC Int. *93*, 1243–1248.

Atallah, Z.K., Bae, J., Jansky, S.H., Rouse, D.I., and Stevenson, W.R. (2007). Multiplex real-time quantitative PCR to detect and quantify Verticillium dahliae colonization in potato lines that differ in response to verticillium wilt. Phytopathology *97*, 865–872.

Bahrdt, C., Krech, A., Wurz, A., and Wulff, D. (2010). Validation of a newly developed hexaplex real-time PCR assay for screening for presence of GMOs in food, feed and seed. Anal. Bioanal. Chem. *396*, 2103–2112.

Basuni, M., Muhi, J., Othman, N., Verweij, J.J., Ahmad, M., Miswan, N., Rahumatullah, A., Aziz, F.A., Zainudin, N.S., and Noordin, R. (2011). A pentaplex real-time polymerase chain reaction assay for detection of four species of soil-transmitted helminths. Am. J. Trop. Med. Hyg. *84*, 338–343.

Bengtsson, M., Karlsson, H.J., Westman, G., and Kubista, M. (2003). A new minor groove binding asymmetric cyanine reporter dye for real-time PCR. Nucleic Acids Res. *31*, e45.

Bentley, H.A., Belloni, D.R., and Tsongalis, G.J. (2005). Parameters involved in the conversion of real-time PCR assays from the ABI prism 7700 to the Cepheid SmartCycler II. Clin. Biochem. *38*, 183–186.

Biotools. Retrieved 3/11/2011, 2011. Available at: http://www.biotools.eu/our_genomic.html.

Braasch, D.A., and Corey, D.R. (2001). Locked nucleic acid (LNA): fine-tuning the recognition of DNA and RNA. Chem. Biol. *8*, 1–7.

Broude, N.E. (2002). Stem-loop oligonucleotides: a robust tool for molecular biology and biotechnology. Trends Biotechnol. *20*, 249–256.

Buh Gasparic, M., Cankar, K., Zel, J., and Gruden, K. (2008). Comparison of different real-time PCR chemistries and their suitability for detection and quantification of genetically modified organisms. BMC Biotechnol. *8*, 26.

Buh Gasparic, M., Tengs, T., La Paz, J.L., Holst-Jensen, A., Pla, M., Esteve, T., Zel, J., and Gruden, K. (2010). Comparison of nine different real-time PCR chemistries for qualitative and quantitative applications in GMO detection. Anal. Bioanal. Chem. *396*, 2023–2029.

Burucoa, C., Garnier, M., Silvain, C., and Fauchere, J.L. (2008). Quadruplex real-time PCR assay using allele-specific scorpion primers for detection of mutations conferring clarithromycin resistance to Helicobacter pylori. J. Clin. Microbiol. *46*, 2320–2326.

Cepheid. SmartNotes Retrieved 3/11/2011, 2011, from http://www.cepheid.com/resources-and-support/support/smartnotes/

Chui, L., Couturier, M.R., Chiu, T., Wang, G., Olson, A.B., McDonald, R.R., Antonishyn, N.A., Horsman, G., and Gilmour, M.W. (2010). Comparison of Shiga toxin-producing *Escherichia coli* detection methods using clinical stool samples. J. Mol. Diag. (JMD) *12*, 469–475.

Ciglenecki, U.J., Grom, J., Toplak, I., Jemersic, L., and Barlic-Maganja, D. (2008). Real-time RT-PCR assay for rapid and specific detection of classical swine fever virus: Comparison of SYBR Green and TaqMan MGB detection methods using novel MGB probes. J. Virol. Methods. *147*, 257–264.

Cirino, N.M., Tavakoli, N.P., Madison-Anenucci, S., and Egan, C. (2007). Multiplex rtPCR in microbiology. In Real-Time PCR in Microbiology, Mackay, I.M., ed (Caister Academic Press, Norfolk, UK), pp. 183–192.

Costa, J.M., Ernault, P., Olivi, M., Gaillon, T., and Arar, K. (2004). Chimeric LNA/DNA probes as a detection system for real-time PCR. Clin. Biochem. *37*, 930–932.

Donald, C., Qureshi, F., Burns, M., Holden, M., Blasic, J., and Woolford, A. (2005). An inter-platform repeatability study investigating real-time amplification of plasmid DNA. BMC Biotechnol. *5*, 15.

Duck, P., Alvarado-Urbina, G., Burdick, B., and Collier, B. (1990). Probe amplifier system based on chimeric cycling oligonucleotides. Biotechniques. *9*, 142–148.

DxS Diagnostics. Molecular diagnostic technologies. Scorpions reaction. Available at: http://www.dxsdiagnostics.com/(S(x1w3v045o5urvn45skrh2b45))/site/content/dxs/TheScorpionsReaction.aspx?Z=18

Egholm, M., Buchardt, O., Nielsen, P.E., and Berg, R.H. (1992). Peptide nucleic acids (PNA). Oligonucleotide analogs with an achiral peptide backbone. J. Am. Chem. Soc. *114*, 1895–1897.

Fiandaca, M.J., Hyldig-Nielsen, J.J., Gildea, B.D., and Coull, J.M. (2001). Self-reporting PNA/DNA primers for PCR analysis. Genome Res. *11*, 609–613.

Frackman, S., Ekenberg, S., Hoffmann, K., Krenke, B., Sprecher, C., and Storts, D. (2005). Plexor™ technology: a new chemistry for real-time PCR. Promega Notes *90*, 2.

French, D.J., Archard, C.L., Brown, T., and McDowell, D.G. (2001). HyBeacon probes: a new tool for DNA sequence detection and allele discrimination. Mol. Cell. Probes. *15*, 363–374.

Fuhrman, L.E., Shianna, K.V., and Aballay, A. (2008). High-throughput isolation and mapping of *C. elegans* mutants susceptible to pathogen infection. PLoS ONE *3*, e2882.

Giancola, S., McKhann, H.I., Berard, A., Camilleri, C., Durand, S., Libeau, P., Roux, F., Reboud, X., Gut, I.G., and Brunel, D. (2006). Utilization of the three high-throughput SNP genotyping methods, the GOOD assay, Amplifluor and TaqMan, in diploid and polyploid plants. Theor. Appl. Gen. *112*, 1115–1124.

Giglio, S., Monis, P.T., and Saint, C.P. (2003). Demonstration of preferential binding of SYBR Green I to specific DNA fragments in real-time multiplex PCR. Nucleic Acids Res. *31*, e136.

Haakensen, M., Dobson, C.M., Deneer, H., and Ziola, B. (2008). Real-time PCR detection of bacteria belonging to the Firmicutes Phylum. Int. J. Food Microbiol. *125*, 236–241.

Heid, C.A., Stevens, J., Livak, K.J., and Williams, P.M. (1996). Real-time quantitative PCR. Genome Res. *6*, 986–994.

Herrmann, M.G., Durtschi, J.D., Voelkerding, K.V., and Wittwer, C.T. (2006). Instrument comparison for DNA genotyping by amplicon melting. J. Assoc. Lab. Autom. *11*, 273–277.

Herrmann, M.G., Durtschi, J.D., Wittwer, C.T., and Voelkerding, K.V. (2007). Expanded instrument comparison of amplicon DNA melting analysis for mutation scanning and genotyping. Clin. Chem. *53*, 1544–1548.

Higuchi, R., Dollinger, G., Walsh, P.S., and Griffith, R. (1992). Simultaneous amplification and detection of specific DNA sequences. Biotechniques *10*, 413–417.

Hilscher, C., Vahrson, W., and Dittmer, D.P. (2005). Faster quantitative real-time PCR protocols may lose sensitivity and show increased variability. Nucleic Acids Res. *33*, e182.

Hoorfar, J., Cook, N., Malorny, B., Wagner, M., De Medici, D., Abdulmawjood, A., and Fach, P. (2003). Making internal amplification control mandatory for diagnostic PCR. J. Clin. Microbiol. *41*, 5835.

Hopkins, M.J., Moorcroft, J.F., Correia, J.B., and Hart, I.J. (2011). Using the full spectral capacity (six channels) of a real-time PCR instrument can simplify diagnostic laboratory screening and typing protocols for pandemic H1N1 influenza. Influenza Other Respi. Viruses 5, 110–114.

Huang, Q., Zheng, L., Zhu, Y., Zhang, J., Wen, H., Huang, J., Niu, J., Zhao, X., and Li, Q. (2011). Multicolor combinatorial probe coding for real-time PCR. PLoS One *6*, e16033.

Ishiguro, T., Saitoh, J., Yawata, H., Yamagishi, H., Iwasaki, S., and Mitoma, Y. (1995). Homogeneous quantitative assay of hepatitis C virus RNA by polymerase chain reaction in the presence of a fluorescent intercalater. Anal. Biochem. *229*, 207–213.

Javorschi-Miller, S., and Delgado Orlic, I. (2011). Real-time PCR instrumentation: an instrument selection guide. In PCR troubleshooting and optimization: The essential guide, Kennedy, S., and Oswald, N., eds (Caister Academic Press, Norfolk), pp. 119–138.

Josefsen, M.H., Jacobsen, N.R., and Hoorfar, J. (2004). Enrichment followed by quantitative PCR both for rapid detection and as a tool for quantitative risk assessment of food-borne thermotolerant campylobacters. Appl. Environ. Microbiol. *70*, 3588–3592.

Josefsen, M.H., Lofstrom, C., Sommer, H.M., and Hoorfar, J. (2009). Diagnostic PCR: comparative sensitivity of four probe chemistries. Mol. Cell. Probes. *23*, 201–203.

Khan, Z. (2011). Setup of a PCR Laboratory. In PCR Protocols, Park, D.J., ed. (Humana Press, New York), pp. 3–14.

Khripin, Y. (2006). High-throughput genotyping with energy transfer-labeled primers. Meth. Mol. Biol. *335*, 215–240.

Kim, Y.H., Yang, I., Bae, Y.S., and Park, S.R. (2008). Performance evaluation of thermal cyclers for PCR in a rapid cycling condition. Biotechniques *44*, 495–500.

Kirchner, S., Kramer, K.M., Schulze, M., Pauly, D., Jacob, D., Gessler, F., Nitsche, A., Dorner, B.G., and Dorner, M.B. (2010). Pentaplexed quantitative real-time PCR assay for the simultaneous detection and quantification of botulinum neurotoxin-producing clostridia in food and clinical samples. Appl. Environ. Microbiol. *76*, 4387–4395.

Krämer, N., Löfström, C., Vigre, H., Hoorfar, J., Bunge, C., and Malorny, B. (2011). A novel strategy to obtain quantitative data for modelling: Combined enrichment and real-time PCR for enumeration of salmonellae from pig carcasses. Int. J. Food Microbiol. *145*, S86–S96.

Kubista, M. (2010). Real-time PCR – where we are and where are we heading. Drug Discov. World Spring, 59–65.

Kumar, R., Singh, S.K., Koshkin, A.A., Rajwanshi, V.K., Meldgaard, M., and Wengel, J. (1998). The first analogues of LNA (locked nucleic acids): phosphorothioate-LNA and 2′-thio-LNA. Bioorg. Med. Chem. Lett. *8*, 2219–2222.

Kutyavin, I.V., Afonina, I.A., Mills, A., Gorn, V.V., Lukhtanov, E.A., Belousov, E.S., Singer, M.J., Walburger, D.K., Lokhov, S.G., Gall, A.A., *et al.* (2000). 3′-minor groove binder-DNA probes increase sequence specificity at PCR extension temperatures. Nucleic Acids Res. *28*, 655–661.

Lasalle, H.E., Duncan, G., and McCord, B. (2011). An analysis of single and multi-copy methods for DNA quantitation by real-time polymerase chain reaction. Forensic Sci. Int. Genet. *5*, 185–193.

Lee, D., Chen, P.J., and Lee, G.B. (2010). The evolution of real-time PCR machines to real-time PCR chips. Biosens. Bioelectron. *25*, 1820–1824.

Lee, M.A., Squirrell, D.J., Leslie, D.L., and Brown, T. (2009). Homogeneous fluorescent chemistries for real-time PCR. In Real-time PCR. Current technology and applications, Logan, J., Edwards, K., and Saunders, N., eds (Caister Academic Press, Norfolk), pp. 23–47.

Li, P., Ruel, T., Fujimoto, K., Hatano, H., Yukl, S., Eller, L.A., Liegler, T., Kamya, M., Gassasira, A., Dorsey, G., *et al.* (2010). Novel application of Locked Nucleic Acid chemistry for a TaqMan assay for measuring diverse human immunodeficiency virus type 1 subtypes. J. Virol. Methods *170*, 115–120.

Li, Q., Luan, G., Guo, Q., and Liang, J. (2002). A new class of homogeneous nucleic acid probes based on specific displacement hybridization. Nucleic Acids Res. *30*, E5.

Lind, K., Stalberg, A., Zoric, N., and Kubista, M. (2006). Combining sequence-specific probes and DNA binding dyes in real-time PCR for specific nucleic acid quantification and melting curve analysis. Biotechniques *40*, 315–319.

Lo, Y.M., and Chan, K.C. (2006). Introduction to the polymerase chain reaction. Meth. Mol. Biol. *336*, 1–10.

Logan, J., and Edwards, K., 2009. An overview of PCR platforms. In Real-time PCR: Current Technology and Applications, Logan, J., Edwards K., and Saunders, N., eds (Caister Academic Press, Norfolk, UK), pp. 7–22.

Lu, S., Smith, A.P., Moore, D., and Lee, N.M. (2010). Different real-time PCR systems yield different gene expression values. Mol. Cell. Probes *24*, 315–320.

McCrea, J.K., Liu, C., Ng, L.K., and Wang, G. (2007). Detection of the *Escherichia coli* pathogenic gene eae with three real-time polymerase chain reaction methods. Can. J. Microbiol. *53*, 398–403.

Mackay, J., and Landt, O. (2007). Real-time PCR fluorescent chemistries. Meth. Mol. Biol. *353*, 237–261.

Malinen, E., Kassinen, A., Rinttila, T., and Palva, A. (2003). Comparison of real-time PCR with SYBR Green I or 5′-nuclease assays and dot-blot hybridization with rDNA-targeted oligonucleotide probes in quantification of selected faecal bacteria. Microbiology *149*, 269–277.

Mao, F., Leung, W.Y., and Xin, X. (2007). Characterization of EvaGreen and the implication of its physicochemical properties for qPCR applications. BMC Biotechnol. *7*, 76.

Markey, A.L., Mohr, S., and Day, P.J. (2010). High-throughput droplet PCR. Methods *50*, 277–281.

Marras, S.A. (2006). Selection of fluorophore and quencher pairs for fluorescent nucleic acid hybridization probes. Meth. Mol. Biol. *335*, 3–16.

Matero, P., Hemmilä H., Tomaso, H., Piiparinen, H., Rantakokko-Jalava, K., Nuotio, L., and Nikkari, S. (2011). Rapid field detection assays for *Bacillus anthracis*, *Brucella* spp., *Francisella tularensis* and *Yersinia pestis*. Clin. Microbiol. Infect. *17*, 34–43.

May, J.P., Brown, L.J., van Delft, I., Thelwell, N., Harley, K., and Brown, T. (2005). Synthesis and evaluation of a new non-fluorescent quencher in fluorogenic oligonucleotide probes for real-time PCR. Org. Biomol. Chem. *3*, 2534–2542.

Molenkamp, R., van der Ham, A., Schinkel, J., and Beld, M. (2007). Simultaneous detection of five different DNA targets by real-time TaqMan PCR using the Roche LightCycler480: Application in viral molecular diagnostics. J. Virol. Methods. *141*, 205–211.

Monis, P.T., Giglio, S., and Saint, C.P. (2005). Comparison of SYT09 and SYBR Green I for real-time polymerase chain reaction and investigation of the effect of dye concentration on amplification and DNA melting curve analysis. Anal. Biochem. *340*, 24–34.

Morrison, L.E. (2010). Basic principles of fluorescence and energy transfer applied to real-time PCR. Mol. Biotechnol. *44*, 168–176.

Nazarenko, I., Lowe, B., Darfler, M., Ikonomi, P., Schuster, D., and Rashtchian, A. (2002). Multiplex quantitative PCR using self-quenched primers labeled with a single fluorophore. Nucleic Acids Res. *30*, e37.

Nazarenko, I.A., Bhatnagar, S.K., and Hohman, R.J. (1997). A closed tube format for amplification and detection of DNA based on energy transfer. Nucleic Acids Res. *25*, 2516–2521.

Neuzil, P., Zhang, C., Pipper, J., Oh, S., and Zhuo, L. (2006). Ultra fast miniaturized real-time PCR: 40 cycles in less than six minutes. Nucleic Acids Res. *34*, e77.

Nordgren, J., Bucardo, F., Dienus, O., Svensson, L., and Lindgren, P.E. (2008). Novel light-upon-extension real-time PCR assays for detection and quantification of genogroup I and II noroviruses in clinical specimens. J. Clin. Microbiol. *46*, 164–170.

Osterman, H.L. (2009). The next step in near infrared fluorescence: IRDye® QC-1 dark quencher, LI-COR Biosciences, Lincoln, NE 68504.

Peng, X., Chen, H., Draney, D.R., Volcheck, W., Schutz-Geschwender, A., and Olive, D.M. (2009). A nonfluorescent, broad-range quencher dye for Förster resonance energy transfer assays. Anal. Biochem. *388*, 220–228.

Petersen, K., Vogel, U., Rockenbauer, E., Nielsen, K.V., Kolvraa, S., Bolund, L., and Nexo, B. (2004). Short PNA Molecular beacons for real-time PCR allelic discrimination of single nucleotide polymorphisms. Mol. Cell. Probes. *18*, 117–122.

Pornprasert, S., Phusua, A., Suanta, S., Saetung, R., and Sanguansermsri, T. (2008). Detection of alpha-thalassemia-1 Southeast Asian type using real-time gap-PCR with SYBR Green1 and high resolution melting analysis. Eur. J. Haematol. *80*, 510–514.

Price, E.P., Smith, H., Huygens, F., and Giffard, P.M. (2007). High-resolution DNA melt curve analysis of the clustered, regularly interspaced short-palindromic-repeat locus of *Campylobacter jejuni*. Appl. Environ. Microbiol. *73*, 3431–3436.

Reed, G.H., Kent, J.O., and Wittwer, C.T. (2007). High-resolution DNA melting analysis for simple and efficient molecular diagnostics. Pharmacogenomics *8*, 597–608.

Reynisson, E., Josefsen, M.H., Krause, M., and Hoorfar, J. (2006). Evaluation of probe chemistries and platforms to improve the detection limit of real-time PCR. J. Microbiol. Methods *66*, 206–216.

Ririe, K.M., Rasmussen, R.P., and Wittwer, C.T. (1997). Product differentiation by analysis of DNA melting curves during the polymerase chain reaction. Anal. Biochem. *245*, 154–160.

Sahoo, R., Harini, V.V., Babu, V.C., V Patil Okaly, G., Rao, S., Nargund, A., Venkataswamy, E., Rao, R., and Kumar, B.S. (2011). Screening for EGFR mutations in lung cancer, a report from India. Lung Cancer. *73*, 316–319.

Saito, K., Miyake, S., Moriya, H., Yamazaki, M., Itoh, F., Imai, K., Kurosawa, N., Owada, E., and Miyamoto, A. (2003). Detection of the four sequence variations of MDR1 gene using TaqMan MGB probe based real-time PCR and haplotype analysis in healthy Japanese subjects. Clin. Biochem. *36*, 511–518.

Saunders, G.C., Dukes, J., Parkes, H.C., and Cornett, J.H. (2001). Interlaboratory study on thermal cycler performance in controlled PCR and random amplified polymorphic DNA analyses. Clin. Chem. *47*, 47–55.

Schlesinger, J., Tönjes, M., Schueler, M., Zhang, Q., Dunkel, I., and Sperling, S.R. (2010). Evaluation of the LightCycler™ 1536 Instrument for high-throughput quantitative real-time PCR. Methods *50*, S19-S22.

Schoder, D., Schmalwieser, A., Schauberger, G., Hoorfar, J., Kuhn, M., and Wagner, M. (2005). Novel approach for assessing performance of PCR cyclers used for diagnostic testing. J. Clin. Microbiol. *43*, 2724–2728.

Shen, F., Du, W., Kreutz, J.E., Fok, A., and Ismagilov, R.F. (2010). Digital PCR on a SlipChip. Lab Chip. *10*, 2666–2672.

Solinas, A., Thelwell, N., and Brown, T. (2002). Intramolecular TaqMan probes for genetic analysis. Chem. Comm. *19*, 2272–2273.

Stephens, K.W., Hutchins, R.J., and Dauphin, L.A. (2010). Cross-platform evaluation of commercial real-time reverse transcription PCR master mix kits using a quantitative 5'nuclease assay for Ebola virus. Mol. Cell. Probes. *24*, 370–375.

Sundberg, S.O., Wittwer, C.T., Gao, C., and Gale, B.K. (2010). Spinning disk platform for microfluidic digital polymerase chain reaction. Anal. Chem. *82*, 1546–1550.

Svanvik, N., Westman, G., Wang, D., and Kubista, M. (2000). Light-up probes: thiazole orange-conjugated peptide nucleic acid for detection of target nucleic acid in homogeneous solution. Anal. Biochem. *281*, 26–35.

Sykes, P.J., Neoh, S.H., Brisco, M.J., Hughes, E., Condon, J., and Morley, A.A. (1992). Quantitation of targets for PCR by use of limiting dilution. Biotechniques *13*, 444–449.

Teo, I.A., Choi, J.W., Morlese, J., Taylor, G., and Shaunak, S. (2002). LightCycler qPCR optimisation for low copy number target DNA. J. Immunol. Methods *270*, 119–133.

Tyagi, S., and Kramer, F.R.,1996. Molecular beacons: probes that fluoresce upon hybridization. Nat. Biotechnol. *14*, 303–308.

Vogelstein, B., and Kinzler, K.W. (1999). Digital PCR. Proc. Natl. Acad. Sci. U.S.A. *96*, 9236–9241.

Whitcombe, D., Theaker, J., Guy, S.P., Brown, T., and Little, S. (1999). Detection of PCR products using self-probing amplicons and fluorescence. Nat. Biotechnol. *17*, 804–807.

Wittwer, C.T., Ririe, K.M., Andrew, R.V., David, D.A., Gundry, R.A., and Balis, U.J. (1997a). The LightCycler: A microvolume multisample fluorimeter with rapid temperature control. Biotechniques *22*, 176–181.

Wittwer, C.T., Herrmann, M.G., Moss, A.A., and Rasmussen, R.P. (1997b). Continuous fluorescence monitoring of rapid cycle DNA amplification. Biotechniques *22*, 130–138.

Wittwer, C.T., Reed, G.H., Gundry, C.N., Vandersteen, J.G., and Pryor, R.J. (2003). High-resolution genotyping by amplicon melting analysis using LCGreen. Clin. Chem. *49*, 853–860.

Wolff, D.J., Heaney, D.L., Neuwald, P.D., Stellrecht, K.A., and Press, R.D. (2009). Multi-site PCR-based CMV viral load assessment-assays demonstrate linearity and precision, but lack numeric standardization: a report of the association for molecular pathology. J. Mol. Diag. *11*, 87–92.

Yang, G.P., Erdman, D.D., Tondella, M.L., and Fields, B.S. (2009). Evaluation of tetramethylrhodamine and black hole quencher 1 labeled probes and five commercial amplification mixes in TaqMan real-time RT-PCR assays for respiratory pathogens. J. Virol. Methods. *162*, 288–290.

Yang, I., Kim, Y.H., Byun, J.Y., and Park, S.R. (2005). Use of multiplex polymerase chain reactions to indicate the accuracy of the annealing temperature of thermal cycling. Anal. Biochem. *338*, 192–200.

Zhang, Y., and Ozdemir, P. (2009). Microfluidic DNA amplification – A review. Anal. Chim. Acta *638*, 115–125.

Zhao, J.R., Bai, Y.J., Zhang, Q.H., Wan, Y., Li, D., and Yan, X.J. (2005). Detection of hepatitis B virus DNA by real-time PCR using TaqMan-MGB probe technology. World J. Gastroenterol. *11*, 508–510.

Quantification Strategies in Real-time Polymerase Chain Reaction

3

Michael W. Pfaffl

Abstract

The present chapter describes the quantification strategies used in real-time RT-PCR (RT-qPCR), focusing on the main elements that are essential to fulfil the MIQE guidelines. The necessity of initial proper data adjustment and background correction is discussed to allow reliable quantification. The advantages and disadvantages of the absolute and relative quantification approaches are also described. In conjunction with relative quantification, the importance of an amplification efficiency correction is shown, and software tools that are available to calculate relative expression changes are presented.

Introduction

In general, two quantification strategies can be applied to analyse real-time RT-PCR (RT-qPCR) expression data. The levels of expressed genes may be measured by absolute quantification, using a standard curve, or relative quantification, using validated reference genes in RT-qPCR.

Absolute quantification relates the PCR signal to input copy number using a calibration curve with known concentrations of the amplified gene of interest (Fig. 3.1). Relative quantification instead measures the expression changes between two or more analysed experimental groups. The reliability of an absolute RT-qPCR assay depends mainly on the validity of the 'calibration curve' used and the 'identical' amplification efficiencies for both the native target in the biological sample and the artificial standard material used in the calibration curve, in the RT and qPCR reactions (Souaze *et al.*, 1996; Bustin *et al.*, 2005).

From the handling perspective, the relative quantification approach seems easier to perform than the absolute one, because a calibration curve is not needed. But it bears other essential problems, such as the changing amplification efficiencies of the analysed target and reference genes, and lacks of amplification variability at low transcript concentrations. It is based on the expression levels of a gene of interest versus one (or more) reference gene(s), whereby the reference gene selection is not trivial (Vandesompele *et al.*, 2002, 2009). To guarantee optimal quantification results, the recently published MIQE guidelines – Minimum Information for Publication of Quantitative Real-Time PCR Experiments – are highly recommended and a prerequisite for any quantitative PCR study (Bustin *et al.*, 2009).

Data adjustment

Prior to any quantification, a proper data adjustment is necessary so that the biological samples and the standards are comparable. Between reaction vessel differences are obvious, which are introduced by the operator, by the handling errors or are derived from the complex enzymatic reaction steps. Furthermore, some instruments excite all reaction vials individually and detected the fluorescence independently, which also may introduce some variability. To overcome this potential variation between reaction set-ups and replicates, it is possible to incorporate a passive reference dye, such as ROX or Fluorescein (FL), in each reaction mix. The passive reference signal is monitored during the course of the amplification, and each

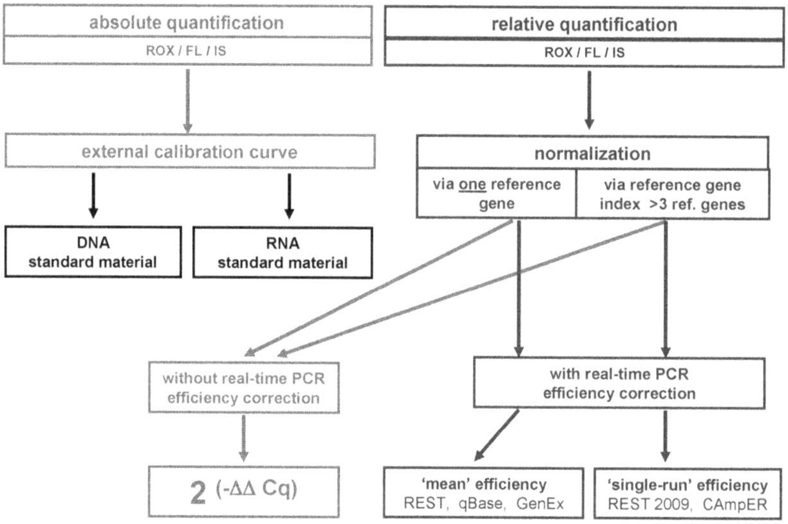

Figure 3.1 Quantification strategies in quantitative real-time RT-PCR.

raw data fluorescence reading is then corrected and equalized using the unchanging reference signal (ROX or FL). An advantage of the data normalization strategy is that the impact of any optical variability across the instrument block will be minimized, but as an additional fluorescence channel is introduced, this adds additional variability in the quantification workflow. The use of a passive reference dye may however help to remove background fluorescence and reduce the noise in the background phase from the measurements. This can be done by the software using the 'automatic background correction' option. An optimal fluorescence history shows low initial background signal, a significant and exponential fluorescence increase resulting from a stable PCR amplification, and a stable plateau (Keer, 2008; Pfaffl *et al.*, 2009).

After proper data adjustment, all samples under investigation are comparable and can be analysed dependent on the applied quantification strategy selected according to the scientific or analytical question. Either an absolute quantification approach, which results in defined units, e.g. copies per defined nucleic acid concentration, or a relative quantification, leading to gene expression ratios between different treatment groups, is used. In this later case, absolute units remain unknown.

Absolute quantification

The so-called 'absolute quantification' is misleading because the method is never 'absolute', as such, it is relative to the created calibration or standard curve. Hence everything depends on the reliability and validity of the standard curve and the used standard material. In particular, the external calibration curve model has to be thoroughly validated as the fidelity of 'absolute quantification' in real-time RT-qPCR depends entirely on the accuracy and quality of the standard used. It should be pure, not degraded, stable, and for this reason yield highly reproducible results over time and the entire quantification range. The standard curve model should allow the generation of highly specific, sensitive, and reproducible quantification data (Reischl and Kochanowsky, 1995; Bustin, 2000; Pfaffl and Hageleit, 2001; Reiter *et al.*, 2011). The standard composition, sequence design, purity, determination of the exact concentration, and stability over long storage time is not straightforward and can be problematic. In addition, the matrix of the biological samples is important. Some biological samples under investigation bear inhibitors for the enzymatic RT and/or qPCR reaction. Therefore, the standard curve itself should reflect as well the biological sample which is the focus for quantification. What is the real benefit of a suboptimal quantification model if the standard curve and the biological do not show the same amplification behaviour? Often the standard curve shows perfect properties with pure standard material and a perfect amplification history whereas the biological samples demonstrate high background, low amplification efficiency and seem to be inhibited by any unknown substance. Very importantly, both the standard curve dilutions and the unknown biological samples have to amplify in the same manner and with the same amplification characteristics. Elsewise the quantification is NOT reliable and will result in wrong quantification findings (Pfaffl, 2004).

What are the characteristics of a good standard curve? The dynamic range of the performed calibration curve should be as broad as possible. It can be up to nine orders of magnitude, starting at one theoretical molecule in the lowest, up to billions of molecules in the highest concentration. If plotted, all standard points and their replicates should be equally spread over the quantification range. All standard points will be summarized in a linear regression which should be linear with low variation, expressed in the confidence interval with a high correlation coefficient (r^2), and a slope, ideally –3.33, which corresponds to an amplification efficiency of 100% or a perfect doubling in each cycle.

The quantification range depends much on the applied standard material stability whereas the biggest difference appears between RNA and DNA (Pfaffl and Hageleit, 2001; Reiter *et al.*, 2011). The analytical question determines the type of the standard material, either DNA or RNA. If the focus is on the quantification of DNA, e.g. quantification of microorganisms in a sample, the calibration curve should be made out of DNA. If the interest is on the quantification of expressed transcripts or to measure RNA viruses concentrations, the standard substance must be RNA based.

For a DNA standard curve, the type of standard material is always based on known concentrations of double- or single-stranded DNA molecules. In most cases, serial dilutions with 10-fold dilution steps are sufficient (Fig. 3.2). Furthermore, the stability and reproducibility of the absolute quantification curve depends on the type of standard material applied, e.g. genomic DNA, RT-PCR product, recombinant plasmid DNA, commercially synthesized oligo-nucleotide or artificial genes (Morrison *et al.*, 1998, Pfaffl, 2004). Cloned

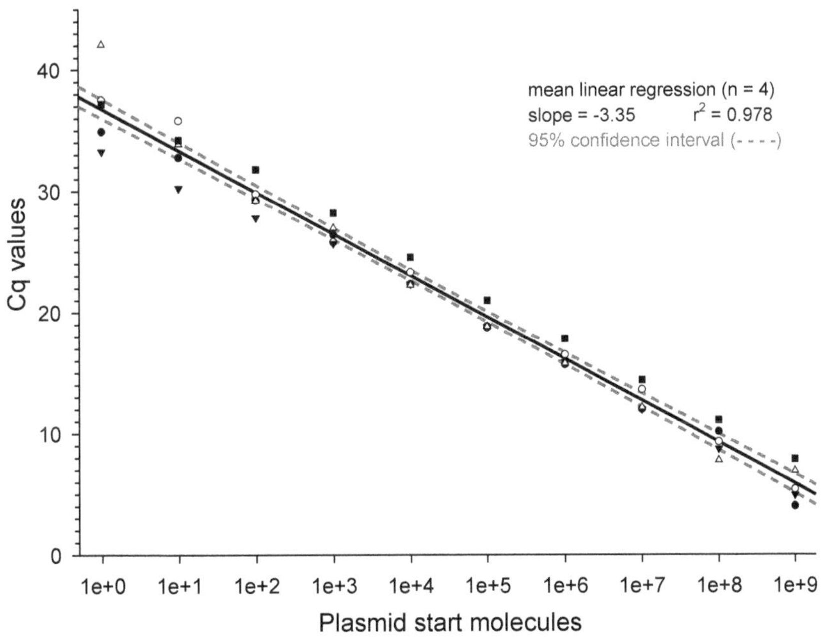

Figure 3.2 DNA standard curve.

recombinant DNA, artificial genes, and genomic DNA are very stable and generate highly reproducible standard curves even after a long storage time up to several years in a –20°C freezer (own findings).

RNA standards are much more sensitive to RNAse activity and degradation, especially at low concentration. Owing to the stability problem, it will introduce variability in the quantification of low abundant transcripts, and low concentration cannot be measured with high confidence. Furthermore, the longer templates derived from artificial genes mimic the average native length of messenger RNA of roughly 2 kb better than shorter templates or oligo-nucleotides. These longer standards have longer overhanging ends and are hence more resistant against unspecific enzymatic cleavage during RT or PCR reaction (Rasmussen, 2004).

A major problem with DNA-based calibration curves is that they can only be applied to quantify at the DNA level, and they are not suitable to quantify any RNA. DNA is subject to the PCR step only, unlike the unknown mRNA samples that must first be reverse transcribed to convert them to complementary DNA (cDNA). The RT step efficiency is usually around 30–40% and bears high variability between technical RT replicates (additional 30%) (Stahlberg *et al.*, 2004a,b). This decreases the amount of synthesized cDNA dramatically; pushes the variability in the quantitative RT-qPCR study and the derived quantification results may not reflect the reality of RNA initially present in the biological sample. If RNA is in the focus of the quantification, a recombinant RNA standard material *in vitro* transcribed from a plasmid, or artificially synthesized is mandatory (Pfaffl *et al.*, 2001; Fronhoffs *et al.*, 2002; Reiter *et al.*, 2011).

Independent of the standard material used, an absolute quantification model requires careful optimization of its precision. The reverse transcription efficiencies for both, standard curve material and native biological sample, must be tested thoroughly and should be confirmed by multiple technical RT replicates (Tichopad *et al.*, 2003b). We recommend performing replicates in the same RT reaction run, to get information about the RT assay variability, and replicates in separate runs, to get the inter-assay variation or reproducibility. This will help to find the detection and quantification limitations within the given absolute quantification procedure, and will show exactly the technical variance (Pfaffl and Hageleit, 2001; Tichopad *et al.*, 2009).

A special focus should be on the existence of inhibitors in the biological sample, which hampers either RT and/or PCR reactions. Inhibitors may be present in the biological sample but not in the standard dilutions used in the calibration curve. Known and unknown biological matrix compounds interact with the enzymatic reaction and therefore impact the validity and trueness. If present, a strong bias is introduced, leading to an overestimation of the standard curve and an underestimation of the unknown sample concentration. This will lead to completely wrong quantification results. The presence of inhibitors in the biological matrix should always be tested first. Many factors present in samples as well as exogenous contaminants have been reported to inhibit enzymatic reactions, either RT or PCR. For example, the presence of humic acids, haemoglobin, fatty acids, glycogen, cell components, calcium ions, and nucleic acid binding proteins are known inhibiting compounds (Rossen *et al.*, 1992; Wilson, 1997; Tichopad *et al.*, 2004). Additionally, exogenous contaminants such as glove powder, DNA or RNA degrading decontamination chemicals, and ethanol or phenolic compounds from the extraction process will also have an inhibitory effect (Rossen *et al.*, 1992; Wilson, 1997). Furthermore, low-quality and impure plasticware with high nucleic acid-binding capacity interact with the quantification procedure and should be avoided. To test for such inhibitors, a SPUD assay can be performed, which allows determining the degree of inhibition in the biological sample (Nolan *et al.*, 2006).

To get the most correct and valid results in an absolute quantification approach, either the user must take actions to get rid of any interfering compounds in the unknown biological samples during nucleic acid purification, or the calibration curve must be adapted. This means that the calibration curve amplification behaviour and the overall RT and qPCR efficiency must be brought to the same level as the biological sample material. In the end, both the standard curve and the biological samples should show most comparable amplification behaviour. The so called 'biological matrix background effect' in the standard curve model should be obtained from the sample under investigation. Eventually, identical background levels in the calibration curve and the biological samples should result in comparable levels of inhibition and therefore leads to a more realistic and valid quantification (Pfaffl, 2004).

Regardless on how accurate the concentration of the standard material is determined, the final result is always reported compared with a defined unit of interest, e.g. copies per defined RNA or DNA concentration, copies per genome, copies per solitaire-cell, copies per tissue unit, copies per millilitre blood, copies per gram of soil, etc. If absolute changes in copy number are important then the denominator must still be shown to be fully stable across the entire study. The applicability depends strongly on the adherence to the MIQE guidelines, the general 'good laboratory practice' and the personal skills of the researcher (Bustin *et al.*, 2009; Health Protection Agency, 2010). Failure to comply with the mentioned rules may introduce enormous additional variability to the 'absolute' quantification system and will result in artificial quantification results.

Relative quantification

Relative quantification determines the relative changes in steady-state messenger RNA (mRNA) levels of an expressed gene across multiple biological samples. Today the use of unregulated and stable expressed reference genes for normalization of transcript levels is certainly the gold standard (Vandesompele *et al.*, 2002). It is adequate for most purposes to investigate even low expression changes of minute amounts of mRNA or microRNA. The units used to express relative quantities are irrelevant, and the relative quantities can be compared across multiple real-time RT-qPCR experiments (Vandesompele *et al.*, 2009). These controls are internally grown, named reference genes (formerly known as housekeeping genes), and can be co-amplified in a multiplex assay or can be amplified in separate tubes (Wittwer *et al.*, 2001). Therefore, relative quantification does not require standards with known concentrations. The controls can be one or more reference gene(s), which can be any transcript as long as it is expressed in all biological samples and properly validated according to the MIQE guidelines (Bustin *et al.*, 2009).

To calculate the expression difference of a target gene in relation to an adequate reference gene, two general mathematical calculation models are established in the scientific qPCR community. The so called 'delta-delta Cq model', developed by Livak and Schmittgen (2001) or the efficiency corrected model, developed by Pfaffl (2001). Both calculations models are based on the comparison of the distinct cycle of quantification (Cq). The most applied and common calculation model is the 'delta-delta Cq method' which is assuming optimal amplification efficiency of 100% in any cycle, over the entire PCR cycles, with all target transcripts. It does not take any suboptimal PCR efficiency or its variation into account, derived from PCR inhibition or handling errors, either for the target or reference genes. As described in the previous 'absolute quantification' section, inhibitors or suboptimal reaction conditions may interfere with the reaction performance and hence the reaction efficiency is never exactly 100%. Therefore, a model was modified by Pfaffl and co-workers in 2001 by adding a real-time PCR efficiency correction. This correction is applied separately for each gene studied, e.g. the target gene (E_{target}) and the reference gene (E_{ref}), as shown in the equation:

$$\text{ratio} = \frac{(E_{target})^{\Delta CP_{target}(control - sample)}}{(E_{ref})^{\Delta CP_{ref}(control - sample)}} \tag{3.1}$$

This efficiency corrected relative quantification model was further developed, driven by innovations from other groups. It was expanded by adding multiple reference genes for normalization (summarized by Vandesompele *et al.*, 2009), by applying the model to group measurements (Pfaffl *et al.*, 2002), or by adding new statistical analysis tools to get more significant valid relative expression results (Qiagen, 2009).

We now know that reference genes have to be carefully selected and validated, which can be done using free available software applets, e.g. geNorm (Vandesompele, 2002), BestKeeper (Pfaffl *et al.*, 2004) or NormFinder (Andersen *et al.*, 2004). The main goal of the normalization procedure is to remove experimental induced variation, each with their own advantages and considerations (Huggett *et al.*, 2005). While most of these methods cannot completely reduce all sources of variation, they have been shown to be very important to control most sources of variation along the entire workflow of PCR-based gene expression analysis. If one does not meticulously try to standardize each step, variation can and will be introduced in the results which cannot be eliminated by applying a final normalization (Stahlberg *et al.*, 2004b). It is thus recommended to ensure that similar biological sample sizes and extraction volumes are used, and also to standardize the workflow. The amount of time invested in DNase treatment and reverse transcription optimization is important and can impact the cDNA conversion efficiency. Furthermore, artificial RNA molecules can be spiked into the sample prior to extraction or to the RNA extract prior to reverse transcription (Smith *et al.*, 2003; Gilsbach *et al.*, 2006). This will give an indication of the RT and qPCR efficiencies, and will reveal any inhibition. We however need to keep in mind that the spiked artificial RNA may not be extracted with the same yield as natural mRNAs and will not fully control for the final amount of input material in the reaction (Stahlberg *et al.*, 2004a).

Summarizing the qPCR literature from the last decade, it has been agreed that the reference gene concept is the currently preferred way of normalizing real-time PCR data for relative expression profiling (Huggett *et al.*, 2005; Vandesompele *et al.*, 2009). The reference gene concept is particularly attractive because the reference genes are grown internal controls that are affected by all sources of variation during the experimental workflow, from sampling, storage, extraction, nucleic acid stabilization, and enzymatic reaction (RT, DNAse digestion, and PCR) in the same way as the genes of interest. Furthermore, PCR based quantification findings for a gene of interest are best normalized using a factor that is measured using the same quantitative methodology set-up and chemicals.

After selecting the right references, the relative expression difference of the target genes can be calculated by multiple software tools. Either a software application is already implemented in the real-time cycler software package or the data can be exported and further analysed using proper quantification tools. The oldest available software is the REST (Relative Expression Software Tool), established by Pfaffl and colleagues (Pfaffl *et al.*, 2002; http://REST.gene-quantification.info), qBase and qBase plus, developed around the group of Vandesompele and co-workers (Hellemans *et al.*, 2007; http://www.biogazelle.com), or GeneEx set up by Kubista and colleagues (Bergkvist *et al.*, 2010; http://www.multid.se). In these software, the target gene expression is normalized by multiple validated non-regulated reference gene expressions, and even more advanced applications are implemented to evaluate if the target gene is expressed as part of a gene cluster, via cluster analysis, principal component analysis or neural networks (summarized in Pfaffl *et al.*, 2009; Bergkvist *et al.*, 2010). This way, relative quantification can generate useful and biologically relevant information when used appropriately according to the MIQE guidelines (Bustin *et al.*, 2009).

PCR amplification efficiency

Constant PCR amplification efficiency in all investigated biological samples is one important criterion for reliable comparison between samples and genes. Individual reaction set-ups can however generate different fluorescence histories in quantitative real-time PCR. The shapes of amplification curves differ in the background fluorescence level, the steepness of fluorescence increase, and in the absolute fluorescence levels at plateau. PCR efficiency has a major impact at each cycle and finally in the overall cycle fidelity or accuracy of the assay. Therefore, efficiency evaluation is an essential and requested criterion in gene quantification procedure (Tichopad *et al.*, 2003a; Bustin *et al.*, 2009). Some quantification models assume that the analysed genes amplify optimally with 100% efficiency (Livak and Schmittgen, 2001) or at least with similar efficiency. But this is often not the case. As discussed, sporadic RT and PCR inhibitors, sample impurity in the reaction set-up, or nucleic acid degradation may reduce the efficiency of the RT or the PCR reaction (Stahlberg *et al.*, 2004b; Becker *et al.*, 2010). To determine the level of inhibition, a dilution series can be performed and the inhibitory factors can often be diluted out, causing a non-linear standard curve (Morrison *et al.*, 1998; Rasmussen, 2001).

A correction for PCR efficiency and its variation, as performed in efficiency corrected mathematically models, is strongly recommended and is one of the key element of the MIQE guidelines (Bustin *et al.*, 2009). It results in a more reliable estimation of the 'real expression ratio' compared with no efficiency correction in the relative quantification model (Pfaffl, 2001). Small efficiency differences between target and reference genes, or between biological samples, generate false expression ratio and results in wrong scientific findings. Overall, the researcher over- or under-estimates the 'real and initial' mRNA amount present in the biological sample (LightCycler Relative Quantification Software; Roche Applied Science, 2001). This is why efficiency quantification corrections should be included in the calculation procedure, and is a major goal to implement it in any real-time qPCR cycler relative quantification software.

The most applied method to evaluate the PCR efficiency is from the slopes of the calibration curve or dilution row from a pool of biological samples. From the regressed expression data versus the concentration, the PCR efficiency (E) can be calculated according to the equation $E = 10^{[-1/\text{slope}]}$, as described earlier (Higuchi *et al.*, 1993; Rasmussen, 2001). Efficiency determination by the dilution method can be performed sample by sample, but this is very time consuming and needs lots of RNA. If not available, the efficiency can be evaluated using a pool of all starting total RNAs to accumulate all possible positive and negative 'impacts' on kinetic PCR efficiency estimation. Usually, this so called 'mean PCR efficiency' (Fig. 3.2) vary around the ideal regression line, behave linearly ($0.980 > r^2 > 0.999$), and range from $E = 1.75$ to maximal values up to $E = 2.10$ for cDNA input ranges over at least three orders of magnitude (Wittwer *et al.*, 1991; Pfaffl *et al.*, 2002). Typically, the relationship between Cq and the logarithm of the starting copy number of the target sequence should remain linear in the calibration curve as well as in the native sample RNA (Muller *et al.*, 2002).

Over the years further methods were established, for a more realistic estimation of the real-time PCR efficiency. These models are based on the analysis of all available data points in the fluorescence history. Raw fluorescence data reading points were fitted according to a given mathematical model. Often, sigmoidal, logistic or exponential curve model were used in different approaches and even multiple fitting models were established (Liu and Saint, 2002a,b; Tichopad *et al.*, 2003a; Zhao and Fernald, 2005). The advantage of such models is that all data points will be included in the calculation process and the 'single run' efficiency can be estimated from each biological sample individually. No work consuming and cost intensive dilution row has to be performed, and no background subtraction is necessary (Raeymaekers, 2000; Tichopad *et al.*, 2004). It can be measured at different points, either at the point of inflexion or at the absolute maximum fluorescence increase which equals the 'first derivate maximum' point. Alternatively, it can be measured earlier than or at the so called 'second derivate maximum' point, where the optimal reaction point is defined. However, the derived slope parameters generated by such more advanced models are in the eyes of a small group of researchers more reliable and 'directly' comparable with the real PCR efficiency (Liu and Saint, 2002a,b; Tichopad *et al.*, 2003a). These 'single run' efficiency methods can be fully automated, are easier to perform for the operator, but bear the problem of higher variability in a set of biological samples. Thus, PCR efficiency is different according to biological samples and has a variability which should be taken into account in the quantification model. To cope with the knowledge of individual 'single-run' PCR efficiencies in each biological sample, new software tools were established, e.g. REST 2009 (Pfaffl *et al.*, 2002; Qiagen, 2009) and Champer (Blom, 2006).

Conclusion

Any calculation model, any applied algorithm, and any quantification strategy are making assumptions which try to describe the nature of the qPCR reaction at its best. Which standard curve material is the most stable? Which biological background best mimics the biological reality? Which normalization approach is the most reliable? Which PCR efficiency calculation method is 'the most correct one'? In the end, the quantification model that shows the most valid expression results is still unknown. With the latest progress, we are however in a good position to find it in the near future.

References

Andersen, C.L., Jensen, J.L., Ørntoft, T.F. (2004) Normalization of real-time quantitative reverse transcription-PCR data: a model-based variance estimation approach to identify genes suited for normalization, applied to bladder and colon cancer data sets. Cancer Res. *64*, 5245–5250.

Becker, C., Hammerle-Fickinger, A., Riedmaier, I, and Pfaffl, M.W. (2010) mRNA and microRNA quality control for RT-qPCR analysis. Methods *50*, 237–243.

Bergkvist, A., Rusnakova, V., Sindelka, R., Garda, J.M., Sjögreen, B., Lindh, D., Forootan, A., and Kubista, M. (2010) Gene expression profiling – Clusters of possibilities. Methods *50*, 323–335.

Blom, J. (2006) Champer software http://camper.cebitec.uni-bielefeld.de.

Bustin, S.A. (2000) Absolute quantification of mRNA using real-time reverse transcription polymerase chain reaction assays. J. Mol. Endocrinol. *25*, 169–193.

Bustin, S.A., Benes, V., Nolan, T., and Pfaffl, M.W. (2005) Review: quantitative real-time reverse transcription polymerase chain reaction – A perspective. J. Mol. Endocrinol. *34*, 597–601.

Bustin, S.A., Benes, V., Garson, J.A., Hellemans, J., Huggett, J., Kubista M., Mueller R., Nolan, T., Pfaffl, M.W., Shipley, G.L., *et al.* (2009) The MIQE guidelines: minimum information for publication of quantitative real-time PCR experiments. Clin. Chem. *55*, 611–622.

Fronhoffs, S., Totzke, G., Stier, S., Wernert, N., Rothe, M., Bruning, T., Koch, B., Sachinidis, A., Vetter, H., and Ko, Y., (2002) A method for the rapid construction of cRNA standard curves in quantitative real-time reverse transcription polymerase chain reaction. Mol. Cell Probes *16*, 99–110.

Gilsbach, R., Kouta, M., Bönisch, H., and Brüss, M. (2006) Comparison of *in vitro* and *in vivo* reference genes for internal standardization of real-time PCR data. Biotechniques *40*, 173–177.

Health Protection Agency (2010). Good laboratory practice when performing molecular amplification assays. National Standard Method QSOP 38 Issue 4.

Hellemans, J., Mortier, G., De Paepe, A., Speleman, F., and Vandesompele, J. (2007) qBase relative quantification framework and software for management and automated analysis of real-time quantitative PCR data. Genome Biol. *8*, R19.

Higuchi, R., Fockler, C., Dollinger, G., and Watson, R. (1993) Kinetic PCR analysis: Real-time monitoring of DNA amplification reactions. Biotechnology *11*, 1026–1030.

Huggett, J., Dheda, K., Bustin, S., and Zumla A. (2005) Real-time RT-PCR normalization – strategies and considerations. Genes Immun. *6*, 279–284.

Keer, J.T. (2008) Essentials of Nucleic Acid Analysis: A Robust Approach (Royal Society of Chemistry, Cambridge, UK), pp. 132–166.

Liu, W., and Saint, D.A. (2002a), Validation of a quantitative method for real time PCR kinetics. Biochem. Biophys. Res. Commun. *294*, 347–353.

Liu, W., and Saint, D.A., (2002b) A new quantitative method of real time reverse transcription polymerase chain reaction assay based on simulation of polymerase chain reaction kinetics. Anal. Biochem. *302*, 52–59.

Livak, K.J., and Schmittgen, T.D. (2001) Analysis of relative gene expression data using real-time quantitative PCR and the 2^[-delta delta C(T)] method. Methods *25*, 402–408.

Morrison, T.B., Weis, J.J., and Wittwer, C.T. (1998) Quantification of low-copy transcripts by continuous SYBR Green I monitoring during amplification. Biotechniques *24*, 954–962.

Muller, P.Y., Janovjak, H., Miserez, A.R., and Dobbie, Z. (2002) Processing of gene expression data generated by quantitative real-time RT-PCR. Biotechniques *32*, 1372–1378.

Nolan T., Hands R.E., Ogunkolade W., and Bustin S.A. (2006) SPUD: a quantitative PCR assay for the detection of inhibitors in nucleic acid preparations. Anal. Biochem. *351*, 308–310.

Pfaffl, M.W. (2001) A new mathematical model for relative quantification in real-time RT-PCR. Nucleic Acids Res. *29*, e45.

Pfaffl, M.W. (2004) Quantification strategies in real-time RT-PCR. In The Real-time PCR Encyclopaedia A–Z of Quantitative PCR (1st edition), Bustin, S.A., ed (International University Line, La Jolla, CA), pp. 87–120

Pfaffl, M.W., and Hageleit, M. (2001) Validities of mRNA quantification using recombinant RNA and recombinant DNA external calibration curves in real-time RT-PCR. Biotechnol. Lett. *23*, 275–282.

Pfaffl, M.W., Lange, I.G., Daxenberger, A., and Meyer, H.H.D. (2001) Tissue-specific expression pattern of oestrogen receptors (ER): quantification of ER alpha and ER beta mRNA with real-time RT-PCR. APMIS 109(5), 345–355.

Pfaffl, M.W., Horgan, G.W., and Dempfle, L. (2002) Relative expression software tool (REST) for group-wise comparison and statistical analysis of relative expression results in real-time PCR. Nucleic Acids Res. 30(9), e36.

Pfaffl M.W., Tichopad A., Prgomet C., and Neuvians T.P. (2004) Determination of stable housekeeping genes, differentially regulated target genes and sample integrity: BestKeeper – Excel-based tool using pair-wise correlations. Biotechnol. Lett. *26*, 509–515.

Pfaffl, M.W., Vandesompele, J., and Kubista, M. (2009) Real-time PCR technology. In Data Analysis Software, Logan, J., Edwards, K., and Saunders, N., eds (Caister Academic Press, Norwich), pp. 65–83.

Qiagen (2009) REST 2009 Software – for gene expression analysis using real-time PCR data from the Rotor-Gene Q and other cyclers. REST 2009 Software. Available at: http://www.REST.de.com

Raeymaekers, L., (2000) Basic principles of quantitative PCR. Mol Biotechnol. *15*, 115–122.

Rasmussen, R. (2001) Quantification on the LightCycler. In Rapid Cycle Real-time PCR, Methods and Applications, Meuer, S., Wittwer, C., and Nakagawara, K., eds (Springer Press, Heidelberg), pp. 21–34.

Reischl, U., and Kochanowski, B. (1995) Quantitative PCR. A survey of the present technology. Mol. Biotechnol. *3*, 55–71.

Reiter, M., Kirchner, B., Müller, H., Holzhauer, C., Mann, W., and Pfaffl, M.W. (2011) Quantification noise in single cell experiments. Nucleic Acids Res. *39*, e124.

Rossen, L., Norskov, P., Holmstrom, K., and Rasmussen, F.O. (1992). Inhibition of PCR by components of food sample, microbial diagnostic assay and DNA-extraction solutions. Int. J. Food Microbiol. *17*, 37–45.

Smith, A.B., Mock, V., Melear, R., Colarusso, P., and Willis, D.E. (2003) Rapid detection of influenza A and B viruses in clinical specimens by Light Cycler real time RT-PCR. J. Clin. Virol. *28*, 51–58.

Souaze, F., Ntodou-Thome, A., Tran, C.Y., Rostene, W., and Forgez, P. (1996) Quantitative RT-PCR: limits and accuracy. Biotechniques *21*, 280–285.

Stahlberg A., Kubista M., and Pfaffl M.W. (2004a) Comparison of reverse transcriptases in absolute and relative gene expression analysis. Clin. Chem. *50*, 1678–1680.

Stahlberg, A., Håkansson, J., Xian, X., Semb, H., and Kubista, M. (2004b) Properties of the reverse transcription reaction in mRNA quantification. Clin. Chem. *50*, 509–515.

Tichopad, A., Dilger, M., Schwarz, G., and Pfaffl, M.W. (2003a) Standardized determination of real-time PCR efficiency from a single reaction setup. Nucl. Acids Res. *31*, e122.

Tichopad, A., Dzidic, A., and Pfaffl, M.W. (2003b) Improving quantitative real-time RT-PCR reproducibility by boosting primer-linked amplification efficiency. Biotechnol. Lett. *24*, 2053–2056.

Tichopad, A., Didier, A., and Pfaffl, M.W. (2004) Inhibition of real-time RT-PCR quantification due to tissue specific contaminants. Mol. Cell. Probes *18*, 45–50.

Tichopad, A., Kitchen, R., Riedmaier, I., Becker, C., Ståhlberg, A., and Kubista M. (2009) Design and optimization of reverse-transcription quantitative PCR experiments. Clin. Chem. *55*, 1816–1823.

Vandesompele, J., De Preter, K., Pattyn, F., Poppe, B., Van Roy, N., De Paepe, A., and Speleman, F. (2002) Accurate normalization of real-time quantitative RT-PCR data by geometric averaging of multiple internal control genes. Genome Bio. 3(7), RESEARCH0034.

Vandesompele, J., Kubista, M., and Pfaffl, M.W. (2009) Real-Time PCR Technology. In Reference Gene Validation Software for Improved Normalization, Logan, J., Edwards, K., and Saunders, N., eds (Caister Academic Press, London), pp. 47–64.

Wilson, I.G. (1997). Inhibition and facilitation of nucleic acid amplification. Appl. Environ. Microbiol. *63*, 3741–3751.

Wittwer, C.T., and Garling, D.J. (1991) Rapid cycle DNA amplification: Time and temperature optimization. Biotechniques *10*, 76–83.

Wittwer, C.T., Herrmann, M.G., Gundry, C.N., and Elenitoba-Johnson, K.S. (2001) Real-time multiplex PCR assays. Methods *25*, 430–442.

Zhao, S., and Fernald, R.D. (2005) Comprehensive algorithm for quantitative real-time polymerase chain reaction. J. Comput. Biol. *12*, 1047–1064.

GenEx: Data Analysis Software

4

Mikael Kubista, Vendula Rusnakova, David Svec, Björn Sjögreen and Ales Tichopad

Abstract

As the qPCR field advances, the design of experiments and the analysis of data are becoming more important and more challenging. Calculation of relative expression of a reporter gene to a reference gene in pairs of samples using the ΔΔCq method is no longer sufficient. Studies are now designed using multiple markers, nested levels, exploring or confirming the effect of multiple factors, occasionally in paired designs, etc. Proper handling of such data requires software that support the planning and design of experiments, and data analysis. Several software with these capacities are emerging. This chapter describes some of the features of one of the most powerful of those: GenEx from MultiD Analysis.

Introduction

In the early days of quantitative real-time PCR (qPCR) the experiments performed were rather uncomplicated, involving only small number of samples and analysis could be performed using a standard spread sheet such as Microsoft Excel. However, as the technology develops, more advanced experimental designs become common with large data sets, often involving multiple plates or runs on high-throughput instruments. These analyses require more information on how the experiment was set up to handle references, standards, and controls appropriately, and correctly account for the variance and covariance in the measured data. On some high throughput instruments, even the transfer of data from the instrument to the analysis software is challenging, requiring user intervention with obvious risk to accidently change or lose some data. In parallel with the instruments becoming more advanced and powerful, qPCR data analysis software are being developed. Today, this development is heavily supported by the instrument companies, who have realized their customers' needs. The main qPCR software on the market today are GenEx from MultiD (www.multid.se), which is supporting all leading qPCR instruments, Statminer from Integromics (www.integromics.com), which supports qPCR instruments from Life Technologies, and qBase plus from Biogazelle (www.biogazelle.com), which supports Bio-Rad instruments. Since GenEx is currently the only generic platform and that it is also among the most developed software, we will focus this chapter on it. Much of what is said, though, is general for qPCR analysis and applies also to most other software on the market.

Data arrangement

When groups are compared, data are classically arranged with the measured Cq (explained variables) in columns headed with the experimental group label (explanatory variables). This arrangement provides easy overview of the data (Fig. 4.1). However, it is not practical for advanced studies that may include more than one nominal factor or covariate (variable of metric character, e.g. time, age, dose, etc.), multiple markers, replicate measurement, repeated measurements (same subject sampled repeatedly), multiplate measurements, etc. A more flexible approach is to arrange the data with samples as rows and all variables in columns (Fig. 4.1). This is standard arrangement in most statistical software today.

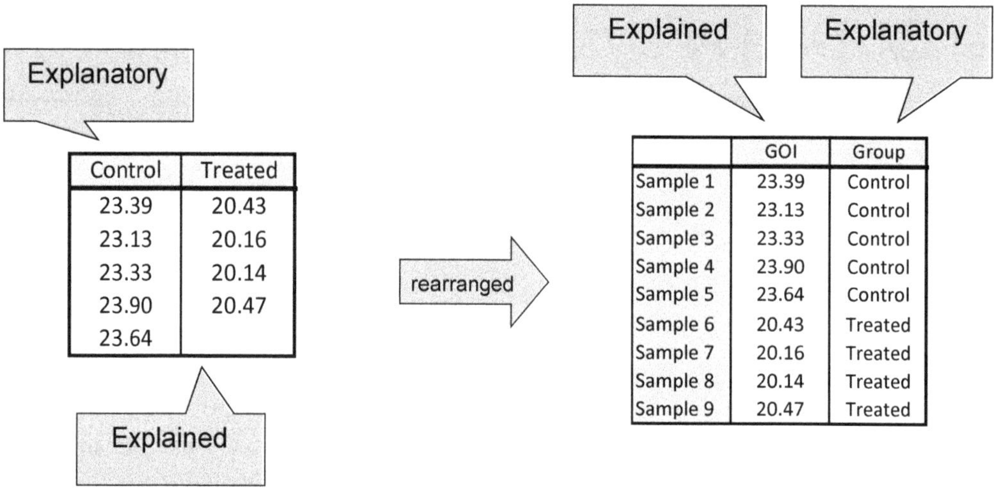

Figure 4.1 Traditional arrangement of data with experimental group labels in column heads (left) and modern, generic arrangement of data with one variable per column (right).

The format is readily generalized to any number of markers and additional columns and rows can be added that specify the experimental design, index samples, markers, plates etc. In GenEx, these are referred to as classification columns and classification rows and have labels starting with #. In the example shown in Fig. 4.2, #Repeat indexes qPCR technical replicates (samples with the same index are replicates on the qPCR level). These are expected to be highly similar and shall be averaged during data pre-processing. #Treatment indexes treatment groups that eventually shall be compared using a statistical test. Finally, the study is paired, meaning that each subject received both treatments and a sample was collected after each treatment. Paired study designs are more powerful, because the pairing reduces confounding variation. This elevates the power of the test and the experiment requires fewer subjects. It can be, for example, samples collected from all subjects before treatment and a second set collected after treatment. It can also be positive and negative samples collected from the same subject or genetically similar individuals such as siblings, identical twins or clones. A special type of paired study design is repeated samplings, used in more

	MG	RG	#Repeat	#Treatment	#Pairing
S1	22.22481...	16.71773...	1	1	1
S2	17.93896...	16.59702...	1	1	2
S3	19.76137...	16.60889...	2	1	3
S4	17.23068...	16.23785...	2	1	4
S5	16.57810...	16.89222...	3	1	5
S6	18.25507...	16.54029...	3	1	6
S7	29.47699...	16.86087...	1	2	1
S8	18.68488...	16.08867...	1	2	2
S9	22.91167...	16.67638...	2	2	3
S10	17.80890...	16.15741...	2	2	4
S11	16.80731...	16.05589...	3	2	5
S12	19.77758...	16.19985...	3	2	6
#Gene	0	1			

Figure 4.2 Example of data arrangement in GenEx. The first column list the samples. Second and third columns are measured Cq values. Fourth column indexes technical replicates, fiftth column indexes treatment groups and sixth column indexes paired samples. Bottom row identifies marker and reference genes.

than two subsequent measurements. In general, the word 'paired' is replaced by 'repeated'. Specialized statistical procedures are available to analyse repeated samplings.

Data import

The experimental design is defined in part by deciding on the experimental factors and covariates involved in the experiment and in part when the samples and assays are mixed while dispensed into the qPCR containers. This information is critical for proper analysis and mining of the measured data. This has been realized by several of the leading qPCR instruments and assay providers. The Roche LC480 software for Real-Time Ready custom and focus panels, for example, names all genes and samples, indicates reference genes, and specifies technical and biological replicates at various levels. This information is transferred to GenEx and arranged appropriately for downstream analysis using a wizard. A similar high level user friendly solution is provided by Exiqon, who offers a customized version of GenEx with a powerful wizard to read their miRCURY LNA™ Universal RT microRNA PCR platform (www.exiqon.com/qpcr-software). On the BIOMARK microfluidic platform from Fluidigm, technical and biological replicates are indicated by the naming of assays and samples, and appropriate classification columns are created automatically. Data generated on other qPCR instruments can also be read by GenEx, including the Stratagene MX300X from Agilent, Realplex from Eppendorf, CFX96/384 from Bio-Rad, Eco from Illumina, and the many different qPCR platforms from Life Technologies. These efforts from the instrument manufacturers to transfer experimental design information automatically or at least semi-automatically into GenEx substantially simplify the pre-processing needed to prepare qPCR data for statistical analysis

Data pre-processing

The days when the ΔΔCq method was sufficient to analyse qPCR data have past. Not that we are calculating differently, the studies have become larger and the experiments more complex. In fact, for most studies performed today, it is not even possible to write a closed form expression to calculate the resulting expression response. Rather the measured data must be processed sequentially to account for the various aspects of the experiment. In particular, it is essential to correctly define the statistical unit (often referred to as a subject when organisms are used). Each unit should be associated with a single value for each variable to use common statistical methods. This value must, however, frequently be assembled from various measurements, i.e. responses of target and reference genes, estimated amplification efficiencies, etc. In order to integrate the process into a logical workflow, GenEx provides an intuitive wizard with the following sequential operations:

1 *Interplate calibration*. Many studies cannot be fitted in a single experimental run or for practical reasons have to be extended over time. qPCR instruments perform base-line correction and set threshold separately for each run, which introduces a bias between the Cqs measured in different runs. This bias can be compensated for by performing a common amplification in all plates, where the same sample is analysed for a given assay. This sample is called Inter-Plate Calibrator, IPC. Any variation in the measured Cqs of the IPC among runs reflects systematic variation due to instrument factors and should be compensated for. It is sufficient to run a single IPC for each channel in the instrument if a common threshold and base-line correction is used. It is not recommended to perform separate inter-plate calibrations for each target. Since every correction adds confounding variation to the data, unimportant corrections shall be avoided, as they may impair data quality rather than improving it. For the same reason, it is a good strategy to use a robust sample for IPC and analyse it in replicates. In multiplate experiments, the runs and the inter-plate calibrators shall be indexed in classification columns.

2 *Efficiency correction*. If PCR efficiency has been estimated, the measured Cq values can be corrected to account for suboptimal amplification. Typically, PCR efficiencies are

estimated from serial dilutions run separately. The PCR efficiencies may then be listed in a classification row for automatic correction in GenEx.

3 *Normalize using spiking.* PCR efficiency depends on the sample matrix. Usually it is assumed that the sample matrix and thus the PCR efficiency is constant. But occasionally there are variations, which can be tested for using an exogenous spike added to the samples. Differential expression of the spike between the test and a standard sample reflects the sample's specific inhibition and can to some degree be accounted for.

4 *Normalize to sample amount.* Measured Cq values depend on the sample input. This can be the sample volume processed, amount of RNA used for reverse transcription, or cell count. If sample input varies, data may have to be normalized. The sample input shall be indicated in a classification column.

5 *Average qPCR replicates.* If qPCR replicates are available they shall be indexed in a classification column and their Cq values shall be averaged.

6 *Correct for genomic DNA background.* When quantifying RNA levels using RT-qPCR, the assays may also amplify genomic copies of the target if the DNase treatment used is insufficient. The amount of genomic background can be assessed by measuring either RT^- controls or by using the ValidPrime approach. The contribution to Cq from the genomic background can be calculated and the Cqs corrected.

7 *Normalize with reference genes.* In expression studies normalization to endogenous controls, such as stably expressed reference genes, is popular. With GenEx, we can normalize to any number of reference genes; we can even normalize sets of reporter genes to sets of reference genes to match the genes' properties such as expression levels, stabilities, distribution in tissues, etc. It is also possible to normalize to the mean expression of all the genes (global normalization). Optionally, reference genes can be indexed in a classification row for automatic processing. Normalization to the expression of reference genes corresponds to calculating ΔCq in the classical approach.

8 *Average technical replicates.* If additional technical replicates are available, such as RT, extraction, and sampling replicates, they shall be indexed in classification columns and averaged.

9 *Normalize with reference sample(s).* In some paired designs, systematic variation can be reduced by normalizing to the paired sample during pre-processing.

10 *Relative quantities.* An arbitrary reference level is selected (which corresponds to ΔΔCq in the classical approach) and data are converted to linear scale ($2^{-\Delta\Delta Cq}$ in the classical approach). The reference level can be the most expressed sample, the least expressed sample, mean expression of all the samples, mean expression of a group of samples, or percentage (sum of the expression in all samples set to 1). It is also possible to convert the ΔCq values to an arbitrary linear scale ($2^{-\Delta Cq}$).

11 *Convert to log scale.* For statistical analysis with parametric methods, the data shall be converted to logarithmic scale. Available options are $\log_2$, $\log_{10}$, ln, and $\log(X+1)$.

All the steps in the workflow are not needed, since some cancel the effect of others. The appropriate steps depend on the experimental design, the controls and references that are available, and the analysis that will be performed.

In addition to the pre-processing work flow, GenEx has correction for missing data. GenEx recognizes two types of missing data, random missing (failed experiments) and non-random missing (off-scale data). There is a built-in handling of random missing data among technical replicates, which are replaced based on available information, in the course of the pre-processing. This is very useful, since the missing information can be ignored and is automatically accounted for. There are also several means to handle non-random missing or off scale data that are due to too low target amounts, which may bias the biological effect and invalidate the statistical inference in the majority of the statistical tests employed. Outliers in the data can be tested for based on standard deviation and the Grubb's outlier test. The pre-processing of data is logged and stored in a log file.

Screening by correlation

Several companies including Roche, Exiqon, Life Technologies, Lonza, Qiagen and TATAA Biocenter, offer pre-plated assays for smooth expression profiling and screening purposes. Data from those plates are readily read into GenEx. Rarely are all assays relevant for every study and a strategy is to analyse a few representative samples of each kind in a pilot study to identify differentially expressed genes to be used in a larger downstream study. This is readily done using the GenEx scatter plot (Fig. 4.3). Replicated measurements can be compared with test the reproducibility between plates (top left), or screen for differentially expressed genes under two conditions (top right, bottom).

Correlations between genes' expressions can be quantified by calculating the Pearson or Spearman correlation coefficients. This is typically applied to larger number of samples and has, for example, been used to reveal correlations between genes expressed in individual cells (Stahlberg *et al.*, 2011).

Preparing the data for analysis

Groups for comparison are created using the GenEx Data Manager. Treatment groups or treatment factors in multifactorial studies, such as studies of the effect of gender or covariates such as age, time or drug load, can be indexed in classification columns and used to assign subjects into groups automatically. The groups are assigned colours and symbols for plotting. A neat feature is that colours and symbols can be set independently, which makes it possible to assign subjects to multiple groups and identify these in plots by the shape, size, and colour of the symbol. Even shades of colours can be used creatively to indicate various levels of covariates at ordinary scale (e.g. darker shade indicates higher drug load).

In the data manager, subjects and genes can be removed temporarily from analysis to compare results based on analyses of subsets of data. Data can also be mean centred (subtraction of the mean value) or autoscaled (subtraction of mean followed by division with the standard deviation) to change the weights of the genes/samples in analyses. This is particularly useful in expression profiling analysis, where genes having different expression levels can be assigned equal weights. For analyses that apply models based on measured data, such as the standard curve, reverse calibration, neural networks, self organized map, potential curves etc., samples (and genes) can be assigned either training or test. Training data are used to create the model, which is applied to classify the test samples. The various analyses available in GenEx are listed in Table 4.1.

Standard curve and reverse calibration

Amounts of pathogens in field samples can be quantified using qPCR by comparing the measured Cqs of the field samples with those of standard samples by means of a standard curve. Representative data arrangement is shown in Fig. 4.4. In addition to the measured data, the concentration of the standards is given in a classification column. Additional classification columns can be used to index replicates and to identify the standard and test samples. Any technical replicates shall be averaged during pre-processing and the averaged Cq shall be considered a single data point. Independently prepared standards are treated as different data points, while replicate measurements of field samples are averaged and used as a single more precise estimate. Groups are created in the GenEx Data Manager and assigned either test or training status. Samples can also be reversibly removed from analysis (Fig. 4.5). A confidence level is set for the analysis.

The standard curve is the best straight line fit of the Cqs measured for the standard samples to their concentration in logarithmic scale (Fig. 4.6). It is calculated using linear regression and defines the intercept, which is the Cq expected for a sample containing a single template molecule, and the slope. From the slope, the PCR efficiency is estimated. GenEx also calculates the uncertainties in the estimates of the slope and the intercept, which are reflected by the dashed lines in the plot as the Working-Hotelling area, and the

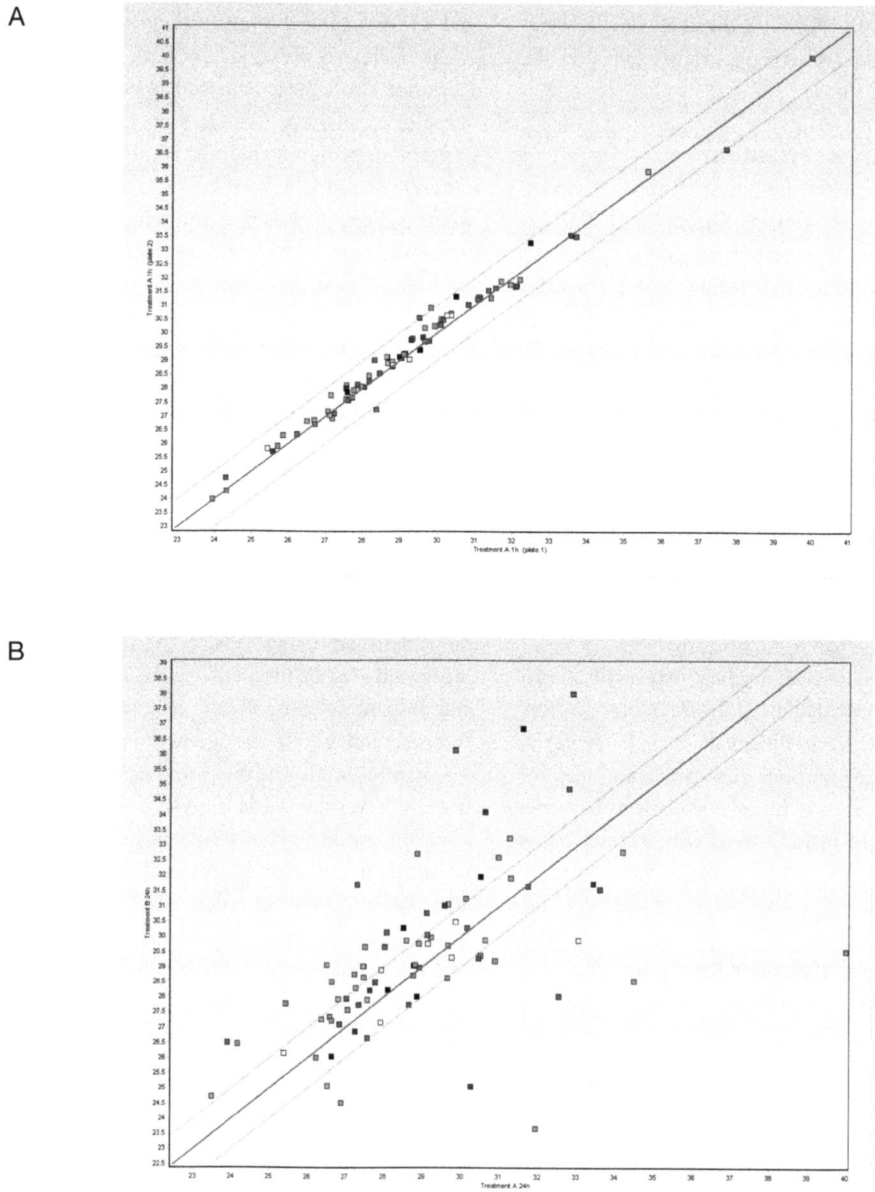

Figure 4.3 Scatter plot comparing replicate measurements performed in separate plates (A) and two different conditions (B). Differentially expressed genes are tabulated (C).

confidence interval for the PCR efficiency (Fig. 4.6). It is essential to calculate the confidence information since it reflects the precision of the estimated efficiency. In this example, the precision of the estimated efficiency is quite high, because a large number (21) of standards was used and a wide concentration range was covered (6 logs). In the literature, we frequently see standard curves based on a substantially lower number of standards. The PCR efficiencies estimated from such standard curves are highly uncertain and any corrections made are unreliable. There are even cases

C

	A	B	C	D	E	F	G	H	I
1	Over-expres...					Under-expre...			
2		Treatment B...	Treatment A...	Diff			Treatment B...	Treatment A...	Diff
3									
4	gene 95	36.13	29.83	6.3		gene 24	29.54	40	-10.46
5	gene 42	36.86	31.58	5.28		gene 55	23.67	31.94	-8.27
6	gene 37	38.01	32.89	5.12		gene 12	28.54	34.49	-5.95
7	gene 83	31.7	27.29	4.41		gene 78	25.05	30.25	-5.2
8	gene 32	32.71	28.86	3.85		gene 8	28.03	32.53	-4.5
9	gene 61	34.12	30.61	3.51		gene 93	29.89	33.04	-3.15
10	gene 49	26.5	23.89	2.61		gene 85	24.52	26.88	-2.36
11	gene 29	29.06	26.5	2.56		gene 81	31.56	33.6	-2.04
12	gene 16	27.78	25.42	2.36		gene 34	31.74	33.43	-1.69
13	gene 9	26.45	24.15	2.3		gene 84	29.21	30.88	-1.67
14	gene 96	29.65	27.5	2.15		gene 70	25.08	26.52	-1.44
15	gene 65	34.87	32.79	2.08		gene 60	32.79	34.19	-1.4
16	gene 2	30.13	28.07	2.06		gene 63	29.28	30.45	-1.17
17	gene 71	33.24	31.25	1.99		gene 44	29.37	30.5	-1.13
18	gene 39	28.5	26.63	1.87		gene 51	28.62	29.64	-1.02
19	gene 73	30.28	28.5	1.78					
20	gene 94	30.77	29.1	1.67					
21	gene 53	32.61	30.96	1.65					
22	gene 15	29.65	28.03	1.62					
23	gene 54	29.01	27.46	1.55					
24	gene 5	29.01	27.47	1.54					
25	gene 59	28.75	27.24	1.51					
26	gene 1	31.97	30.5	1.47					
27	gene 18	31.01	29.57	1.44					

in literature when separate standard curves are measured for the same assay in different plates and the estimated PCR efficiencies are used to adjust the two sets of data. Such adjustment is nonsense since any difference between the estimates of efficiency is caused by random variation in the measurements rather than any systematic difference between the runs.

The residual plot shows the deviations of the standard samples' measured Cq and their predicted Cq by the standard curve (Fig. 4.6). If the straight line standard curve is adequate to model the data, residuals should fluctuate randomly. If the model is inadequate, runs of positive and negative residuals will be seen. GenEx performs a statistical test for the number of runs, and if they are too few, warns that the linear standard curve may be an inadequate model. Outliers are readily identified visually in the residual plot and GenEx further uses the Grubb's test to support the removal of outliers. In general, no more than one outlier should be removed from a standard curve (Anonymous, 2003). If multiple outliers are indicated, the approach used is likely to be unstable and should be overseen. When replicates are available, the residual plot also reveals if noise increases at low concentrations.

A reliable standard curve is critical for accurate estimation of the concentrations of field samples, which in GenEx are referred to as test samples. The estimates improve if the field samples are available in replicates that can be averaged to reduce confounding variation. Concentrations of the unknown samples are estimated by entering the standard curve at the measured Cq and reading out the log of the concentration on the x-axis (Fig. 4.7). The Working-Hotelling area, which reflects the prediction uncertainty, is wider than before because of the additional error contribution from the measured Cq. GenEx calculates confidence intervals for the estimated concentrations. The confidence intervals are symmetric around the mean in logarithmic scale, while they are asymmetric around the mean in linear scale (Fig. 4.7). The uncertainty in the estimates is larger than most people think. Even though the standard curve in the example is based on seven concentrations of standards covering six logs, each measured in triplicates for a total of 21 readings, and the assay has 96% efficiency, the uncertainty in the estimated concentrations of the unknowns is substantial. For example, for 'Test 1', estimated concentration is 46,700 copies, with the 95% confidence interval: 31,000–61,000 copies! With less accurate standard curve, the precision in the estimated concentration would be even worse.

Table 4.1 Genex versions

GenEx version comparison	Standard	Pro	Enterprise
Pre-processing of data			
Interplate calibration	×	×	×
PCR efficiency correction	×		×
Normalize to sample amount	×		×
Normalize to reference genes/samples	×		×
Normalize to spike	×	×	×
Missing data handling and primer dimer correction	×	×	×
Relative quantities and fold changes	×	×	×
Finding optimal reference genes			
geNorm	×	×	×
NormFinder	×	×	×
Plots			
Scatterplots	×	×	×
Line plots	×	×	×
Bar plots	×	×	×
Box and whiskers plot	×	×	×
Principal component analysis			
PCA		×	×
P-curve			×
Cluster analysis			
Hierarchical clustering/dendogram	×	×	×
Heatmap analysis		×	
Networks			
Self-organizing map (SOM)		×	×
Artificial neural networks (ANN)			×
Support vector machine (SVM)			×
Regression analysis			
Standard curve	×	×	×
Reverse calibration		×	×
Limit of detection (LOD)		×	×
Partial least square (PLS)			×
Three-way analysis			
Trilinear decomposition			×
Statistics			
Descriptive statistics	×	×	×
Parametric *t*-test	×	×	×
Non-parametric tests	×	×	×
One-way ANOVA	×	×	×
Two-way ANOVA		×	×
Nested ANOVA		×	×

GenEx version comparison	Standard	Pro	Enterprise
Correlation			
Spearman rank correlation coefficient	×	×	×
Pearson correlation coefficient	×		×
Experimental design			
Sample size		×	×
Experimental design optimization			×

Limit of detection

The limit of detection (LOD) is 'the lowest amount of analyte in a sample that can be detected with (stated) probability, although perhaps not quantified as an exact value', with analyte here referring to the targeted nucleic acid (Anonymous, 1995, 2004). For classical tests, when a signal is measured against a background, LOD is estimated from the standard deviation of the blank readout at the standard curve intercept. This approach is, however, not applicable to qPCR, which, due to its real-time readout, gives no reading for a negative sample (Cq for a blank sample is formally infinity). Instead, for an analytical process that involves qPCR, LOD can be estimated from multiple standard curves (Burns and Valdivia, 2008). A minimum of six is recommended and concentrations around the expected LOD should be assessed. The measured data are transferred to binary format indicating positive and negative PCR's and the fraction of positive calls at each concentration is calculated. LOD is the concentration at which replicates are positive at the stated rates (e.g. 95%). GenEx fits the measured positive rates at different concentrations to estimate the LOD (Fig. 4.8).

Selecting reference genes

With qPCR, the amount of target molecules in a sample is measured rather than their concentration. A large sample is expected to contain more target molecules than a small one and to compensate for the effect of size, normalization must be applied. There are several options to normalize. A popular option in gene expression analysis is to normalize with reference genes, since this should not only compensate for variation in sample amount, but also for variations in extraction yield, reverse transcription efficiency, and RNA quality. In the early days of PCR, genes needed for basic housekeeping functions were thought to have stable expression and could serve as references. Experience has shown this is not always true, and before a gene is used as a reference, this assumption should be validated. Criteria for a good reference gene is that it has a stable expression among samples and that its expression is invariant of the treatment applied. Stability of expression is reflected by the standard deviation (SD) of biological replicates. However, we cannot just take a set of samples, measure the Cqs and calculate the SD's, because we do not know how to normalize

	A	B	C	D	E
1		Gene	#conc	#repeats	#test&train
2	Calibration 1	27.63	1	1	1
3	Calibration 2	25.31	2	2	1
4	Calibration 3	21.42	3	3	1
5	Calibration 4	17.82	4	4	1
6	Calibration 5	14.65	5	5	1
7	Calibration 6	10.63	6	6	1
8	Calibration 7	7.95	7	7	1
9	Calibration 1	28.79	1	1	1
10	Calibration 2	24.61	2	2	1
11	Calibration 3	21.38	3	3	1
12	Calibration 4	17.87	4	4	1
13	Calibration 5	14.27	5	5	1
14	Calibration 6	10.7	6	6	1
15	Calibration 7	8.02	7	7	1
16	Calibration 1	27.91	1	1	1
17	Calibration 2	24.68	2	2	1
18	Calibration 3	21.42	3	3	1
19	Calibration 4	17.91	4	4	1
20	Calibration 5	14.45	5	5	1
21	Calibration 6	10.35	6	6	1
22	Calibration 7	8.14	7	7	1
23	Test 1	15.76	0	22	2
24	Test 2	18.22	0	23	2
25	Test 3	20.74	0	24	2

Figure 4.4 Data prepared for analysis using a standard curve. Classification columns indicate concentrations of standards, sample replicates, and indexes test and training samples.

A

B

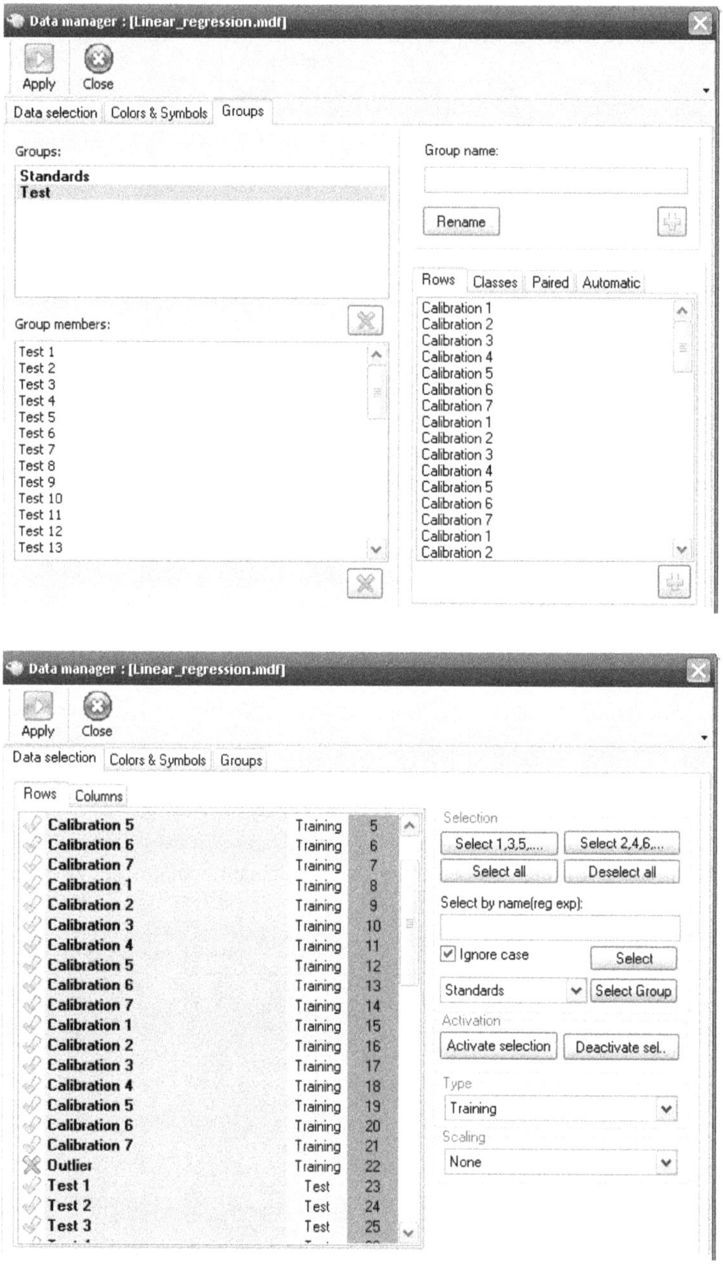

Figure 4.5 Groups are created in the GenEx data manager optionally using information in classification columns (A). Samples are assigned training or test status, and may be reversibly inactivated (B).

the samples for this exercise. Of course, we can use the same amount of RNA in the analyses, but then, how do we know that all samples were extracted and reverse transcribed with the same yield, and that they have the same mRNA/total RNA ratio and the same RNA quality? Furthermore, if we were to assume that we can evaluate genes' expression stabilities based on samples normalized to the same amount of RNA, then we would have already decided that total RNA is the best norm. The gene selected based on minimum SD measured on samples having the same amount of

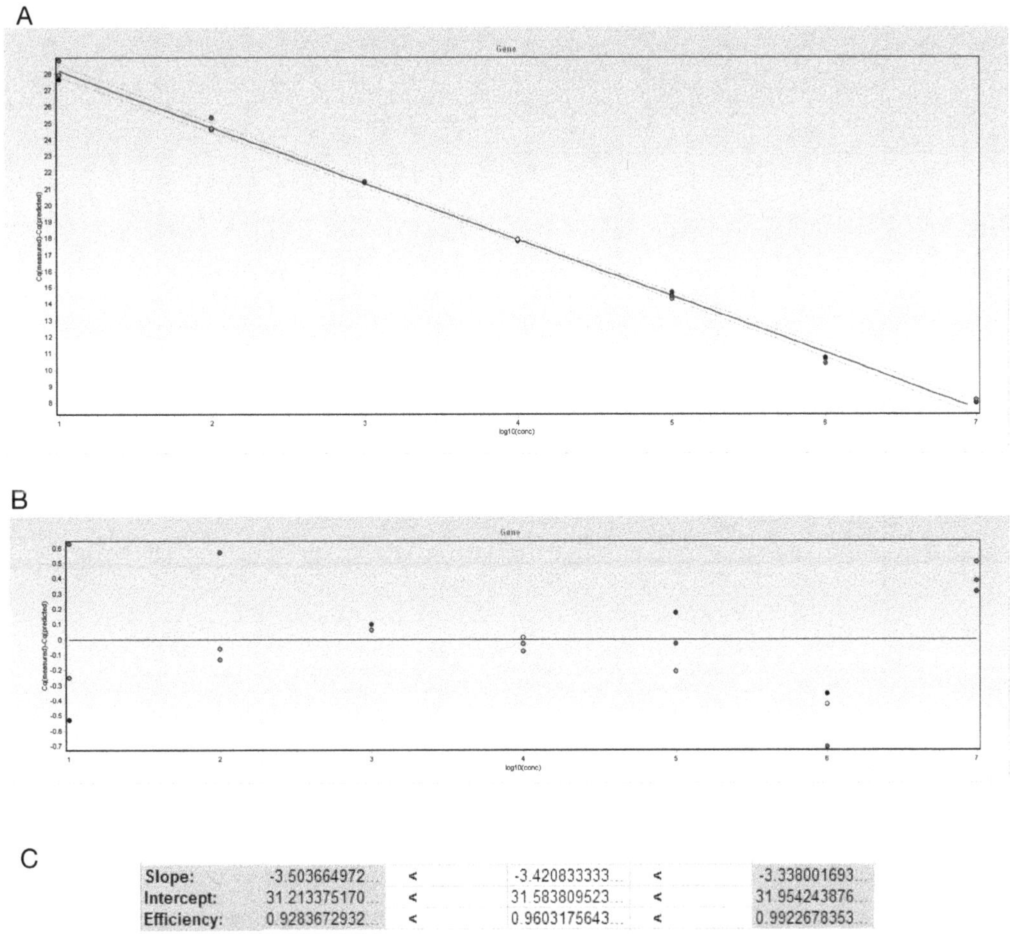

Figure 4.6 (A) qPCR standard curve (line) with uncertainty indicated (dashed lines). (B) residual plot showing the deviations of the standard points from the fitted standard curve. (C) estimated slope, intercept and PCR efficiency for the standard curve including confidence intervals.

RNA will be the gene that shows a variation that correlates the most with that of total amount of RNA, and we may then as well normalize to the amount of RNA directly. If we suspect that the total amount of RNA is not the best norm, we have to identify optimum reference genes using different strategy.

An appropriate approach to select reference genes is a special form of analysis of variance, which in qPCR literature is best known as using the tool NormFinder (Andersen *et al.*, 2004). NormFinder is applied to a panel of candidate reference genes that is analysed in a set of representative samples. In essence, NormFinder calculates a global average expression of all the genes in all the samples, to which the individual genes are compared. Based on this comparison, SD for each candidate reference gene is estimated. Furthermore, if the samples are from different treatment groups, NormFinder separates the variation into an intragroup and an intergroup contribution. Fig. 4.9 shows an example where reference genes were sought for an obesity study in mice, where wild-type mice and an obese strain were compared. The genes were selected from the TATAA reference gene panel (www.tataa.com/products/Gene-Expression-Assays-and-Panels/View-all-products.html), which was measured on

A

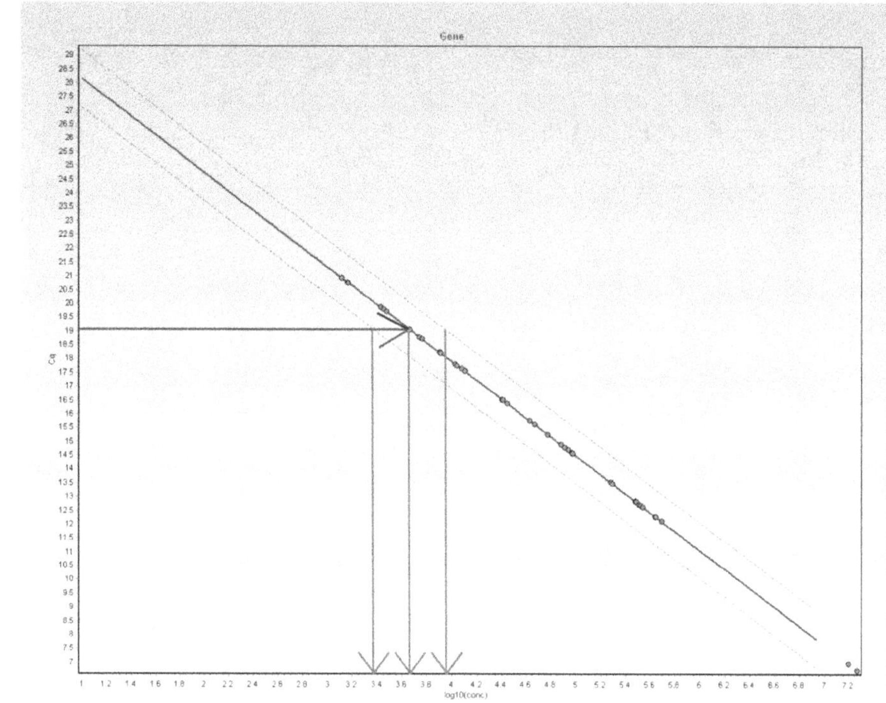

B

	Log(Conc) Low	Log(Conc)	Log(Conc) Up	Repeats	Conc. Low	Conc.	Conc. Up
Test 1	4.4934490538...	4.6403340873...	4.7872191208...	3	31149.354679...	43685.175758...	61265.942768...
Test 2	3.8124309780...	3.9494692883...	4.0865075985...	3	6492.7843427...	8901.6248683...	12204.151734...
Test 3	2.9954390885...	3.1494692883...	3.30349948811...	3	989.55306551...	1410.8124655...	2011.4048274...
Test 4	4.2768800219...	4.4181659996...	4.5594519773...	3	18918.209126...	26191.839432...	36262.018684...
Test 5	4.8034314462...	4.9618931616...	5.1203548770...	3	63596.240975...	91599.512357...	131933.43718...
Test 6	3.3072282896...	3.4515399338...	3.5958515781...	3	2028.7488638...	2828.3941731...	3943.2251776...
Test 7	4.7014554271...	4.8556812249...	5.0099070228...	3	50286.965231...	71726.761884...	102307.39410...
Test 8	5.3415332445...	5.5280320167...	5.7145307888...	3	219549.90047...	337312.17479...	518239.83075...
Test 9	3.9399952514...	4.0771184966...	4.2142417417...	3	8709.5406689...	11943.139268...	16377.278779...
Test 10	5.11376236059...	5.2873499216...	5.4609374828...	3	129945.83394...	193798.28152...	289026.37955...
Test 11	3.5855139776...	3.7243779363...	3.8632418949...	3	3850.4720715...	5301.2457380...	7298.6391935...
Test 12	5.301050398549	5.4851574734...	5.6692645483...	3	200009.39609...	305602.90163...	466943.73021...
Test 13	6.9432878254...	7.2449625891...	7.5466373528...	3	8775822.3962...	17577721.901...	35207675.507...
Test 14	5.4682718668...	5.6625021750...	5.8567324832...	3	293948.91864...	459729.29095...	719005.94817...

Figure 4.7 (A) Calibration by means of the standard curve to estimate the test sample's concentration and its confidence interval. (B) summary of estimated concentrations including confidence intervals in logarithmic (left) and linear (right) scales.

seven representative mice from each strain. The intragroup variation estimated is the SD of the genes in the different treatment groups, while the intergroup variation is differential expression and sums to zero for every gene over all the groups. Good reference genes shall have low intergroup variation in all groups and negligible intergroup variation. Good strategy using NormFinder is to inspect the calculated intra and intergroup variations to identify any genes that appear regulated or exceedingly unstable and remove them from the data set using the GenEx Data Manager. NormFinder analysis is then repeated without considering the groups, since the remaining genes are not regulated. This produces more robust result with a single SD estimate for each gene, based on which the genes are ranked (Fig. 4.10). The gene with lowest SD is the optimum reference gene. GenEx also calculates the accumulated SD expected if multiple reference genes are used for

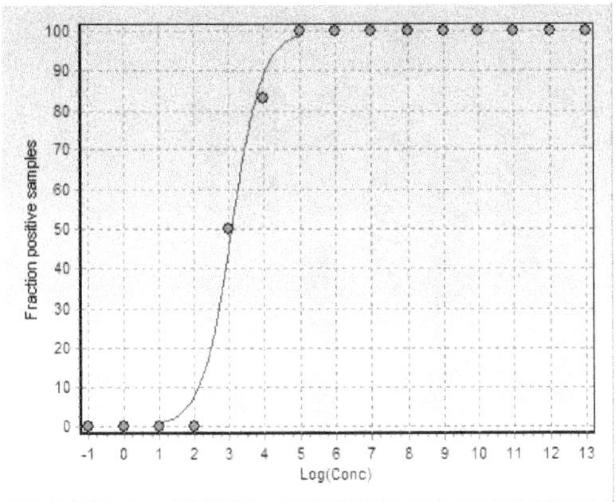

Figure 4.8 Estimation of LOD by fitting the rate of positive calls as function of concentration.

A

NormFinder (Intergroup variation)

	A	B	C
1	Gene Name	1	2
2	PPIA	-0.0349	0.0349
3	HPRT	0.0572	-0.0572
4	YWHAZ	-0.0507	0.0507
5	TUBB5	-0.0178	0.0178
6	B2M	0.1122	-0.1122
7	ACTB	0.0029	-0.0029
8	TBP	0.0658	-0.0658
9	GAPDH	0.0086	-0.0086
10	PGK1	-0.0542	0.0542
11	18S rRNA	-0.0471	0.0471
12	GUSB	-0.0099	0.0099
13	ARBP	-0.0321	0.0321

B

NormFinder (Intragroup variation)

	A	B	C
1	Gene Name	1	2
2	PPIA	0.0008	0.0021
3	HPRT	0.0242	0.0062
4	YWHAZ	0.0190	0.0168
5	TUBB5	0.0086	0.0043
6	B2M	0.0139	0.0057
7	ACTB	0.0158	0.0017
8	TBP	0.0154	0.0157
9	GAPDH	0.0035	0.0139
10	PGK1	0.0055	0.0013
11	18S rRNA	0.0184	0.0152
12	GUSB	0.0072	0.0006
13	ARBP	0.0092	0.0048

Figure 4.9 Estimated intergroup and intragroup variations with NormFinder of genes from the TATAA reference panels in representative brain samples of wild-type and obese mice.

normalization. If we use larger number of reference genes, random variation among the genes' expression partially cancel reducing the SD. Comparing the SD contributed from different number of reference genes selected based on stability, a minimum in the accumulated SD plot is obtained, indicating the number of reference genes that give the lowest SD (Fig. 4.10). However, analysing more genes cost time and money, and one should consider the degree of improvement and the overall noise contributed by the reference genes when making a decision. In the example, the largest improvement is observed when including the second reference gene; including additional reference genes only slightly improves the result. Furthermore, the noise contribution from the best reference gene is only 0.05 cycles and as little as 0.04 cycles when combining the two best reference genes. Considering that the repeatability of a qPCR instrument is rarely less than 0.1 cycle (estimated as SD of technical replicates), using more than one reference gene, and definitely more than two, will in this study not improve the quality of the data appreciably.

Using NormFinder, normalization with reference genes can be compared with normalization with total RNA by adding an extra column in the data sheet with the RNA concentrations per analysed sample in logarithmic scale (Fig. 4.11). The algorithm is ignorant of the nature of the variables, and will compare their variation. For the data in our example, normalization with total RNA is essentially as stable as normalization with PPIA, which is the single optimum reference gene here. In this study the samples analysed were flash-frozen biopsies from mouse brains, from which RNA of very high quality (RIN 8–9) was extracted. Our experience is that for samples with high quality RNA, normalization to total amount of RNA is often as good as normalizing with a single reference gene. In samples of poor RNA integrity, or when expression may have been induced, normalization with reference genes is preferred.

An older method to identify good reference genes that still is being used is geNorm. It uses the same input data as NormFinder, but it does not consider groups; all samples are treated as being from a single population. geNorm sequentially

A

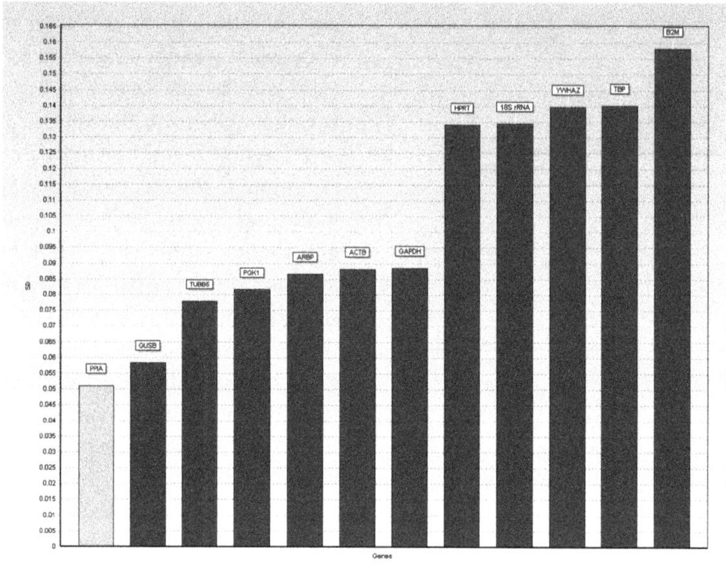

B

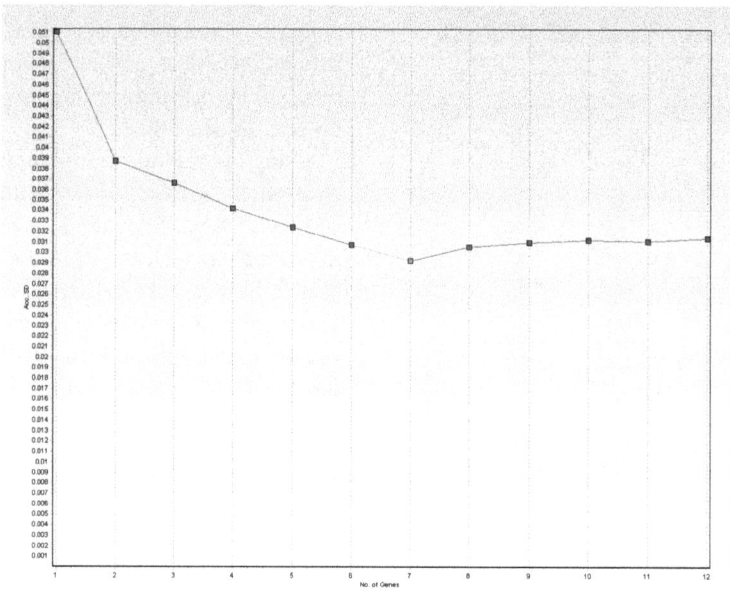

Figure 4.10 SD (A) and accumulated SD (B) for the reference gene candidates in the TATAA reference gene panel in 14 brain samples from mice estimated using Normfider.

eliminates the gene that shows the highest variation relative to all the other genes based on paired expression values in all the studied samples. The variability is reflected by a so-called M-value (Fig. 4.12). Because of the elimination process, geNorm cannot identify an optimum reference gene, and ends up by suggesting a pair of genes that shows high correlation and should be suitable for normalization. The M-value is related to the SD, but as calculated, the M-values for the genes are based on different sample sizes and are therefore not strictly comparable. Furthermore,

A

	A	K	L	M	N	O
1		18S rRNA	GUSB	ARBP	#group	log(RNA)
2	NC1	7.48	24.83	20.77	1	2
3	NC2	7.61	24.88	20.49	1	2
4	NC3	7.72	24.83	20.32	1	2
5	NC4	7.78	25.03	20.62	1	2
6	NC5	7.7	25.12	20.6	1	2
7	NC6	7.68	25	20.55	1	2
8	NC7	7.7	24.97	20.49	1	2
9	HFD1	7.58	24.92	20.7	2	2
10	HFD2	7.66	24.97	20.64	2	2
11	HFD3	7.63	24.94	20.6	2	2
12	HFD4	7.98	25	20.65	2	2
13	HFD5	7.8	24.96	20.57	2	2
14	HFD6	7.78	24.91	20.51	2	2
15	HFD7	7.79	24.99	20.51	2	2

B

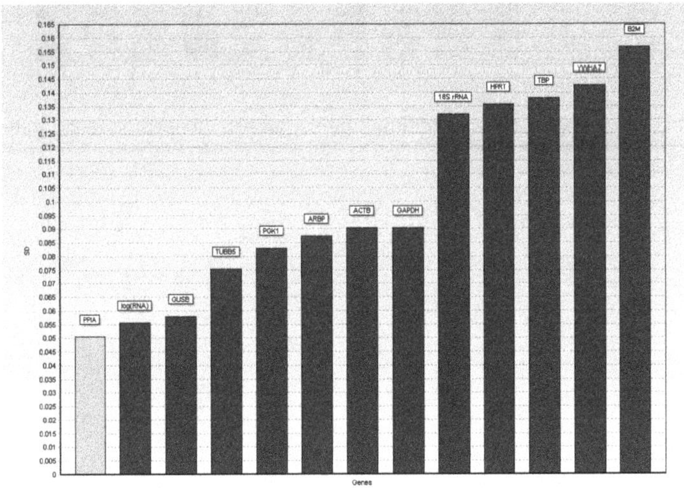

Figure 4.11 (A) Input data for comparison of normalization with reference genes to normalization with total amount of RNA using Normfider. A column indexing the used total RNA concentrations in logarithmic scale has been added. (B) Output indicating that for these particular samples, total RNA normalization is almost as stable as normalization with the single best reference gene.

as the comparison of any individual candidate gene is performed towards a plurality of genes, assumed to resemble most closely the anticipated stable behaviour, it is prone to systematic failure where group of co-regulated instable genes may be involved in the analysis. Any such co-regulated complex of instable genes may dominate over the stable genes and hence point at deviant genes as candidates. Usually, the gene rankings by geNorm and NormFinder are similar, which is reassuring. Should the rankings differ, there would be a reason to suspect the selection to include one or more regulated genes, and the result should be interpreted with caution.

Relative quantification

Treatment groups are readily compared visually in bar graphs using descriptive statistics (Fig. 4.13) and statistical comparison is made using ANOVA (one factor, two or more levels) or two-way ANOVA (two factors, two or more levels each) or, in the case of two groups, with either t-test (paired/unpaired, one-tail/two-tail) or non-parametric tests (Mann–Whitney, Wilcoxon). The difference between the groups is shown in either linear or logarithmic scale and the confidence interval is indicated (Fig. 4.14). Note that the confidence interval of the differential expression is asymmetric

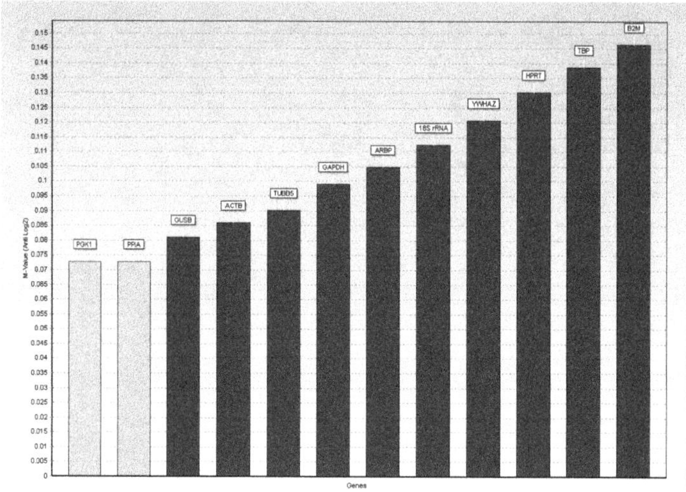

Figure 4.12 Ranking of reference gene candidates by sequential elimination using geNorm. The final two genes cannot be compared further.

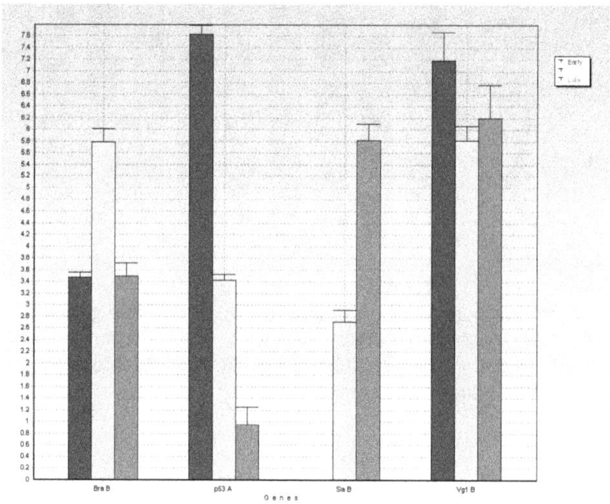

Figure 4.13 Comparison of the expression of four genes in three groups. Bars indicate mean expression and the error bar either SD, SE or CI.

when data are presented in linear scale. When expression of many genes is compared, GenEx offers several means to control for the false discovery rate due to multiple testing, including Bonferroni, Benjamini–Hochberg, Westfall and Young, and Benjamini–Yekutieli (Fig. 4.15).

Expression profiling

t-Test and ANOVA are univariate methods that analyse the expression of every gene separately, effectively assuming that the genes are expressed independently of each other. This is rarely the case; genes' expressions tend to be correlated. This correlation can be exploited in the analyses using multivariate statistical methods. GenEx

A

	A	B	C	D	E
1		Bra B (Early)	Bra B (Late)	p53 A (Early)	p53 A (Late)
2		3.1947	3.1021	8.14879	0
3		3.50028	3.94476	7.96359	1.0186
4		3.7503	3.20396	7.7784	1.4816
5		3.6114	3.44472	7.408	1.1112
6		3.40768	4.69482	7.2228	1.9446
7		3.30582	3.07432	7.2228	2.0372
8			2.69466		0
9			3.75956		0
10	**KS**	0.11224	0.17524	0.20745	0.23979
11	**Norm. Dist.**	Passed	Passed	Passed	Passed
12	**KS P-Value**	>0.1	>0.1	>0.1	>0.1
13	**Count**	6	8	6	8
14	**Mean**	3.46169	3.48986	7.62406	0.94915
15	**STDEV**	0.20283	0.62914	0.39577	0.86123
16	**df**		12		12
17	**SD^2**		0.24804		0.49793
18	**t**		0.10472		17.5153
19	**P-Value**		0.918329628		1E-8
20	**Confidence level**		0.95		0.95
21	**SEM**		0.23735		0.3447
22	**t***		2.17881		2.17881
23	**Difference (log scale)**		-0.02817		6.67491
24	**CI start/end (log)**	-0.5453	0.48897	5.92387	7.42595
25	**Fold change**		-1.01971		102.17599
26	**CI start/end (linear)**	-1.45933	1.40344	60.71033	171.96305

B

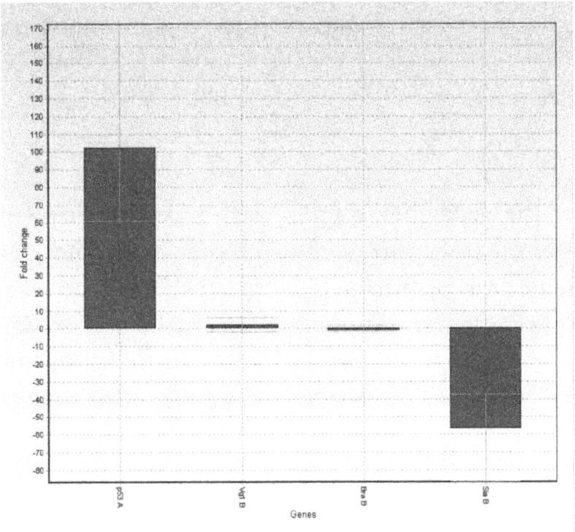

Figure 4.14 Comparison with *t*-test for the expression of four genes between two groups. (A) Descriptive statistics include normality test and *P*-values. (B) Bar graph showing the differential expression in linear scale. Note, the indicated confidence intervals are asymmetric around the means.

offers several unsupervised as well as a selection of supervised methods to classify samples and categorize genes based on expression profiles. Unsupervised methods classify samples and genes based on the measured profiles only. They include classical hierarchical clustering combined with heat map, which can be based on various clustering schemes including the Ward's algorithm and several distance measures including the Euclidian distance and the magnitude of the Pearson correlation. While the Euclidian distance clusters genes based on similarities and consider up-regulation and down-regulation being opposite, hence anti-correlated, distance based on the magnitude of the Pearson correlation considers up-regulation and down-regulation to be correlated. The latter is useful to classify, for example, genes that show the same temporal response to

	A	B	D	E	F	G	H	I
1	(Early) vs (Late)	Normality ...	Fold change	Difference (log...	P-Value	P (Benjamini-Hochb...	P (Westfall & Young)	P (Benjamini-Yekutieli)
2	Xcad2	Passed	4.25835	2.09029	6.6E-8	0.00003	0	0.00015
3	Pax6	Passed	-4.22654	-2.07948	1.445E-6	0.00002	0	0.00009
4	An1	Passed	3.34994	1.74414	2.069E-6	0.00006	0	0.00029
5	Oct-60	Passed	3.9197	1.97074	2.852E-6	0.01028	0	0.05169
6	VegT	Passed	4.25014	2.08751	3.858E-6	0.01028	0	0.05169
7	Xdazl	Passed	3.94797	1.98111	7.068E-6	0.01076	0	0.05411
8	Vg1	Passed	4.07709	2.02754	8.66E-6	0.01157	0	0.05816
9	HNF-3beta	Passed	-3.95275	-1.98286	2.1577E-5	0.00032	0	0.00161
10	follistatin	Passed	-4.10137	-2.03611	9.174E-5	0.00074	0	0.00372
11	cytokeratin	Passed	-3.45001	-1.7866	0.000140734	0.01235	0.01376	0.06208
12	An2	Passed	-3.67176	-1.87647	0.00041507	0.03288	0.01615	0.16528
13	Est-1	Passed	2.22441	1.15342	0.000942705	0.02423	0.02698	0.12182
14	U3-snoRNA	Failed	-3.61566	-1.85426	0.00102422	0.00287	0.01376	0.01445
15	Xmam1	Failed	2.28333	1.19114	0.001689097	0.03097	0.03872	0.15572
16	GSK-3beta	Passed	-2.48805	-1.31501	0.001826109	0.03288	0.06069	0.16528
17	N-CAM	Passed	-2.96568	-1.56836	0.003911296	0.07654	0.1341	0.38479
18	Otx-1	Passed	2.61282	1.38561	0.005112953	0.02423	0.10217	0.12182
19	Wnt11	Passed	3.24284	1.69726	0.00888664	0.0757	0.16567	0.38058
20	dishevelled	Passed	1.77314	0.82631	0.033430491	0.17244	0.37813	0.86689
21	ubiquitin	Passed	-2.20924	-1.14355	0.054665413	0.21805	0.47228	> 0.99
22	alpha-tubulin	Passed	1.61367	0.69035	0.06707434	0.13149	0.40531	0.66102
23	goosecoid	Passed	-2.01217	-1.00875	0.134073733	0.1493	0.46947	0.75056
24	Stat3	Passed	-1.43598	-0.52203	0.152717812	0.31918	0.71231	> 0.99

Figure 4.15 Comparison of the expression of multiple genes between two groups corrected for the high false discovery rate due to multiple testing. Columns 2 and 3 indicate results of normality test and column 4 the differential expression in log scale. Column 5 indicates *P*-values calculated with *t*-test. Grey indicates *P*-values considered significant based on Bonferroni correction, light grey are *P*-values below stipulated uncorrected confidence level that are not significant with Bonferroni correction, and dark grey are *P*-values above confidence threshold. The three last columns indicate *P*-values calculated using Benjamini-Hochberg, Westfall and Young, and Benjamini–Yekutieli corrections for the high false discovery rate.

treatment independently of the genes being up or down-regulated. The clustering is visualized in a dendrogram, which in GenEx can be mirrored in every node. Mirroring in a node changes the visual appearance of the dendrogram, producing an equivalent mathematical solution. A small self-organized map (SOM) can be used to force classification into a defined small number of groups based on expression similarities. SOM can also be used to validate a classification model, based on the distribution of samples/genes in a large map. Principal component analysis (PCA)

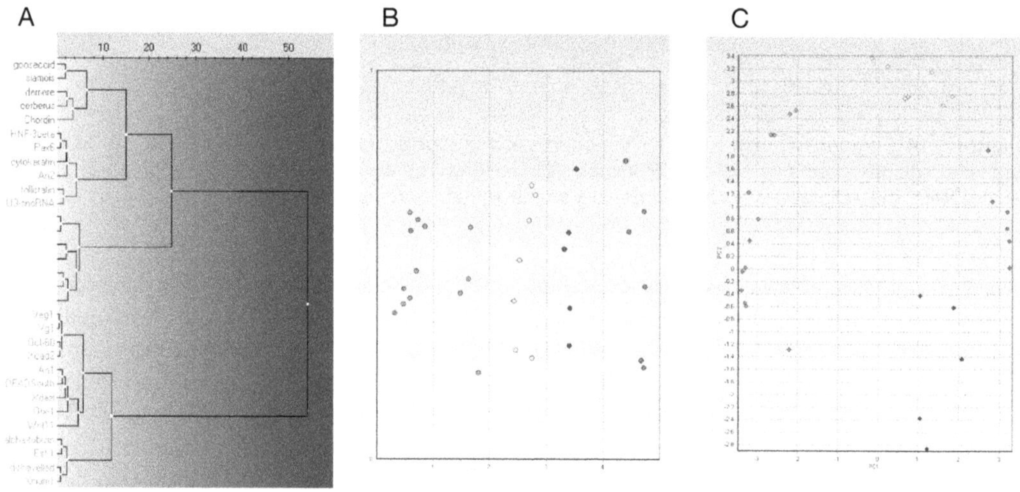

Figure 4.16 Classification of genes expressed during early development of *Xenopus laevis* based on hierarchical clustering (A), SOM (B), and PCA (C).

groups samples/genes based on correlated expression in reduced space. Fig. 4.16 shows example of hierarchical clustering, SOM, and PCA of genes expressed during the development of the African claw frog *Xenopus laevis* from the oocyte to tadpole stage (Bergkvist *et al.*, 2010). Hierarchical clustering of sample and genes can be combined showing also the measured intensities in a heat map (Fig. 4.17). The appearance of the heat map can be changed in GenEx within equivalent mathematical solutions by mirroring the dendrograms' in nodes.

Supervised methods require a training set of samples with known classification; for example, negative and positive samples, or short, medium and long term survivals. A model is developed based on the training set that can be used to classify new data (in GenEx called test data). The procedure is similar to a regression based on a standard curve, but here it is based on multiple genes and the model does not have to be linear. Supervised methods available in GenEx include partial least squares (PLA), which is used to calculate a single standard curve based on the expression of multiple genes to predict concentrations or other measures of test samples. Potential curves is a variant for prediction of new data based on PCA, and artificial neural network (ANN) and support vector machines (SVM) are multivariate non-linear methods to classify samples. Logistic regression, Probit, receiver operating characteristics (ROC), and survival analysis will soon also be available.

Experimental design

Designing experiments is more difficult than analysing results. In a well designed experiment, confounding variation is minimized and the number of subject is sufficient to obtain conclusive results. A good strategy is to perform a fully nested pilot study before specifying the test protocol for a larger study (Fig. 4.18; Tichopad *et al.*, 2009). Fig. 4.19 shows the result of a nested pilot study wherein three heifers were studied by collecting three blood samples of each that were reverse transcribed in triplicates, and each cDNA was analysed in triplicate using qPCR. Using a

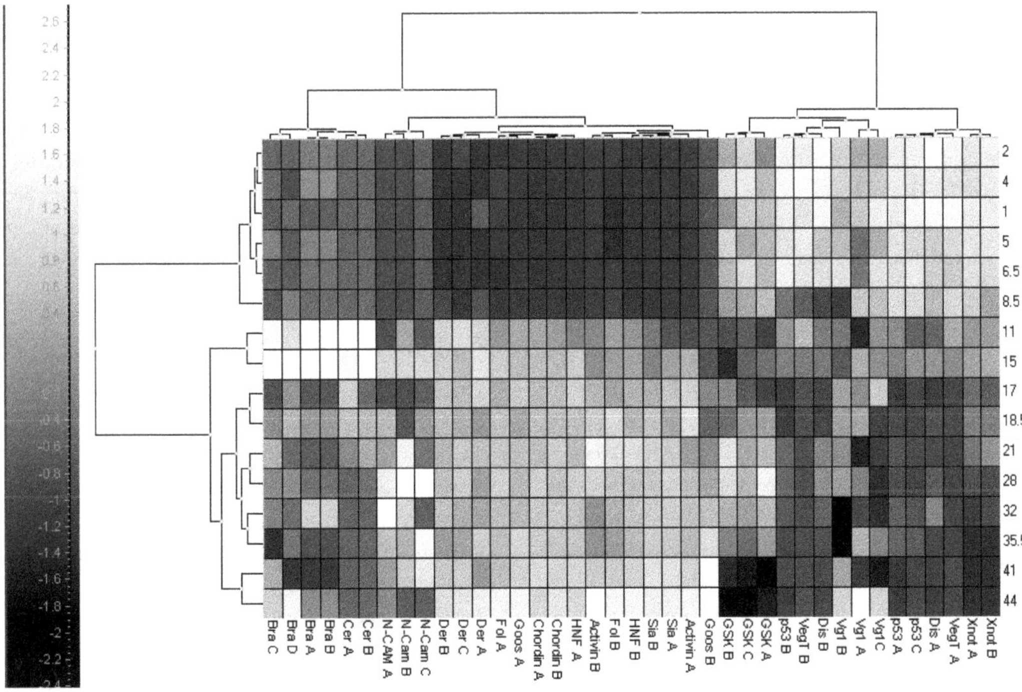

Figure 4.17 Classification of genes and samples visualized with dendrograms and a heat map indicating expression levels. The heat map appearance can be changed by mirroring in any of the dendrograms' nodes.

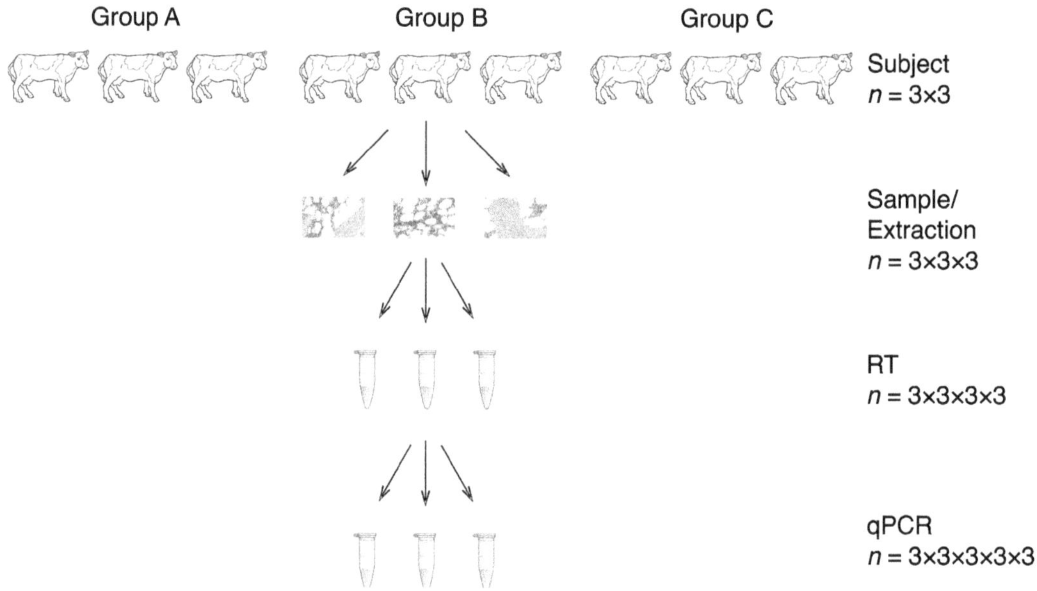

Figure 4.18 Nested experimental design. Three subjects (heifers) are tested; three samples are collected from each subject and extracted; the extracted material is reverse transcribed in triplicates. Each cDNA is analyzed in triplicates for qPCR.

nested ANOVA, the variation arising from the different experimental steps can be estimated and expressed either as standard deviations or as variance contributions (Kitchen *et al.*, 2010). While β-actin and caspase-3 show generally low standard deviations in all steps, interleukin 1β and interferon-levels varied substantially among the heifers. For liver samples, the picture was different, with the data evidencing large variation in the sampling step (Trichopad *et al.*, 2009). Knowing the costs associated with the different experimental steps, the follow up study can be cost optimized. For example, for genes exhibiting SD = 0.1 cycle for the qPCR and RT steps, SD = 0.2 cycles for the sampling/extraction step, and SD = 1 cycle for the variation among the animals, and assuming a cost of 1 unit for the qPCR, 3 units for the RT, 10 units for sampling/extraction, and 100 unit for each animal, with a total budget of 1000 units, the best we can do is to analyse eight animals, sample each animal once, perform RT in triplicates and qPCR in duplicates. The total standard error (SE) for this study is expected to be about 0.36 cycles (Fig. 4.20). Using the same tool, we also estimated the SD among animals analysed with a single sample collected from each animal, RT performed in triplicates and qPCR in duplicate to 1.02 cycles. This can then be fed into a Power analysis to estimate the number of animals needed to ensure a particular difference with certain confidence and power. If we accept 5% false-positive rate (95% confidence) and 5% false-negative rate (95% power) we construct a graph showing how many subject are needed to measure a particular difference due to treatment, For example, to measure a 2-fold difference (ΔCq of 1) we require under these criteria 15 animals (Fig. 4.20).

Conclusions

GenEx is the first professional software that offers the full palette of methods for qPCR data analysis and qPCR experimental design. The user interface is attractive and the graphical output is of high quality. The help file is adequate and the audiovisual tutorials based on example data are excellent (www.multid.se/tutorials.php). A big plus is the free support forum, where hundreds of GenEx users share experiences and advises (www.qpcrforum.com). The important support from virtually all leading qPCR instrument manufacturers is also most valuable.

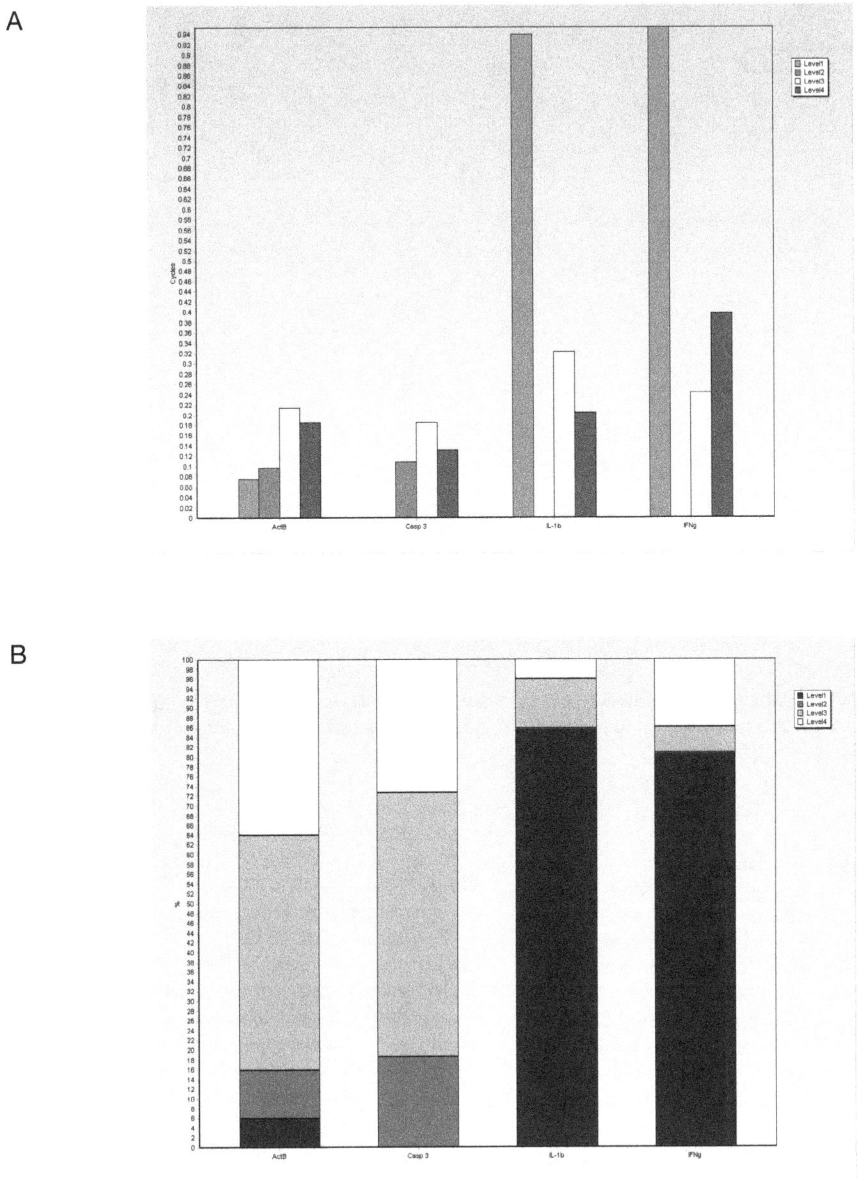

Figure 4.19 Decomposition of the variance in a four-level nested study of blood samples collected from heifers. (A) estimated standard deviations (SD) among heifers (grey), among replicate blood samplings (dark grey), among reverse transcription replicates (white), and among qPCR replicates (black). (B) Same data presented as variance contributions expressed in percentages.

Future trends

qPCR is rapidly becoming a broadly accepted platform, which leads to needs for standardization and quality control. This will require future software to be GLP, ISO, and Title 21 CFR Part 11 compliant. We also expect followers to Roche and Exiqon that already offer excellent wizards for transferring data and experimental design information to qPCR analysis software. Integration with Laboratory Information Management Systems (LIMS) is also expected.

A

	Level 1	Level 2	Level 3	Level 4	Tot. Var.	SD	Cost
Design Total	10	6	1	1	0.1055	0.3248076353...	980
Design Subject	1	6	1	1	1.055	1.0271319292...	98

B

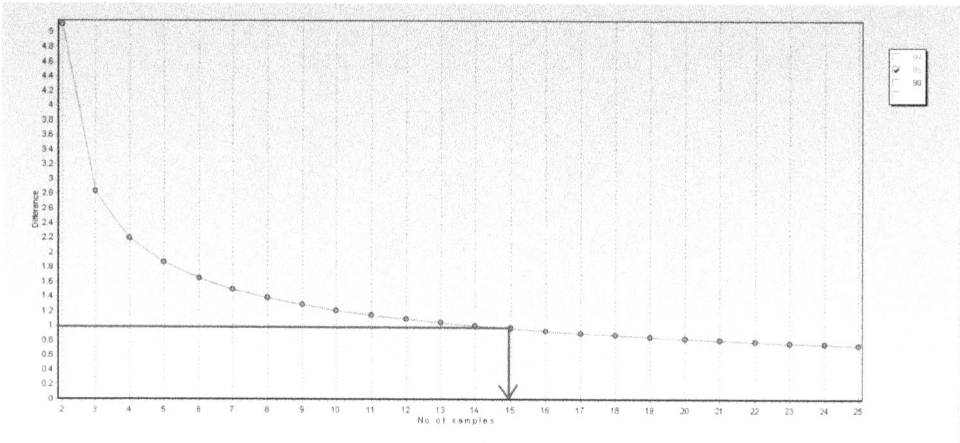

Figure 4.20 Power analysis assuming SD = 1.055 cycles and 5% false-positive and 5% false-negative calls. Straight line and arrow indicate that 15 subjects are needed to measure a difference of one cycle owing to treatment under those conditions.

References

Andersen, C.L., Jensen, J.L., and Ørntoft, T.F. (2004). Normalization of real-time quantitative reverse transcription-PCR data: a model-based variance estimation approach to identify genes suited for normalization, applied to bladder and colon cancer data sets. Cancer Res. *64*, 5245–5250.

Anonymous (1995). Glossary of Terms for Biological Substances Used for Texts of the Requirements. WHO Expert Committee on Biological Standardization. Document WHO-BS/95.173. Geneva, World Health Organization.

Anonymous (2003). Approved Guideline NCCLS. EP6-A, Vol. 23, no. 16.

Anonymous (2004). Approved Guideline NCCLS. EP17-A, Vol. 24, no. 10.

Bergkvist, A., Rusnakova, V., Sindelka, R., Garda, J.M.A., Sjögreen, B., Lindh, D., Forootan, A., and Kubista, M. (2010). Gene expression profiling. Clusters of possibilities. Methods *50*, 323–335.

Burns, M., and Valdivia, H. (2008). Modelling the limit of detection in real-time quantitative PCR. Eur. Food Res. Technol. *226*, 1513–1524.

Kitchen, R.R., Kubista, M., and Tichopad, A. (2010). Statistical aspects of quantitative real-time PCR experiment design. Methods *50*, 231–236.

Stahlberg, A., Andersson, D., Aurelius, J., Faiz, M., Pekna, M., Kubista, M., and Pekny, M. (2011). Defining cell populations with single-cell gene expression profiling: correlations and identification of astrocyte subpopulations. Nucl. Acids Res. *39*, e24.

Tichopad, A., Kitchen, R., Riedmaier, I., Becker, C., Stahlberg, A., and Kubista, M. (2009). Design and optimization of reverse-transcription quantitative PCR experiments. Clin. Chem. *55*, 1816–1823.

Part II

qPCR for Quantifying Microorganisms

Part II

qPCR for Quantifying Microorganisms

Quantification of Microorganisms Targeting Conserved Genes in Complex Environmental Samples using Quantitative Polymerase Chain Reaction

5

Claudia Goyer and Catherine E. Dandie

Abstract

Development of quantitative PCR (qPCR) has facilitated major advances in assessment of microbial community abundances in complex environmental samples including water, soil, sediments, compost and manure and in our understanding of factors influencing community sizes *in situ*. qPCR has increasingly been used in environmental studies due to its sensitivity, ease of use, and the capacity to run large numbers of samples. However, qPCR has some limitations, which are specifically caused by the nature of environmental samples, including the variability in microorganism distribution, the efficiency of DNA recovery and purification, and the amount and type of PCR inhibitors co-extracted with the target nucleic acids. The heterogeneity of the templates amplified by qPCR can generate PCR biases and artefacts. Accuracy of the quantification of broad groups of microorganisms is influenced by the number of gene copies per genome of the selected marker. In this review, we will discuss the main experimental considerations for using qPCR in environmental studies, including the factors affecting key steps in the process of performing quantification of microorganisms in environmental samples. Although quantification of microorganisms is challenging, it is possible to reliably quantify microorganisms in complex environmental samples using qPCR. We will also briefly review the findings of studies which have used qPCR to quantify microorganisms from complex matrices.

Introduction

Development of quantitative PCR (qPCR) and its application in the field of microbiology has facilitated major advances in our understanding of microbial community abundance by targeting conserved genes including ribosomal genes (16S, 18S rRNA genes or internal transcribed spacer (ITS)) (Takai and Horikoshi, 2000) and functional genes (Dandie *et al.*, 2007; Smith and Osborn, 2009; Henderson *et al.*, 2010). qPCR quantifies the number of copies of targeted microbial genes in environmental samples thus providing valuable information on total microbial population size (Smith and Osborn, 2009) and on the abundances of microbial functional groups (Dandie *et al.*, 2008; Attard *et al.*, 2010; Nettmann *et al.*, 2010).

The first reports of the use of qPCR in environmental studies appeared in the literature early in the year 2000 with several studies quantifying bacteria or archaea using small subunit ribosomal genes in DNA extracts from marine water (Takai and Horikoshi, 2000; Suzuki *et al.*, 2000), lake water (Becker *et al.*, 2000) and groundwater (Hristova *et al.*, 2001). Quantifications of bacteria involved in the nitrogen cycle (Gruntzig *et al.*, 2001; Hermansson and Lindgren, 2001), and capable of pollutant degradation (Stults *et al.*, 2001) were successfully performed in complex matrices including soil and sediment. Microorganisms were quantified using qPCR from other complex matrices including pathogens (Higgins *et al.*, 2001) and antibiotic-resistant bacteria (Yu *et al.*, 2005) from manures. In subsequent years, several studies have utilized qPCR targeting conserved genes to quantify microorganisms for a large variety of purposes including understanding the distribution of microorganisms in the environment (Ochsenreiter *et al.*, 2003; Cassler *et al.*, 2008), quantification of changes in abundance of microorganisms involved in biogeochemical processes (Lee *et al.*, 2008; Attard *et al.*, 2010), pathology (Wielgoss *et al.*, 2010), and bioremediation (Taş *et al.*, 2009) or assessment of soil microbial community structure at broad taxonomic levels (Philippot *et al.*, 2009; Wessén *et al.*, 2009).

qPCR has several advantages for quantification of microbes in environmental samples. It is a culture-independent methodology, a distinct advantage compared with classical culture-based microbiology techniques considering that most microorganisms in soils are yet not culturable in the laboratory using traditional conditions (Kellenberger, 2001). In recent years, qPCR has increasingly been used in environmental studies due to its sensitivity in detecting low copy number target templates. In addition to the ease of performing qPCR reactions, the technique has the capacity to rapidly run large numbers of samples, an important feature given that the spatial heterogeneity of environmental samples requires large numbers of replicates to palliate the data variability. Moreover, the development of a variety of commercial kits specifically designed to improve nucleic acid extraction from environmental samples and PCR reaction mix kits to reliably quantify microorganisms has widened the usage of qPCR in environmental studies. However, qPCR has several limitations which are specifically caused by the nature of the environmental samples, including the variability in microorganism distribution, the efficiency of DNA recovery and purification, and the amount and type of PCR inhibitors co-extracted with the target nucleic acids. The heterogeneity of the templates amplified by qPCR can generate PCR biases such as changing the relative frequencies of genes in PCR products, including the moles per cent guanine-plus-cystosine (G + C) content of template DNA, mismatches between primer or probe and template and formation of primer dimer molecules. The choice of the gene selected as a marker to quantify microorganisms also has consequences on the accuracy of the qPCR quantification.

In this review, the main experimental considerations in using qPCR in environmental studies, including the factors affecting key steps in the process of performing PCR quantification of microorganisms in environmental samples will be discussed. This will include discussions on sampling, the recovery of DNA, sources of PCR inhibitors, possible PCR biases and choice of marker gene and data analysis (Fig. 5.1) from complex matrices such as water, soil, manure, compost, and sediment. In addition, an overview of environmental studies that have used qPCR targeting ribosomal genes (16S, 18S rRNA genes or ITS) to quantify microorganisms in complex matrices and their findings will be presented.

qPCR and chemistries used in environmental studies

qPCR uses a similar fundamental approach to conventional PCR. It is based on the exponential amplification of templates using several cycles consisting of (1) DNA denaturation, (2) annealing of primers targeting specific sequences, and (3) extension of a complementary strand from each annealed primer by a thermostable DNA polymerase (Mullis and Faloona, 1987). qPCR differs from conventional PCR by monitoring the accumulation of PCR products in real-time during the cycling process by detection of fluorescent molecules bound to the newly synthesized DNA fragments. In environmental studies, fluorescent dyes, i.e. SYBR Green and TaqMan technology, are the two approaches most commonly used and they will be briefly reviewed in this section, however more details can be found in Chapter 2. The fluorescent dye, SYBR Green, intercalates or binds to the surface of any double stranded DNA (Zipper *et al.*, 2004) without compromising the polymerization reaction (Wittwer *et al.*, 1997). As more PCR products are amplified during the PCR reaction, more SYBR Green dye molecules are adducted or intercalated into the double-stranded DNA resulting in a concomitant increase in fluorescence signal following light excitation. Primers need to be highly specific to the target sequence

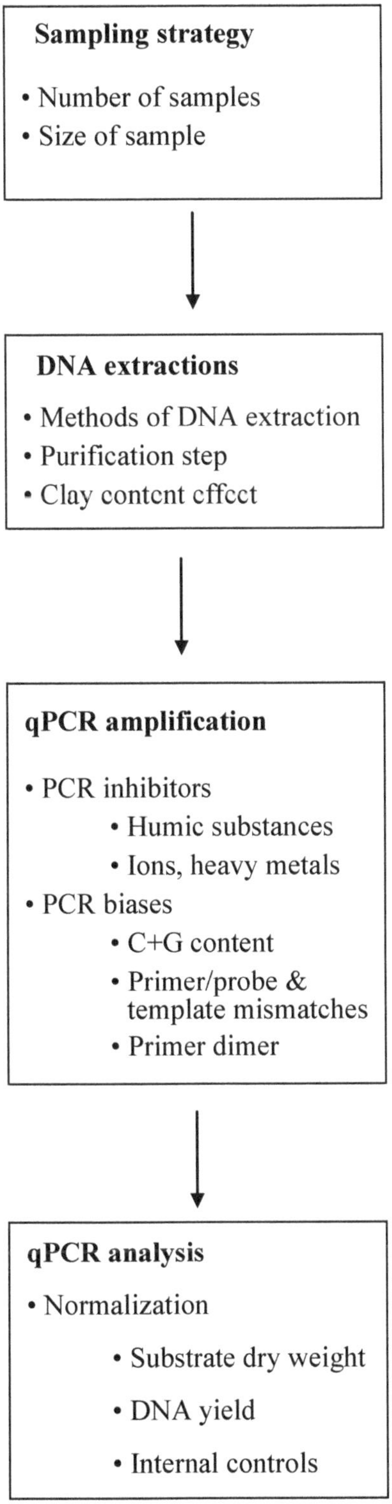

Figure 5.1 Steps required in the quantification of microorganisms using quantitative PCR (qPCR) in environmental studies and key considerations.

when used in SYBR Green qPCR assays since this dye intercalates or adducts any double-stranded DNA templates, thus non-specific PCR products can contribute to the fluorescence signal. Extensive PCR optimization can be required to avoid generation of non-specific PCR products. Primers should be designed to avoid the formation of primer-duplexes that can compete with the targeted specific PCR product for PCR reagents, potentially leading to reagent limitation and reduced amplification efficiency. A dissociation curve or melting curve is required post-PCR to ensure that the fluorescent signal originates from targeted sequences and not from non-specific PCR products or primer dimers. Gel electrophoresis should also be used to visualize PCR products and confirm specificity of primer amplification, along with sequencing of amplified PCR products.

An alternative system uses an oligonucleotide, the TaqMan probe, that is labelled at the 5′ end with a fluorophore whose fluorescence is quenched by a rhodamine derivative linked either a few bases downstream of the 5′ end or at the 3′ end of the probe (Lee *et al.*, 1993; Livak *et al.*, 1995) or using minor groove binding (Kutyavin *et al.*, 2000). The TaqMan probe anneals within the targeted DNA region amplified by a specific set of primers. During the extension phase of the PCR reaction, the TaqMan probe is hydrolysed by the Taq DNA polymerase releasing the fluorophore from the quencher and resulting in an increase in fluorescence in the PCR solution. The TaqMan approach is more specific than the SYBR Green assay as both the primers and the probe anneal to the target template. For that reason, it is recommended to use the TaqMan approach for environmental studies since environmental samples present highly heterogeneous templates. The TaqMan approach can also detect multiple templates in a single reaction by using probes labelled with different fluorophores (Baldwin *et al.*, 2003). Care must be taken with multiplex PCR reactions as they need careful optimization and validation to be accurate. The disadvantages of using TaqMan are that it is more costly to produce labelled probes than to use SYBR Green and that finding conserved gene regions for the primers and probe in a short sequence can be difficult.

For both qPCR approaches, monitoring the

increase in fluorescence in real time, i.e. cycle by cycle, allows calibration by the threshold cycle (Ct) method (Heid *et al.*, 1996) The Ct method is defined as the number of PCR cycles required for fluorescence to reach a preset, threshold, value. This threshold is set based on qualified decisions or a computing procedure. The Ct value for a particular sample will be low if a large amount of template sequences were present initially or high if the PCR started with very few templates. The quantity of a target gene in a sample is calculated using the Ct value obtained during the PCR by using a standard curve of Ct values obtained from a serially diluted standard solution, i.e. target sequence cloned in a vector (absolute quantification). One potential issue in using standards in absolute quantification is that the PCR amplification efficiency can be different between the standards which are purified plasmidic or genomic DNA and DNA extracted from environmental samples that can contain significant amounts of PCR inhibitors.

The key assumptions of the C_T method are identical amplification efficiencies at different target DNA concentrations and a linear relationship between the C_T over a range of target DNA concentrations. However it was shown that this is not always the case in PCR amplification (Liu and Saint, 2002; Gulden *et al.*, 2007). To remediate this problem, an alternative method of quantification named the sigmoidal curve fitting (SCF) method was developed (Liu and Saint, 2002; Goll *et al.*, 2006; Gulden *et al.*, 2007). The SCF method fits an empirical sigmoidal model to the experimental PCR amplification curve leading to the prediction of the amplification efficiency and to the calculation of the initial copy number in the sample. This method eliminates the need for standards to quantify the target. Several studies have reported some inherent problems with quantification using SCF method (Goll *et al.*, 2006; Swillens *et al.*, 2008) probably because this method is based on the assumptions that the sigmoidal model may describe an amplification curve quantitatively even in the portion of the curve where the fluorescence signal is hidden in the noise band. Despite the development of other quantification methods, the Ct method is still the most broadly adopted approach to quantify microorganisms in environmental studies.

Considerations when using qPCR in environmental samples

Sampling from the environment

As with any other environmental measurement, sampling for microbial molecular analyses requires robust sampling protocols, replication and good experimental design to ensure validity of the results obtained (Prosser, 2010). Issues of scale are particularly important in regards to environmental sampling for microorganisms: studies may either focus on small-scale differences in soil structure and associated microbial communities (e.g. Remenant *et al.*, 2009) or on large-scale differences such as plot, field or continent-scale assessment of microbial communities and associated functions. The aim of the study needs to be well defined to reflect the scale of factors being investigated and to choose an appropriate sampling strategy. A detailed discussion of sampling strategies is beyond the scope of this chapter, however we would like to highlight some specific issues related to sampling for quantification of microbes in the environment.

A majority of studies that have evaluated sampling strategies have focused on microbial diversity rather than abundance (Ranjard *et al.*, 2003; Baker *et al.*, 2009; Manter *et al.*, 2010), however the results of these studies should also apply to abundance studies. Environmental samples can be pooled to form a composite sample (Manter *et al.*, 2010) or multiple small subsamples are analysed (Baker *et al.*, 2009). When looking at community diversity, large, pooled samples have been recommended as they appear to capture the dominant members of the community (Manter *et al.*, 2010). Environmental samples are often pooled to alleviate the large heterogeneity in soil microbial communities and to reduce numbers of samples to be processed and associated costs. Depending on the target population, pooling of samples may not be an appropriate strategy. When rare community members are considered, multiple small subsamples are recommended (Baker *et al.*, 2009).

Although a majority of DNA extraction kits are designed for small amounts of soil (~250 mg), soil sample size can also have an influence on the results obtained. A study by Ranjard *et al.* (2003) analysed various sample sizes for bacterial and fungal diversity profiles and found that for bacterial analyses, sample size did not affect the results obtained but for fungal analyses, heterogeneity of samples led to significantly different profiles being obtained when small samples (< 1 g) were extracted. Applying this to quantification of abundance, it is likely that for low abundance targets (i.e. fungal pathogens), larger samples will provide (1) more target template DNA and (2) lower heterogeneity between replicates, whereas for high abundance targets (i.e. total bacteria), small samples are likely to give consistent results.

The above discussion mainly pertains to soil sampling, whereas sampling to quantify microorganisms from water has its own particular challenges. In water, sensitivity can be a major issue, with samples requiring concentration rather than pooling to increase probability of detecting rare targets (i.e. pathogens). Filtration is often used to concentrate microorganisms in water samples prior to DNA extraction, including hollow fibre filters that capture organisms according to their molecular weight (Rajal *et al.*, 2007), carbohydrates or hydroxyapatite columns to capture broad groups of microorganisms (Lindquist, 2011) and immunomagnetic separation using specific antibodies to capture specific organisms (Yang *et al.*, 2011). Robust sampling protocols are again required for water sampling, and temporal sampling effects may also be more important due to highly fluctuating populations in aquatic systems (Pobel *et al.*, 2011).

DNA extraction methods

qPCR-based quantification of microorganisms from water, soil, compost, manure and sediment samples can be strongly influenced by the DNA extraction methods utilized. The efficiency of several DNA extraction methods have been compared in soils (Frostegård *et al.*, 1999; Miller *et al.*, 1999), composts (LaMontagne *et al.*, 2002), manures (Cook and Britt, 2007) and sediments (Miller *et al.*, 1999) and considerable variation in DNA yield and quality was observed. DNA extraction efficiency depends on the lysis and/or physical disruption of cells in complex matrices. Some methods rely on the use of chemical or enzymatic lysis (LaMontagne *et al.*, 2002), however most methods additionally use a physical disruption of cells including freeze–thaw procedures (Zhou *et al.*, 1996; Porteous *et al.*, 1997) or bead beating (Peršoh *et al.*, 1998; Miller *et al.*, 1999; Bürgmann *et al.*, 2001; Martin-Laurent *et al.*, 2001). Bead beating generally results in a higher total cell lysis and hence higher DNA yield compared with other protocols (Miller *et al.*, 1999) and is now widely used in most DNA extraction methods.

Several commercial kits are available, specifically tailored to extract DNA from soils and sediments, and from manure or faeces. Most commercial DNA extraction kits use a bead-beating system. The commercial DNA extraction kits have the advantages of being easy to use and should have less batch to batch variation in solutions. However, they can be costly and have lower DNA yield compared with non-commercial DNA extraction methods. Several automated systems are now available and could be use to extract DNA from soils at very high throughput. The main disadvantages of automated systems are that they utilize expensive equipment (high up-front costs) and require ongoing costly consumables.

Careful preliminary comparison and optimization of the DNA extraction methods for the environmental samples to be included in a study is generally required to obtain satisfactory DNA yield and purity. Because the DNA extraction methods have such an influence on recovery of DNA, it is one of the reasons why direct comparisons of results between studies are often not possible. Although researchers have tested different DNA extraction protocols in attempts to efficiently extract DNA from different soil types (Peršoh *et al.*, 1998; Martin-Laurent *et al.*, 2001), there is still no consensus on a single protocol to be used. To remedy this problem, a standard soil DNA extraction method based on the publication of Martin-Laurent *et al.* (2001) was proposed at the International Standard Operation organization (ISO document number 11,063) and has been approved (Philippot *et al.*, 2010).

Clay content of soils and sediments

An important factor impacting the yield of DNA recovered from environmental samples from soils and sediments is the abundance of clay particles. Adsorption of DNA on clay minerals can seriously reduce the yield of DNA in extracts, especially when soils or sediments possess high clay content (20–70%) (Paget *et al.*, 1992). Clay minerals are divided in several groups including kaolinite, bentonite and montmorillonite. Several studies have investigated the mechanisms of DNA adsorption on clay minerals. Adsorption of DNA on montmorillonite and kaolinite increased with the decrease of pH from 9.0 to 3.0 (Greaves and Wilson, 1969; Khanna and Stotzky, 1992). Divalent and monovalent cations such as Mg^{2+}, Na^{+} and NH_4^{+} can form bridges between phosphate groups of DNA molecules and the negatively charged sites of clays (Paget *et al.*, 1994). The length and tertiary structure of DNA molecules influences the interaction of DNA with soil particles. Linear DNA is adsorbed more readily than supercoiled plasmid (Gallori *et al.*, 1994) and there is an inverse relationship between linear DNA molecule size and adsorption degree (Ogram *et al.*, 1994).

Determination of the soil/sediment texture, i.e. sand, silt and clay content is important in the optimization scheme of the DNA extraction method. Several authors have reported that efficiency of DNA extraction from high clay content soils can be improved by using RNA (Frostegård *et al.*, 1999; Takada-Hoshino *et al.*, 2004), skim milk (Takada-Hoshino *et al.*, 2004) or casein (Ikeda *et al.*, 2008) either as a soil pre-treatment before extraction or in the extraction buffer. Proteins in milk/casein or RNA saturate binding sites on clay minerals thus they reduce DNA adsorption onto clay minerals. However, genomic DNA contamination in skim milk lots was also documented (Ikeda *et al.*, 2008) suggesting that genomic DNA contamination could contribute to qPCR quantification of organisms in environmental samples, especially when universal primers are used to quantify broad microbial communities.

PCR inhibitors

One of the major impediments in quantifying microorganisms from environmental samples is the presence of several potentially PCR-inhibiting factors, including humic substances, heavy metals and ions.

Humic substances

Humic substances are the most commonly reported group of PCR inhibitors in environmental samples and can be found in surface water, soils, sediments, composts and manures (Wilson, 1997). Humic substances are heterogeneous and relatively small molecules derived from the degradation and decomposition of dead biological material (Piccolo, 2002; Schaumann, 2006). The exact molecular weight and structure of humic substances are still poorly understood (Piccolo, 2002; Schaumann, 2006). They can be separated into several groups, humic acids, fulvic acids and humins, based on their solubility properties in the aqueous solutions used as soil extractants (Piccolo, 2002). Humic substances appear as dark-coloured organic compounds in DNA extracts and are resistant to degradation. Concentrations of humic substances co-extracted with soil DNA are generally higher from composts and manures compared with soils and sediments since the former contain high levels of organic matter. Structure and composition of humic substances changes depending on the matrices (e.g. compost vs. soil) (Senesi *et al.*, 2007), on the geography and topography and can be affected by management practices in agricultural soils (Novotny *et al.*, 1999).

Several studies have reported PCR inhibition and ascribed it to the presence of humic acids in DNA extract solutions (Tsai and Olson, 1992a; Tebbe and Vahjen, 1993). The exact mechanism of PCR inhibition by humic substances is still unclear. Some evidence indicates that humic substances inhibit PCR by interfering with DNA polymerase enzymes (Tebbe and Vahjen, 1993; Matheson *et al.*, 2010) and/or with the binding of DNA polymerase to the DNA template (Opel *et al.*, 2010).

Due to the amorphous nature of humic substances, they are notoriously difficult to remove, thus a combination of purification approaches needs to be taken to reduce the load of these substances in DNA extracts. Several approaches have been taken to relieve PCR inhibition either at the DNA extraction, purification or at the PCR amplification steps. Aluminium sulfate was previously added to soil or compost prior to cell lysis to flocculate humic substances, thus reducing co-extraction of humic substances into DNA extracts

(Dong *et al.*, 2006). Additions to the DNA extraction buffer including polyvinylpolypyrrolidone (Picard *et al.*, 1992; Frostegård *et al.*, 1999) or aluminium ammonium sulfate (Braid *et al.*, 2003) were used to bind or flocculate humic substances from the DNA extracts. Differential precipitation of DNA and humic substances using polyethylene glycol (PEG) was shown to be efficient in removing the contamination (La Montagne *et al.*, 2002; Arbeli and Fuentes 2007); however, the concentration of PEG and sodium chloride in the DNA extract solution has an effect on the efficiency in removing humic substances (Arbeli and Fuentes 2007).

Even if chemical additives and differential precipitation are used during the DNA extraction, further DNA purification might still be necessary. Methods that have been developed to remove humic substances from DNA samples to prevent PCR inhibition include gel electrophoresis (Moreira, 1998; Roh *et al.*, 2005), size exclusion chromatography (Tsai and Olson, 1992a), and silica-based DNA binding columns (Tsai and Olson, 1992b; Miller *et al.* 2001). Although further purification may be necessary to remove PCR inhibitors, most purification approaches will result in some additional loss of the DNA template, so care must be taken especially if low concentrations of DNA or low-abundance templates are being quantified.

Some strategies can help relieve inhibition during amplification of DNA by (1) using PCR facilitators, (2) dilution of the DNA templates and/or (3) inclusion of humic acid-tolerant DNA polymerases. Addition of facilitators to the PCR reaction to relieve PCR inhibition is covered by Wilson (1997), thus it will not be covered herein. The most commonly used method to overcome PCR inhibition by humic substances is to dilute the extract, thereby diluting the concentration of the inhibitors to a level where inhibition no longer occurs. However, dilution of the sample also reduces the concentration of DNA which can decrease the efficiency of PCR amplification. It is necessary to strike a balance between diluting out the inhibition whilst still retaining a concentration of DNA template that is within the linear range of amplification efficiency for qPCR. Soil DNA extracts need to be tested for the presence of co-extracted inhibitory substances. There are several approaches that can be taken to detect the presence of PCR inhibitors. A known quantity of a cloned gene fragment is added to serially diluted soil DNA extracts and the cloned gene fragment is amplified by qPCR. The resulting Ct values are compared between spiked soil dilution extracts and the purified cloned gene fragment. Presence of PCR inhibitors will result in greater Ct values for the spiked soil dilution extracts compared with PCR reaction using the cloned gene fragment or in complete failure of PCR amplification if inhibitory substances are at high concentrations. Another approach consists of 10-fold serially diluted soil DNA extracts that are quantified using qPCR. 10-fold dilutions should results in a 3.3 cycle difference in Ct between dilutions. PCR inhibitors will result in larger difference in Ct values obtained between soil DNA dilutions or in absence of PCR amplification. The dilution to be selected for further work is the lowest dilution that has no sign of PCR inhibition.

DNA polymerases have wide variation in their resistance to inhibitors (Al-Soud and Rådström, 1998; Eilert *et al.*, 2009; Matheson *et al.*, 2010). AmpliTaq Gold DNA polymerase, which is the standard for use with a number of commercial PCR kits including qPCR master mix (Applied Biosystems), is one of the most sensitive to inhibition by humic acid (Eilert *et al.*, 2009). Amongst those polymerases tested for humic acid resistance, the most resistant were Pfx50 DNA polymerase (Eilert *et al.*, 2009) and pfu DNA polymerase (Matheson *et al.*, 2010), which showed resistance up to 10 ng/µl and 110 ng/µl of humic acids, respectively, under the conditions tested. Since DNA polymerases included in commercially available qPCR master mixes are not selected based on their resistance to PCR inhibitors, it underscores the importance of DNA extraction and purification methods and highlights opportunities for future improvement of these products.

Ions and heavy metals

Heavy metals (e.g. iron, lead, copper, cadmium, mercury, manganese and zinc) occur naturally in soils or accumulate in soils close to mining sites, in manures and in composts. Heavy metals were reported as a potential source of inhibitors in PCR reactions (Tsai and Olson, 1992b; Straub *et al.*, 1995); however, the mechanisms of PCR inhibition by heavy metals remain unknown. Another study observed no adverse effect of the presence of heavy metals on extraction efficiency and PCR amplification (Kozdrój and van Elsas, 2000), suggesting that the choice of DNA extraction methods or other factors, including the type and concentration of heavy metals, and soil chelating abilities, also influence PCR inhibition. Ions including potassium, calcium, magnesium and sodium can either originate from soils or from the reagents used during the DNA extractions. Commercial DNA polymerases have different resistance to ions with the AmpliGold Taq DNA polymerase and the rTth DNA polymerase being the most sensitive and most resistant, respectively, to monovalent or divalent ions potassium, calcium, magnesium and sodium (Al-Soud and Rådström, 1998).

PCR biases

PCR suffers from several biases that can either change the relative frequency of DNA sequences (per cent guanine-plus-cystosine (G+C) content) or reduce the efficiency of PCR (primer/probe mismatches and primer duplexes). These biases were previously shown to be important in end-point PCR and qPCR can also be similarly impacted.

Guanine-plus-cystosine (G+C) content of template DNA

In environmental studies, the PCR template is a mixture of homologous genes, especially in the case of amplification of genes coding for conserved genes such as the ribosomal RNA (rRNA) genes. The PCR amplification process can lead to a change in the relative frequencies of genes in PCR products from mixed templates, especially with templates showing a wide range mole per cent G+C content (Reysenbach *et al.*, 1992). During the denaturation step of PCR, genes with lower G+C composition dissociate into single stranded DNA more readily than genes with higher G+C composition, resulting in a preferential amplification of templates with a lower G+C content. This result suggests that quantification of microorganisms from mixed templates could underestimate the community size when universal primers are used in qPCR. PCR mix additives including acetamide (Reysenbach *et al.*, 1992), dimethylsulfoxide (DMSO) and betaine (Baskaran *et al.*, 1996; Musso *et al.*, 2006), alternative DNA polymerases (Dutton *et al.*, 1993), high denaturation temperature (98°C) (Dutton *et al.*, 1993; Baskaran *et al.*, 1996), and primers with high melting temperatures (Dutton *et al.*, 1993) were factors that reduced selectivity of PCR amplification from a mix of templates and resulted in efficient amplification of genes with high G+C content.

Primer, probe and template mismatches

Complementarities between primers and template are important for PCR applications because mismatches between primer-probe and template can severely affect PCR efficiency and can ultimately result in complete PCR failure. Mismatches in primer/template are unavoidable, particularly in mixed templates from environmental samples. For example, the longest string of totally conserved bases in the 16S rRNA gene is between positions 788 and 798, and most of the absolutely conserved regions are found in strings of fewer than four bases (Baker *et al.*, 2003). This means that no existing primer targeting the prokaryotic 16S rRNA gene is truly universal and that differential amplification in a mixture of template DNA from a complex microbial community and biases in the microbial community quantification should be expected.

The effects of mismatches depend on numerous factors including the position (Ayyadevara *et al.*, 2000; Bru *et al.*, 2008) and the nature (Ayyadevara *et al.*, 2000; Stadhouders *et al.*, 2010) of the mismatches. Mismatches on the last or on the penultimate nucleotide on the 3′ end of the primer sequence more severely affect PCR amplification, probably due to an inefficient extension of the primer by the DNA polymerase. Primer internal mismatches were previously believed to have less of an effect (Kwok *et al.*, 1990) compared with mismatches on the last nucleotide; however, primer internal mismatches reduce thermal stability of the primer-template duplex (Sipos *et al.*, 2007; Bru *et al.*, 2008), leading to reduced amplification efficiency. Mismatches in the probe were shown to reduce PCR efficiencies more drastically than mismatches in the primers and multiple mismatches to the probe caused the greatest reduction in efficiency (Klein *et al.*, 1999; Smith *et al.*, 2002). The nature of the mismatch has different effects on PCR amplification efficiency with reduction of less than 1.5 Ct for mismatches A–C, C–A, T–G and G–T to more than 7 C_T for mismatches A–A, G–A, A–G and C–C when the forward primer had a single mutation at the 3′ end (Stadhouders *et al.*, 2010). These studies indicate that primer-probe design is an important step and that selection of primer pairs targeting broad groups of microorganisms has inevitable consequences on the qPCR results.

Primer dimers

Primer dimers are produced either by annealing of sequences within a primer (self-complementarity) or between two primers (cross-complementarity). The formation and subsequent amplification of primer dimers reduces the availability of primer to the template molecule resulting in decreased sensitivity and efficiency of the PCR reaction (Löffert *et al.*, 1997). Furthermore, primer dimers are especially problematic when the non-specific SYBR Green dye is used because it will bind to the primer dimers and fluoresce thus leading to erroneous quantification results. It is generally recommended to design primers without complementary sequences to avoid formation of primer dimers (Löffert *et al.*, 1997) and primer design software will reject primer pairs showing high complementarity. However, it can be difficult to find appropriate conserved sequences in a gene present in a large group of microorganisms or when multiplex reactions are necessary, thus in some instances preventing primer dimers is not obvious. When the problem of primer dimer cannot be avoided, several approaches can be used to optimize the PCR reaction conditions to reduce primer dimer formation, including high temperature primer annealing and synthesis steps (Rychlik *et al.*, 1990; Wu *et al.*, 1991), the use of 'hot-start' (D'Aquila *et al.*, 1991; Chou *et al.*, 1992), touch-down PCR (Don *et al.*, 1991) and/or the use of enzyme formulations such as AmpliTaq Gold.

Structural modifications of the primers including addition of a 5′ tag (Brownie *et al.*, 1998) and chimeric primers can reduce or prevent formation of primer dimers. HANDS (Homo-Tag Assisted Non-Dimer System) uses primers tagged with a sequence of nucleotides at the 5′-end that is complementary to the 3′-end of the primer permitting the formation of a stem-loop (Brownie *et al.*, 1998) (Fig. 5.2). The stem–loop structure of the primers excludes annealing of short overlaps between primers but permits annealing to a fully complementary target sequence. Chimeric primers have some DNA bases replaced with RNA bases resulting in lower melting temperatures for two annealed chimeric primers compared with a chimeric primer annealed to the target template (Peleg *et al.*, 2009) (Fig. 5.3). Another approach is to eliminate the contribution of the primer–dimers from the fluorescence signal by performing an additional step in the PCR cycle (Pfaffl, 2004). An additional PCR step after the polymerization step is performed at elevated temperatures (generally in the range of 80–87°C) in order to eliminate the non-specific fluorescence signals derived from primer dimers and ensuring accurate quantification of the desired amplification product (Pfaffl, 2004). This approach is useful as long as there are few primer dimers formed and their inhibitory effect on PCR product synthesis is minor.

Marker genes for microorganism quantification

The marker genes selected to quantify microorganisms may have a significant impact on the quantification results. There are several issues with the use of the bacterial or archaeal 16S rRNA or the fungal 18S rRNA gene to quantify specific species or large groups of microorganisms. A high degree of sequence similarity between closely related species can limit the utility of the bacterial 16S rRNA gene sequences (Gribaldo *et al.*, 1997; Becker *et al.*, 2004), making it difficult if not impossible to design primers of sufficient specificity to track specific guilds. Secondly, the quantification of microorganisms using qPCR assumes that the targeted genes are found in one

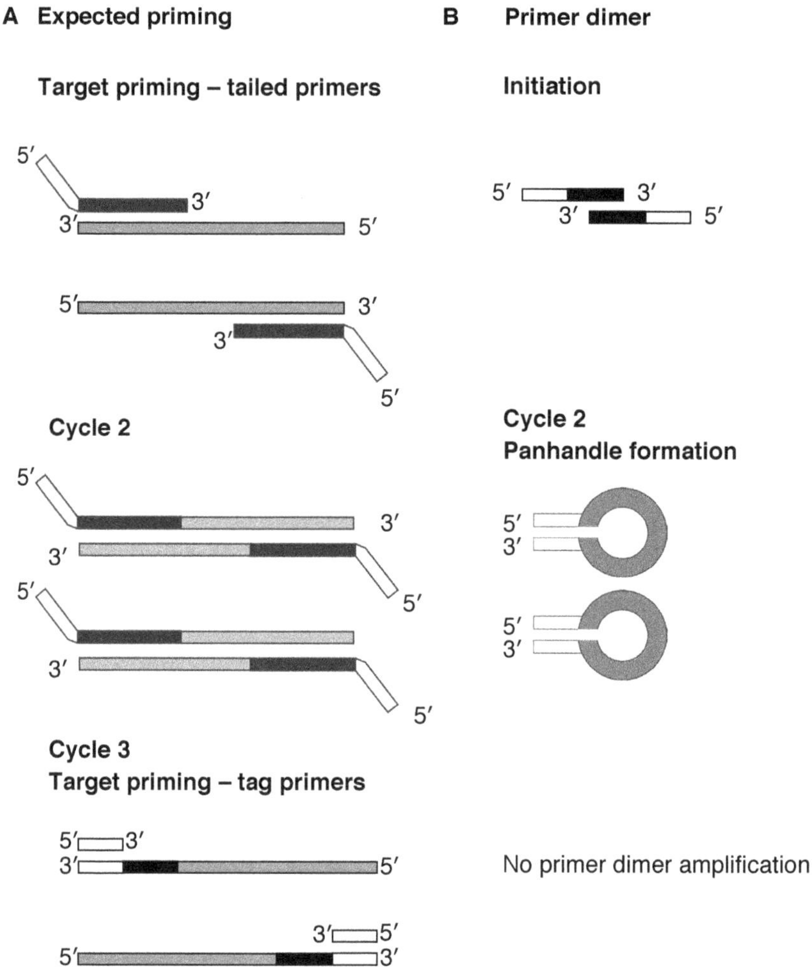

Figure 5.2 Schematic representation of homo-tag assisted no dimer system (HANDS) to prevent formation of primer dimers. Annealing of tailed primer to the target (grey box) (A). Two pairs of primers are included in the reaction: tailed primers consist of a target specific section at their 3′-end (black box) and a tail sequence at their 5′-end (white box) while the tag primer pair is specific to the tail sequence. In early PCR cycles, tailed primers amplify the target and add tail sequences at their 5′-end. In later cycles, tag primers amplify the target using the added tail sequences. Primer dimer formation (B). Amplification of dimers by the tailed primers results in panhandle structure, thus preventing amplification of primer dimers.

copy per genome. However, the number of 16S rRNA and 18S rRNA genes located on chromosomes differs widely from one copy up to 15 copies per genome for bacteria (Klappenbach *et al.*, 2001), one to five copies per genome in archaea (Acinas *et al.*, 2004) and up to 100 copies per genome in fungi (Tsuchiya and Taga, 2001). In fungi, the number of 18S rRNA gene fluctuates with the age and growth stage (Paris *et al.*, 2002) and variation in sequences of the ITS region within species and among strains of a species was reported (Ko *et al.*, 2002). Variability in sequences and in the number of 16S/18S rRNA genes per genome results in under- or over-estimation of the number of individuals from a species or of broad bacterial communities. Hence reporting of microorganisms abundances is generally reported as number of gene copies per g soil or ng DNA rather than number of microorganisms.

Other bacterial conserved genes including the *rpoB* (RNA polymerase subunit beta) and *gyrA* (gyrase subunit A) genes were found to have lower

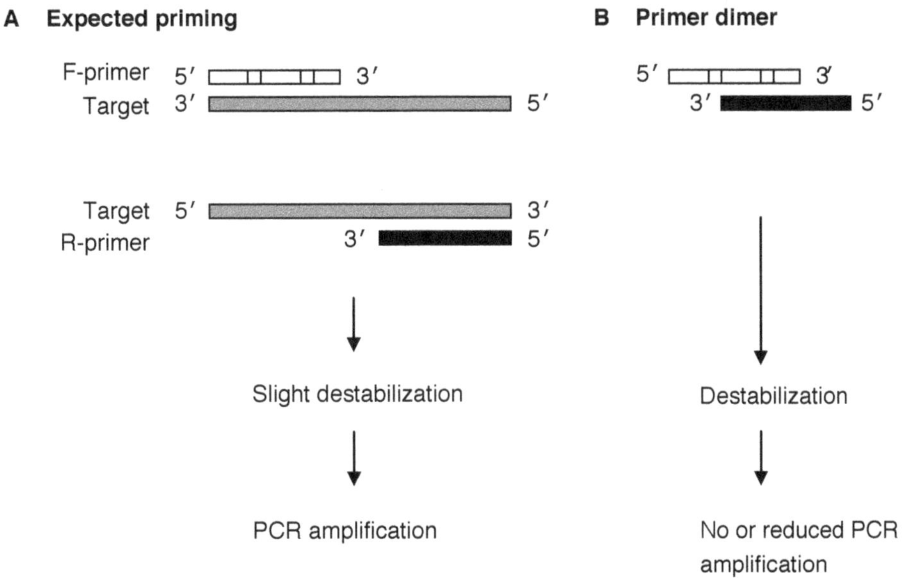

Figure 5.3 Schematic representation of chimeric primers to reduce the formation of primer dimers. Annealing of chimeric forward primer (F-primer, white box) and chimeric reverse primer (R-primer, black box) to target (grey box) (A). When primers anneal to the expected priming sites on the target template, there is a slight destabilization, however PCR amplification is successful. Formation of a primer dimer by the chimeric primers (B). DNA-RNA chimeric primers are constructed by incorporating RNA bases (stippled box) at the 5′ nearest neighbour of position suspected to initiate primer dimers but avoid stretches of RNA bases. Chimeric primer dimers results in an increase in the destabilization, thus eliminating or reducing the amplification of primer dimers.

copy number (1–3 copies) per genome which would make them more suitable for quantification of microorganisms using qPCR (Rodrigues and Tiedje, 2007). However, only few studies have targeted these genes to quantify bacteria from environmental samples (Rodrigues and Tiedje, 2007; Lee and Lee, 2010). Owing to the ubiquity of the 16S rRNA gene and the large databases that have been amassed around the use of this gene as a phylogenetic marker, it will probably remain the gene of choice for quantification of microorganisms for many years to come.

Data analysis

Several approaches have been used to normalize qPCR data in environmental studies including (1) reporting the gene copy number per gram of substrate (e.g. soil) or yield of extracted DNA, (2) using relative abundance of a specific organism to total community, or (3) incorporating the use of internal amplification controls. Most studies normalize qPCR data per gram of substrate. It is, however, important to use the dry weight of the substrate to avoid variability in the water content of solid samples by freeze-drying or oven-drying. The qPCR results can be normalized by reporting the gene copies number per μg or ng of extracted DNA (Töwe *et al.*, 2011). In that case, the variability in DNA extraction efficiency between samples is taken into accounts. Accurate DNA yield measurements from samples with high humic substances are best achieved using intercalating fluorescent dyes including PicoGreen (Invitrogen, California, USA). Relative abundance of microorganisms can be done by using a ratio or percentage that is obtained by comparing the gene copy number of a specific species or taxa to the total bacterial or fungal community (Wessén *et al.*, 2010).

The qPCR results can be normalized to take into account the high variability in the recovery of DNA from complex environmental samples (Lerat *et al.*, 2005; Stoeckel *et al.*, 2009) or the presence of PCR inhibitors (van Doorn *et al.*, 2009) using internal amplification controls. Assessments can be made of the efficiency of DNA extractions from environmental matrices by use of spike-and-recovery controls (Lerat *et al.*, 2005; Stoeckel *et al.*, 2009). In this technique, a known quantity of control bacteria (Stoeckel *et al.*, 2009) or plasmidic DNA (Lerat *et al.*, 2005) is added to the samples prior to DNA extractions and quantified using qPCR. The quantification results of the microorganism of interest are normalized based on the control results which can reduce variability in DNA extraction efficiency. Systematic assessment of PCR inhibition can be done by incorporating internal amplification controls that are added to the extracted DNA of all samples and this technique is particularly well suited to high-throughput qPCR platforms (van Doorn *et al.*, 2009).

Uses of qPCR targeting conserved genes in environmental samples

Although quantification of microorganisms is challenging, it is possible to reliably quantify microorganism in complex environmental samples using qPCR by carefully optimizing each step of the quantification process. Quantification of microorganisms using conserved genes such as the ITS or 16S/18S rRNA gene has been used for different purposes including investigation on (1) the distribution of microorganisms in the environment, (2) studying microorganisms known to be involved in ecological functions such as biogeochemical processes, pathology or bioremediation or (3) assessment of soil microbial community structure at broad taxonomic levels. Examples of studies using qPCR to quantify microorganisms for each of these purposes will be briefly reviewed.

Distribution of microorganisms in the environment

Microbial community structure in environmental samples using sequencing of cloned 16S/18S rRNA gene libraries or high-throughput sequencing revealed that there is a large discrepancy between sequences of culturable microorganisms and microorganisms present in a given environment (Rappé and Giovannoni, 2003). qPCR is often combined with sequencing efforts to quantify dominant groups of microorganisms in a given environment to better understand the distribution of the targeted microorganisms, thus gaining more insight on their potential ecological importance. Frequently, the targeted microorganisms are not yet culturable (Dang *et al.*, 2010; Lefèvre *et al.*, 2010) or few culturable representatives are currently available in culture collections (Cassler *et al.*, 2008). Using qPCR, the studies described briefly in Table 5.1 confirmed that the microorganisms were dominant members of the microbial community of sea sponges (Cassler *et al.*, 2008) while other studies demonstrated a wide distribution of microorganisms in water (Lefèvre *et al.*, 2010) and soil (Ochsenreiter *et al.*, 2003), changes in distribution depending on the environments (da Rocha *et al.*, 2009; Swan *et al.*, 2010) or heterogeneous distribution in sediments (Dang *et al.*, 2010).

Functional groups of microorganisms

Functional groups of microorganisms that are involved in biogeochemical processes, pathology or bioremediation were quantified in environmental samples. Most studies using qPCR targeting conserved genes have focused on abundance of bacteria and to a lesser extent on archaea and fungi. Changes in abundance of the bacterial communities involved in biogeochemical processes including the nitrogen cycle (nitrification and denitrification) and in bioremediation are two subjects that have been extensively studied in bacteria (Table 5.2). For examples, the effect of tillage regime on the abundance of Nitrospira-like bacteria involved in nitrification in soil (Attard *et al.*, 2010) and spatial variation of Dehalococcoides abundance (Taş *et al.*, 2009) that are degraders of hexachlorobenzene pollutant in sediments were investigated (Table 5.2). Several studies have quantified dominant bacterial species or groups in manure or faeces to identify possible biomarkers (e.g. Furet *et al.*, 2009) to track sources of faecal pollution and to quantify biomarkers in water stream and bodies (e.g. Lee *et al.*, 2010) (Table 5.2).

Archaeal quantification in the environment

Table 5.1 Studies investigating the distribution of microorganisms in environments

Environment	Organisms targeted	Comments/results	Reference
Sediment	Anaerobic ammonium oxidizing bacteria (annamox)	'*Candidatus* Scalindua' had a heterogeneous distribution in sediments of the Jiaozhou Bay, China, and its abundance was correlated with nitrite concentration	Dang *et al.* (2010)
Soil	Acidobacteria and Verrucomicrobia	Abundance of Acidobacteria and Verrucomicrobia was higher in rhizospheric soils compared with bulk soils	da Rocha *et al.* (2010)
Soil	Archaea and bacteria	Mesophilic Crenarchaeota were widely distributed in terrestrial environments with a relative abundance of 0.5–3% in the bulk soil sample and 0.16% in the rhizosphere compared with bacteria	Ochsenreiter *et al.* (2003)
Water	Archaea and bacteria	Archaea were numerically dominant at all depths of anoxic sediments of a hypersaline lake and exhibited a lesser response to environmental gradients than bacteria	Swan *et al.* (2010)
Water	Zoosporic fungi	Zoosporic fungi were widely distributed in a lake, suggesting a potential role in microbial food webs	Lefèvre *et al.* (2010)
Water	*Nitrospira*-like microbes and Actinobacteria	The study showed that uncultured groups targeted were more abundant than known cultured bacteria on *Vetulina* sp. sponges	Cassler *et al.* (2008)

has mainly focused around the process of methanogenesis. Methanogenic archaea involved in methane formation in reactors to produce biogas (Nettmann *et al.*, 2010) or in a waste-water treatment reactor (Lee *et al.*, 2008) were quantified. Possible importance of archaea in reducing malodor of manure (Yamada *et al.*, 2007) and distribution of archaeal methanotrophs in several environments other than marine sediments (Miyashita *et al.*, 2008) have been evaluated.

Quantification of fungal communities in the environment has mostly focused on pathogenic fungal species in soil (Kernaghan *et al.*, 2007) or water (Wielgoss *et al.*, 2010) although their abundance in reactor to reduce manure malodor has also been evaluated (Yamada *et al.*, 2009) (Table 5.2).

Soil microbial community structure at broad taxonomic levels

In most studies, the group of microorganisms targeted is a relatively narrow group of microorganisms, i.e. at strain, species or genus levels. A few studies have looked at changes at the higher taxonomic levels of bacteria by targeting different bacterial phyla and classes (Fierer *et al.*, 2007; Philippot *et al.*, 2009; Wessén *et al.*, 2010). Fierer *et al.* (2007) suggested that certain bacterial phyla could be differentiated into copiotrophic and oligotrophic categories (Fierer *et al.*, 2007). This suggested that bacteria within a phylum could share specific ecological traits that could influence their spatial distribution or their responses to changes in environmental conditions and soil physicochemical characteristics. Specific bacterial phyla were demonstrated to followed spatial patterns in a pasture (Philippot *et al.*, 2009), or responded differently to fertilizer regimes (Wessén *et al.*, 2009), suggesting an ecological coherence at broad taxonomic level (Table 5.3).

Conclusions and perspectives

qPCR is a fast and effective technique to quantify microorganisms from environmental samples. It is highly specific and sensitive for targeting templates that are present in a mixed community. Because of the quantitative nature of the results obtained by qPCR, it is important to consider each step of the extraction-quantification process since it has an impact on the quality of the results. DNA extraction method and purification must be tested and optimized for the particular environmental samples and the absence of PCR inhibitors need to be verified, especially for environmental samples where their presence is known to be problematic.

Table 5.2 Quantification of functional groups of microorganisms in environmental samples

Environment	Theme/issue	Organisms	Comments/results	Reference
Compost	Manure malodor	Archaea/ Fungi	An increase in total archaea and fungal-protozoan communities in reactor with hyperthermophilic pre-treatment of manure to reduce malodor was observed compared with reactor using regular process	Yamada *et al.* (2007)
Manure	Biogas	Archaea	*Methanomicrobiales* dominated with 61–99% of total archaeal 16S rRNA gene copies in biogas reactors	Nettmann *et al.* (2010)
	Microbial source tracking	Bacteria	The abundance of three dominant bacterial groups of the human microbiota was different compared with animal species suggesting they could be used as biomarkers for water pollution source tracking	Furet *et al.* (2009)
Sediment	Methane oxidation	Archaea	Two phylogenetic groups of archaeal anaerobic methanotrophs were detected in natural gas fields but not in methanogenic sludges, rice field soils, and lotus field sediments	Miyashita *et al.* (2009)
	Bioremediation	Bacteria	*Dehalococcoides* spp., bacteria degrading hexachlorobenzene (HCB) was significantly more abundant in the upstream river locations than in downstream locations and the abundance was not correlated with HCB pollution	Taş *et al.* (2009)
Soil	Nitrification	Bacteria	*Nitrospira*-like bacteria abundance showed limited changes in soils with different tillage treatments and was not correlated with potential nitrite oxidation suggesting that this species was not the key player in nitrification process in the system studied	Attard *et al.* (2010)
	Pathology	Fungi	The root-rot and replant disease of ginseng pathogen, *Cylindrocarpon destructans* f. sp. *panacis*, was found to be positively correlated with disease severity.	Kernaghan *et al.* (2007)
Water	Waste water treatment	Archaea	Methanogenic archaea community size in a municipal wastewater reactor supplemented with whey permeate increased twice overtime, which corresponded to the two stages of methane formation	Lee *et al.* (2008)
	Microbial source tracking	Bacteria	Bacteroidales was used to quantify general, bovine or human fecal pollution in freshwater streams influenced by bovine and human faeces and persisted in downstream areas	Lee *et al.* (2010)
	Pathology	Fungi	The water-borne pathogen, *Pythium phragmitis* infecting reed was shown to be particularly affected by the stage of reed development and varied temporally	Wielgoss *et al.* (2010)

It is important to keep in mind that there are some unavoidable limitations to this methodology such as the number of conserved genes per genome, which prevent a precise estimation of the number or microorganisms in a given sample, and other PCR biases and artefacts including primer duplex, primer-template mismatches and C + G content when broad groups of microorganisms are targeted. However, with careful considerations of all steps of the quantification process, qPCR gives reliable and reproducible results.

qPCR is particularly informative when changes in microbial community abundance are combined with study of the diversity of microbial communities. In this case, diversity and abundance of microorganisms can be evaluated simultaneously or step-wise, when changes in the diversity of the microbial community in a given environment reveals which dominant groups of microorganisms should be targeted. Furthermore, the combination of changes in abundance of microorganisms and the changes in environmental conditions and

Table 5.3 Quantification of domains and phyla of bacteria and bacterial, archaeal and fungal domains in soil

Theme/problematic	Domains/phyla or classes	Comments/results	Reference
Spatial variability at field level	Acidobacteria, α-Proteobacteria, Actinobacteria, Bacteroidetes, β-Proteobacteria, Gemmatimonadetes, Firmicutes, Verrucomicrobia Total bacteria	Specific phyla or classes followed spatial patterns in a pasture field that correlated to soil properties suggesting that members of a bacterial phyla shared specific ecological traits	Philippot *et al.* (2009)
Agricultural practices – fertilization regime	Actinobacteria, Acidobacteria, Bacteriodetes, Firmicutes, Gemmatimonadetes, Verrucomicrobia, α-Proteobacteria, Crenarchaeota Total bacteria and archaea	The total abundances of phyla or classes were affected in similar ways by different fertilization regimes, however some changes were observed when ratios of the abundance of phyla over the total bacteria community abundance were considered	Wessén *et al.* (2009)
Contrasting soils	Actinobacteria, Acidobacteria, Bacteriodetes, Firmicutes, Gemmatimonadetes, Verrucomicrobia, α-Proteobacteria, Crenarchaeota Total bacteria, Basidiomycota, and fungi	Acidobacteria and Proteobacteria were dominant in prairie, desert and forest soils while Bacteroidetes and Firmicutes phyla were less common. Fungal community was more abundant in forest soils than bacteria	Fierer *et al.* (2005)

physicochemical properties of their environments or changes in biological process rates may allow us to gain a better understanding of what is driving changes in the microbial community in the environment or how these changes affects biological process rates.

In the future, improvement on the accuracy of quantification of microorganisms using qPCR can be expected. For example, the use of more resistant DNA polymerases to PCR inhibitors, one of the major challenges in environmental samples, would greatly improve PCR amplifications. Recently, the use of viable qPCR in environmental samples was proposed (Fittipaldi *et al.*, 2011). In this technique, dyes bind to DNA from dead microorganisms, thus inhibiting PCR amplification while in viable microorganisms, cellular membrane prevents the dye from attaching to DNA. This technique shows great promise in circumventing the problem of quantification of DNA from dead microorganisms.

Acknowledgements

The authors would like to thank Martin Filion for the invitation to write this review and Enrico Tatti and Sophie Wertz for critical reading of the chapter. C. Goyer acknowledges the support of the Sustainable Agriculture Environmental Systems (SAGES) programme of Agriculture and Agri-Food Canada.

References

Al-Soud, W.A., and Rådström, P. (1998). Capacity of nine thermostable DNA polymerases to mediate DNA amplification in the presence of PCR-inhibiting samples. Appl. Environ. Microbiol. *64*, 3748–3753.

Acinas, S.G., Marcelino, L.A., Klepac-Ceraj, V., and Polz, M.F. (2004). Divergence and redundancy of 16S rRNA sequences in genomes with multiple rrn operons. J. Bacteriol. *186*, 2629–2635.

Arbeli, Z., and Fuentes, C.L. (2007). Improved purification and PCR amplification of DNA from environmental samples. FEMS Microbiol. Lett. *272*, 269–275.

Attard, E., Poly, F., Laurent, F., Commeaux, C., Terada, A., Smets, B.F., Recous, S., and Le Roux, X. (2010). Shifts between Nitrospira-and Nitrobacter-like nitrite oxidizers underlie the response of soil potential nitrite oxidation to changes in tillage practices. Environ. Microbiol. *12*, 315–326.

Ayyadevara, S., Thaden, J.J., and Shmookler Reis, R.J. (2000). Discrimination of primer 3′-nucleotide mismatch by Taq DNA polymerase during polymerase chain reaction. Anal. Biochem. *284*, 11–18.

Baker, G.C., Smith, J.J., and Cowan, D.A. (2003). Review and re-analysis of domain-specific 16S primers. J. Microbiol. Methods 55, 541–555.

Baker, K.L., Lagenheder, S., Nicol, G.W., Ricketts, D., Killham, K., Campbell, C.D., and Prosser, J.I. (2009). Environmental and spatial characterisation of bacterial community composition in soil to inform sampling strategies. Soil Biol. Biochem. *41,* 2292–2298.

Baldwin, B.R., Nakatsu, C.H., and Nies, L. (2003). Detection and enumeration of aromatic oxygenase genes by multiplex and real-time PCR. Appl. Environ. Microb. *69,* 3350–3358.

Baskaran N., Kandpal, R.P., Bhargava, A.K., Glynn, M.W., Bale A., and Weissman, S.M. (1996). Uniform amplification of a mixture of deoxyribonucleic acids with varying GC content. Genome Res. *6,* 633–638.

Becker, K., Harmsen, D., Mellmann, A., Meier, C., Schumann, P., Peters, G., and von Eiff, C. (2004). Development and evaluation of a quality-controlled ribosomal sequence database for 16S ribosomal DNA-based identification of *Staphylococcus* species. J. Clin. Microbiol. *42,* 4988–4995.

Becker, S., Böger, P., Oehlmann, R., and Ernst, A. (2000). PCR bias in ecological analysis: a case study for quantitative Taq nuclease assays in analyses of microbial communities. Appl. Environ. Microbiol. *66,* 4945- 4953.

Braid, M.D., Daniels, L.M., and Kitts, C.L. (2003). Removal of PCR inhibitors from soil DNA by chemical flocculation. J. Microbiol. Methods *52,* 389–393.

Brownie, J., Shawcross, S., Theaker, J., Whitcombe, D., Ferrie, R., Newton, C., and Little, S. (1997). The elimination of primer-dimer accumulation in PCR. Nucleic Acids Res. *25,* 3235–3241.

Bru, D., Martin-Laurent, F., and Philippot, L. (2008). Quantification of the detrimental effect of a single primer-template mismatch by real-time PCR using the 16S rRNA gene as an example. Appl. Environ. Microbiol. *74,* 1660–1663.

Bürgmann, H., Pesaro, M., Widmer, F., and Zeyer, J. (2001). A strategy for optimizing quality and quantity of DNA extracted from soil. J. Microbiol. Methods *45,* 7–20.

Cassler, M., Peterson, C., Ledger, A., Pomponi, S., Wright, A., Winegar, R., McCarthy, P, and Lopez, J. (2007). Use of real-time qPCR to quantify members of the unculturable heterotrophic bacterial community in a deep sea marine sponge, *Vetulina* sp. Microbial Ecol. *73,* 6898–6904.

Chou, Q., Russell, M., Birch, D.E., Raymond, J., and Bloch, W. (1992). Prevention of pre-PCR mis-priming and primer dimerization improves low-copy-number amplifications. Nucleic Acids Res. *20,* 1717–1723.

Cook, K.L., and Britt, J.S. (2007). Optimization of methods for detecting *Mycobacterium avium* subsp. paratuberculosis in environmental samples using quantitative, real-time PCR. J. Microbiol. Methods *69,* 154–160.

Dandie, C.E., Burton, D.L., Zebarth, B.J., Henderson, S.L., Trevors, J.T., and Goyer, C. (2008). Changes in bacterial denitrifier community abundance over time in an agricultural field and their relationship with denitrification activity. Appl. Environ. Microbiol. *74,* 5997–6005.

Dandie, C.E., Miller, M.N., Burton., D.L., Zebarth, B.J., Trevors, J.T., and Goyer, C. (2007). Nitric oxide reductase-targeted real-time PCR quantification of denitrifier populations in soil. Appl. Environ. Microbiol. *73,* 4250–4258.

D'Aquila, R.T., Bechtel, L.J., Videler, J.A., Eron, J.J., Gorczyca, P., and Kaplan, J.C. (1991). Maximizing sensitivity and specificity of PCR by pre-amplification heating. Nucleic Acids Res. *19,* 3749.

Dang, H., Chen, R., Wang, L., Guo, L., Chen, P. Tang, Z. Tian, F., Li, S., and Klotz, M.G. (2010). Environmental Factors Shape Sediment Anammox Bacterial Communities in Hypernutrified Jiaozhou Bay, China. Appl. Environ. Microbiol. *76,* 7036–7047.

Don, R.H., Cox, P.T., Wainwright, B.J., Baker, K., and Mattick. J.S. (1991). "Touchdown" PCR to circumvent spurious priming during gene amplification. Nucleic Acids Res. *19,* 4008.

Dong, D., Yan, A., Liu, H., Zhang, X., and Xu, Y. (2006). Removal of humic substances from soil using aluminium sulfate. J. Microbiol. Methods *66,* 217–222.

van Doorn, R., Klerks, M.M., van Gent-Pelzer, M.P.E., Speksnijder, A.G.C.L., Kowalchuk, G.A., and Schoen, C.D. (2009). Accurate quantification of microorganisms in PCR-inhibiting environmental extracts by a novel internal amplification control approach using Biotrove OpenArrays. Appl. Environ. Microbiol. *75,* 7253–7260.

Dutton, C.M., Paynton, C., and Sommer, S. (1993). General method for amplifying regions of very high G + C content. Nucleic Acids Res. *21,* 2953-2954.

Eilert, K.D., and Foran, D.R. (2009). Polymerase resistance to polymerase chain reaction inhibitors in bone. J. Forensic Sci. *54,* 1001–1007.

Fierer, N., Jackson, J.A., Vilgalys, R., and Jackson, R.B. (2005). Assessment of soil microbial community structure by use of taxon-specific quantitative PCR assays. Appl. Environ. Microbiol. *71,* 4117–4120.

Fittipaldi, M., Codony, F., Adrados, B. Camper, A.K., and Morató, J. (2011). Viable real-time PCR in environmental samples: can all data be interpreted directly? Microb. Ecol. *61,* 7–12.

Frostegård, Å., Courtois, S., Ramisse, V., Clerc, S., Bernillon, D., Le Gall, F., Jeannin, P., Nesme, X., and Simonet, P. (1999). Quantification of bias related to the extraction of DNA directly from soils. Appl. Environ. Microbiol. *65,* 5409–5420.

Furet, J.P., Firmesse, O., Gourmelon, M., Bridonneau, C., Tap, J., Mondot, Doré, J., and Corthier, G. (2009). Comparative assessment of human and farm animal faecal microbiota using real-time quantitative PCR. FEMS Microbiol. Ecol. *68,* 351–362.

Gallori, E., Bazzicalupo, M., Dal Canto, L., Fani, R., Nannipieri, P., Vettori, C., and Stotzky, G. (1994). Transformation of *Bacillus subtilis* by DNA bound on clay in nonsterile soil. FEMS Microbiol. Ecol. *15,* 119–126.

Goll, R., Olson, T., Cui, G.L., and Florholmen, J. (2006). Evaluation of absolute quantitation by nonlinear regression in probe-based real-time PCR. BMC Bioinfo. *7,* 107–117.

Greaves, M.P., and Wilson, M.J. (1969). The adsorption of nucleic acids by montmorillonite. Soil Biol. Biochem. *1,* 317–323.

Gribaldo, S., Cookson, B., Saunders, N., Marples, R., and Stanley, J. (1997). Rapid identification by specific PCR of coagulase-negative staphylococcal species important in hospital infection. J. Med. Microbiol. *46*, 45–53.

Gruntzig, V., Nold, S.C., Zhou, J., and Tiedje, J.M. (2001). *Pseudomonas stutzeri* nitrite reductase gene abundance in environmental samples measured by real-time PCR. Appl. Environ. Microbiol. *67*, 760–768.

Gulden, R.H., Levy-Booth, D., Campbell, R., Powell, J.R., Hart, M.M., Dunfield, K., Trevors, J.T., Pauls, K.P., Klironomos, J.N., and Swanton, C.J. (2007). An empirical approach to target environmental samples using real-time polymerase chain reactions. Soil Biol. Biochem. *39*, 1956–1967.

Heid, C.A., Stevens, J., Livak, K.J., and Williams, P.M. (1996). Real time quantitative PCR. Genome Res. *6*, 986–994.

Henderson, S.L., Dandie, C.E., Patten, C.L., Zebarth, B.J., Burton, D.L., Trevors, J.T., and Goyer, C. (2010). Changes in denitrifier abundance, Denitrification gene mRNA levels, nitrous oxide emissions, and denitrification in anoxic soil microcosms amended with glucose and plant residues. Appl. Environ. Microbiol. *76*, 2155–2164.

Hermansson, A., and Lindgren, P.E. (2001). Quantification of ammonia-oxidizing bacteria in arable soil by real-time PCR. Appl. Environ. Microbiol. *67*, 972––976.

Higgins, J.A., Fayer, R., Trout, J.M., Xiao, L., Lal, A.A., Kerby, S., and Jenkins, M.C. (2001). Real-time PCR for the detection of *Cryptosporidium parvum*. J. Microbiol. Methods *47*, 323–337.

Hristova, K.R., Lutenegger, C.M., and Scow, K.M. (2001). Detection and quantification of methyl tert-butyl ether degrading strain PM1 by real-time TaqMan PCR. Appl. Environ. Microbiol. *67*, 5154–5160.

Ikeda, S., Tsurumaru, H., Wakai, S., Noritake, C., Fujishiro, K., Akasaka, M., and Ando, K. (2008). Evaluation of the effects of different additives in improving the DNA extraction yield and quality from Andosol. Microbes Environ. *23*, 159–166.

Kellenberger, E. (2001). Exploring the unknown: the silent revolution of microbiology. EMBO Rep. *2*, 5–7.

Kernaghan, G., Reeleder, R.D., and Hoke, S.M.T. (2007). Quantification of *Cylindrocarpon* destructans f. sp. Panacis in soils by real-time PCR. Plant Pathol. *56*, 508–516.

Khanna, M., and G. Stotzky. (1992). Transformation of *Bacillus subtilis* by DNA bound on montmorillonite and effect of DNase on the transforming ability of bound DNA. Appl. Environ. Microbiol. *58*, 1930–1939.

Klappenbach. J.A., Saxman, P.R., Cole, J.R., and Schmidt. T.M. (2001). rrndb: the ribosomal RNA operon copy number database. Nucl Acids Res. *29*, 181–184.

Klein, D., Janda, P., Steinborn, R., Muller, M., Salmons, B., and Gunzburg, W.H. (1999). Proviral load determination of different feline immunodeficiency virus isolates using real-time polymerase chain reaction: influence of mismatches on quantification. Electrophoresis *20*, 291–299.

Ko, K.S., and Jung, H.S. (2002). Three nonorthologous ITS1 types are present in a polypore fungus Trichaptum abietinum. Mol. Phylogenet. Evol. *23*, 112–122.

Kozdrój J., and van Elsas J.D. (2000). Bacterial community DNA extracted from soils polluted with heavy metals. Polish J. Environ. Studies *9*, 403–407.

Kutyavin, I.V., Afonina, I.A., Mills, A., Gorn, V.V., Lukhtanov, E.A., Belousov, E.S., Singer, M.J., Walburger, D.K., Lokhov, S.G., Gall, A.A., *et al.* (2000). "3′-minor groove binder-DNA probes increase sequence specificity at PCR extension temperatures". Nucleic Acids Res. *28*, 655–661.

Kwok, S., Kellogg, D.E., McKinney, N., Spadic, D., Godal, L., Levenson, C., and Sninsky, J.J. (1990). Effects of primer-template mismatches on the polymerase chain reaction: human immunodeficiency virus type I model studies. Nucleic Acids Res. *18*, 999–1005.

LaMontagne, M.G., Michel, F.C., Holden, P.A., and Reddy, C.A. (2002). Evaluation of extraction and purification methods for obtaining PCR-amplifiable DNA from compost for microbial community analysis. J. Microbiol. Methods *49*, 255–264.

Lee, C., Kim, J., Shin, S.G., and Hwang, S. (2008) Monitoring bacterial and archaeal community shifts in a mesophilic anaerobic batch reactor treating a high-strength organic wastewater. FEMS Microbiol. Ecol. *65*, 544–554.

Lee, C.S., and Lee, J. (2010). Evaluation of new *gyrB*-based real-time PCR system for the detection of *B. fragilis* as an indicator of human specific fecal contamination. J. Microbiol. Methods *82*, 311–318.

Lee, D.-Y., Weir, S.C., and Lee, H., and Trevors, J.T. (2010). Quantitative identification of fecal water pollution sources by TaqMan real-time PCR assays using Bacteroidales 16S rRNA genetic markers. Appl. Microbiol. Biotechnol. *88*, 1373–1383.

Lee, L.G., Connell, C.R., and W. Bloch. (1993). Allelic discrimination by nick-translation PCR with fluorogenic probes. Nucleic Acids Res. *21*, 3761–3766.

Lefèvre, E., Jobard, M., Venisse, J., Bec, A., Kagami, M., Amblard, C., and Sime-Ngando, T. (2010). Development of a real-time PCR assay for quantitative assessment of uncultured freshwater zoosporic fungi. J. Microbiol. Methods *81*, 69–76.

Lerat, S., England, L.S., Vincent, M.L., Pauls, K.P., Swanton, C.J., Klironomos, J.N., and Trevors, J.T. (2005). Real-time polymerase chain reaction quantification of the transgenes for Roundup Ready maize and Roundup Ready soybean in soil samples. J. Agric. Food Chem. *53*, 1337–1342.

Lindquist, H.D.A. (2011). Taking the hay out of the haystack: collecting and processing water samples. In Environmental Microbiology: Current Technology and Water Applications, Sen, K., and Ashbolt, N.J., eds (Caister Academic Press, Norfolk), pp. 39–64.

Liu, W., and Saint, D.A. (2002). A new quantitative method of real time RT-PCR assay based on simulation of PCR kinetics. Anal. Biochem. *302*, 52–59.

Livak, K.J., Flood, S.J., Marmaro, J., Giusti, W., and Deetz, K. (1995). Oligonucleotides with fluorescent dyes at opposite ends provide a quenched probe system useful for detecting PCR product and nucleic acid hybridization. PCR Methods Applic. *4*, 357–362.

Löffert, D., Schaffrath, N., Berkenkopf, M., Stump, S., and Kang, J. (1997). PCR optimization. Qiagen News 2, 1–21.

Manter, D.K., Weir, T.L., and Vivanco, J.M. (2010). Negative effects of sample pooling on PCR-based estimates of soil microbial richness and community structure. Appl. Environ. Microbiol. *76*, 2086–2090.

Martin-Laurent, F., Philippot, L., Hallet, S., Chaussod, R., Germon, J.C., Soulas, G., and Catroux, G. (2001). DNA extraction from soils: old bias for new microbial diversity analysis methods. Appl. Environ. Microbiol. *67*, 2354–2359.

Matheson, C.D., Gurney, C., Esau, N., and Ryan Lehto, R. (2010). Assessing PCR Inhibition from Humic Substances. Open Enz. Inh. J. *3*, 38–45.

Miller, D.N. (2001). Evaluation of gel filtration resins for the removal of PCR-inhibitory substances from soils and sediments. J. Microbiol. Methods *44*, 49–58.

Miller, D.N., Bryant, J.E., Madsen, E.L., and Ghiorse, W.C. (1999). Evaluation and optimization of DNA extraction and purification procedures for soil and sediment samples. Appl. Environ. Microbiol. *65*, 4715–4724.

Miyashita, A., Mochimaru, H., Kazama, H., Ohashi, A., Yamaguchi, T., Nunoura, T., Horikoshi, K., Takai, K., and Imachi, H. (2009). Development of 16S rRNA gene-targeted primers for detection of archaeal anaerobic methanotrophs (ANMEs). FEMS Microbiol. Lett. *297*, 31–374.

Moreira D. (1998). Efficient removal of PCR inhibitors using agarose embedded DNA preparations. Nucleic Acids Res. *26*, 3309–3310.

Mullis, K.B., and Faloona, F.A. (1987). Specific synthesis of DNA *in vitro* via a polymerase-catalyzed chain reaction. Methods Enzymol. *155*, 335–350.

Musso, M., Bocciardi, R., Parodi, S., Ravazzolo, R., and Ceccherini, I. (2006). Betaine, dimethyl sulfoxide, and 7-deaza-dGTP, a powerful mixture for amplification of GC-rich DNA sequences. J. Mol. Diagn. *8*, 544–550.

Nettmann, E., Bergmann, I., Pramschüfer, S., Mundt, K., Plogsties, V., Herrmann, C., and Klocke, M. (2010). Polyphasic analyses of methanogenic archaeal communities in agricultural biogas plants. Appl. Environ. Microbiol. *76*, 2540–2548.

Novotny, E.H., Blum, W.E.H., Gerzabek, M.H., and Mangrich, A.S. (1999). Soil management system effects on size fractionated humic substances. Geoderma 92, 87–109.

Ochsenreiter, T., Selezi, D., Quaiser, A., Bonch-Osmolovskaya, L., and Schleper, C. (2003). Diversity and abundance of Crenarchaeota in terrestrial habitats studied by 16S RNA surveys and real time PCR. Environ. Microbiol. 5, 787–797.

Ogram, A.V., Mathot, M.L., Harsh, J.B., Boyle, J.and Pettigrew, Jr., C.A. (1994). Effects of DNA polymer length on its adsorption to soils. Appl. Environ. Microbiol. *60*, 393–396.

Opel, K.L., Chung, D., and McCord, B.R. (2010). A Study of PCR Inhibition Mechanisms Using Real Time PCR. J. Forensic Sci. 55, 25–33.

Paget, E., Jocteur-Monrozier, L., and Simonet, P. (1992). Adsorption of DNA on clay minerals: protection against DNAse I and influence on gene transfer. FEMS Microbiol. Lett. 97, 31–40.

Paget, E., and Simonet, P. (1994). On the track of natural transformation in soil. FEMS Microbiol. Ecol. *15*, 109–118.

Paris, R., and Lamattina, L. (2002). Increased ratio of mitochondrial rDNA to cytoplasmic rDNA during zoosporic and germinating cyst stages of the life cycle of Phytophthora infestans (Mont.) de Bary. Can. J. Microbiol. *48*, 268–274.

Peleg, O., Baneth, G., Eyal, O., Inbar, J., and Harrus, S. (2009). Use of chimeric DNA–RNA primers in quantitative PCR for detection of Ehrlichia canis and Babesia canis. Appl. Environ. Microbiol. *75*, 6393–6398.

Peršoh, D., Theuer, S., Buscot, F., and Rambold, G. (2008). Towards a universally adaptable method for quantitative extraction of high-purity nucleic acids from soil. J. Microbiol. Methods 75, 19–24.

Pfaffl, M.W. (2004). Quantification strategies in real-time PCR. In A–Z of Quantitative PCR, Bustin, S.A., ed (International University Line Biotechnology Series, La Jolla, CA), pp. 87–112.

Philippot, L., Bru, D., Saby, N.P.A., Cuhel, J., Arrouays, D., Simek, M., and Hallin, S. (2009). Spatial patterns of bacterial taxa in nature reflect ecological traits of deep branches of the 16S rRNA bacterial tree. Environ. Microbiol. *11*, 3096–3104.

Philippot, L., Abbate, C., Bispo, A., Chesnot, T., Hallin, S., Lemanceau, P., Lindström, K., Pandard, P., Romero, E., Schloter, M., *et al.* (2010). Soil microbial diversity: an ISO standard for soil DNA extraction. J. Soils Sediments *10*, 1344–1345.

Picard, C., Ponsonnet, C., Paget, E., Nesme, X., and Simonet, P. (1992). Detection and enumeration of bacteria in soil by direct DNA extraction and polymerase chain reaction. Appl. Environ. Microbiol. *58*, 2717–2722.

Piccolo, A. (2002). The supramolecular structure of humic substances: A novel understanding of humus chemistry and implications in soil science. Adv. Agron. *75*, 57–134.

Pobel, D., Robin, J., and Humbert, J. (2011). Influence of sampling strategies on the monitoring of cyanobacteria in shallow lakes: lessons from a case study in France. Water Res. *45*, 1005–1014.

Porteous, L.A., Seidler, R.J., and Watrud, L.S. (1997). An improved method for purifying DNA from soil for polymerase chain reaction amplification and molecular ecology applications. Molec. Ecol. *6*, 787–791.

Prosser, J.I. (2010). Replicate or lie. Environ. Microbiol. *12*, 1806–1810.

Rajal, V.B., McSwain, B.S., Thompson, D.E., Leutenegger, C.M., Kildare, B.J., and Wuertz, S. (2007). Validation of hollow fiber ultrafiltration and real time PCR using bacteriophage PP7 as surrogate for the quantification of viruses from water samples. Water Res. *41*, 1411–1422.

Ranjard, L., Lejon, D.P.H., Mougel, C., Schehrer, L., Merdinoglu, D., and Chaussod, R. (2003). Sampling strategy in molecular microbial ecology: influence of soil sample size on DNA fingerprinting analysis of fungal and bacterial communities. Environ. Microbiol. *5*, 1111–1120.

Rappé, M.S., and Giovannoni, S.J. (2003). The uncultured microbial majority. Annu. Rev. Microbiol. *57*, 369–394.

Remenant, B., Grundmann G.L., and Jocteur-Monrozier, L. (2009). From the micro-scale to the habitat: assessment of soil bacterial community structure as shown by soil structure directed sampling. Soil Biol. Biochem. *41*, 29–36.

Reysenbach, A.L., Giver, L.J., Wickham, G.S., and Pace, N.R. (1992). Differential amplification of rRNA genes by polymerase chain reaction. Appl. Environ. Microbiol. *58*, 3417–3418.

da Rocha, U.N., van Elsas, J.D., and van Overbeek, L.S. (2010). Real-time PCR detection of Holophagae (Acidobacteria) and Verrucomicrobia subdivision 1 groups in bulk and leek (*Allium porrum*) rhizosphere soils. J. Microbiol. Methods *83*, 141–148.

Rodrigues, D.F., and Tiedje, J.M. (2007). Multi-locus real-time PCR for quantitation of bacteria in the environment reveals Exiguobacterium to be prevalent in permafrost. FEMS Microbiol. Ecol. *59*, 489–499.

Roh, C., Villatte, F., Kim, B.G., and Schmid, R.D. (2005). In-gel patch electrophoresis: a new method for environmental DNA purification. Electrophoresis *2*, 3055–3061.

Rychlik W., Spencer, W.J., and Rhoads, R.E. (1990). Optimization of the annealing temperature for DNA amplification *in vitro*. Nucleic Acids Res. *18*, 6409–6412.

Schaumann, G.E. (2006). Soil organic matter beyond molecular structure. Part I. Macromolecular and supramolecular characteristics, J. Plant Nutr. Soil Sci. *169*, 145–156.

Senesi, N., Plaza, C., Brunetti, G., and Polo, A. (2007). A comparative survey of recent results on humic-like fractions in organic amendments and effects on native soil humic substances. Soil Biol. Biochem. *39*, 1244–1262.

Sipos, R., Székely, A.J., Palatinsky, M., Révész, S., Márialigeti, K., and Nikolausz, M. (2007). Effect of primer mismatch, annealing temperature and PCR cycle number on 16S rRNA gene-targeting bacterial community analysis. FEMS Microbiol. Ecol. *60*, 341–350.

Smith, C.J., and Osborn, A.M. (2009). Advantages and limitations of quantitative PCR (Q-PCR)-based approaches in microbial ecology. FEMS Microbiol. Ecol. *67*, 6–20.

Smith, S., Vigilant, L., and Morin. P.A. (2002). The effects of sequence length and oligonucleotide mismatches on 5′ exonuclease assay efficiency. Nucleic Acids Res. *30*, 111–121.

Stadhouders, R., Pas, S.D., Anber, J., Voermans, J., Mes, T.H.M., and Schutten, M. (2010). The effect of primer-template mismatches on the detection and quantification of nucleic acids using the 5′ nuclease Assay. J. Mol. Diagn. *12*, 109–117.

Stoeckel, D.M., Stelzer, E.A., and Dick, L.K. (2009) Evaluation of two spike-and-recovery controls for assessment of extraction efficiency in microbial source tracking studies. Water Res. *43*, 4820–4827.

Straub, T.M., Pepper, I.L., and Gerba, C.P. (1995). Removal of PCR inhibiting substances in sewage sludge amended soil. Water Sci. Technol. *3*, 311–315.

Stults, J.R., Snoeyenbos-West, O., Methe, B., Lovley, D.R., and Chandler, D.P. (2001). Application of the 52′-fluorogenic exonuclease assay (TaqMan) for quantitative ribosomal DNA and rRNA analysis in sediments. Appl. Environ. Microbiol. *67*, 2781–2789.

Suzuki, M.T., Taylor, L.T., and DeLong, E.F. (2000). Quantitative analysis of small-subunit rRNA genes in mixed microbial populations via 5′-nuclease assays. Appl. Environ. Microbiol. *66*, 4605–4614.

Swan, B.K., Ehrhardt. C.J., Reifel, K.M., Moreno, L.I., and Valentine, D.L. 2010. Archaeal and bacterial communities respond differently to environmental gradients in anoxic sediments of a California hypersaline lake, the Salton Sea. Appl. Environ. Microbiol. *76*, 757–768.

Swillens,.S, Dessars, B., and Housni, H.E. (2008). Revisiting the sigmoidal curve fitting applied to quantitative real-time PCR data. Anal. Biochem. *373*, 370–376.

Takada-Hoshino, Y., and Matsumoto, N. (2004). An improved DNA extraction method using skim milk from soils that strongly adsorb DNA. Microbes Environ. *19*, 13–19.

Takai, K., and Horikoshi, K. (2000). Rapid detection and quantification of members of the archaeal community by quantitative PCR using fluorogenic probes. Appl. Environ. Microb. *66*, 5066–5072.

Taş, N., van Eekert, M.H.A., Schraa, G., Zhou, J., de Vos, W.M., and Smidt, H. (2009). Tracking functional guilds: 'Dehalococcoides' spp. in European river basins contaminated with hexachlorobenzene. Appl. Environ. Microbiol. *75*, 4696–4704.

Tebbe, C.C., and Vahjen,W. (1993). Interference of humic acids and DNA extracted directly from soil in detection and transformation of recombinant DNA from bacteria and a yeast. Appl. Environ. Microbiol. *59*, 2657–2665.

Töwe, S., Wallisch, S., Bannert, A., Fischer, D., Hai, B., Haesler, F., Kleineidam, K., and Schloter, M. (2011). Improved protocol for the simultaneous extraction and column-based separation of DNA and RNA from different soils. J. Microbiol. Methods *84*, 406–412.

Tsai, Y.L., and Olson, B.H. (1992a). Detection of low numbers of bacterial cells in soils and sediments by polymerase chain reaction. Appl. Environ. Microbiol. *58*, 754–757.

Tsai, Y.L., and Olson, B.H. (1992b). Rapid method for separation of bacterial DNA from humic substances in sediments for polymerase chain reaction. Appl. Environ. Microbiol. *58*, 2292–2295.

Tsuchiya, D., and Taga, M. (2001). Application of fibre-FISH (fluorescence *in situ* hybridization) to filamentous fungi: visualization of the rRNA gene cluster of the ascomycete *Cochliobolus heterostrophus*. Microbiology *147*, 1183–1187.

Wessén E., Hallin, S., and Philippot, L. (2010). Differential responses of bacterial and archaeal groups at high taxonomical ranks to soil management. Soil Biol. Biochem. *42*, 59–1765.

Wielgoss, A., Nechwatal, J., Bogs, C., and Mendgen, K. (2009). Host plant development, water level and water parameters shape Phragmites australis-associated oomycete communities and determine reed pathogen dynamics in a large lake. FEMS Microbiol. Ecol. *69*, 55–65.

Wilson, I.G. (1997). Inhibition and facilitation of nucleic acid amplification. Appl. Environ. Microbiol. *63*, 3741–3751.

Wittwer, C.T., Herrmann, M.G., Moss, A.A., and Rasmussen, R.P. (1997). Continuous fluorescence monitoring of rapid cycle DNA amplification. Biotechniques 22, 130–138.

Wu, D.Y., Ugozzoli, L., Pal, B.K., Qian, J., Wallace, R.B. (1991). The effect of temperature and oligonucleotide primer length on specificity and efficiency of amplification by polymerase chain reaction. DNA Cell Biol. *10*, 233–238.

Yang, W., Gu, A.Z., Zeng, S., Li, D. He, M., and Shi, H. (2011). Development o a combined immunomagnetic separation and quantitative reverse transcription-PCR assay for sensitive detection of infectious rotavirus in water samples. J. Microbiol. Methods *84*, 447–453.

Yamada, T., Miyauchi, K., Ueda, H., Ueda, Y., Sugawara, H., Nakai, Y., and Endo, G. (2007). Composting cattle dung wastes by using a hyperthermophilic pre-treatment process: characterization by physicochemical and molecular biological analysis. J. Biosci. Bioeng. *104*, 408–415.

Yu, Z., Michel, Jr. F.C., Hansen, G., Wittum, T., and Morrison, M. (2005). Development and application of real-time PCR assays for quantification of genes encoding tetracycline resistance. Appl. Environ. Microbiol. *71*, 6926–6933.

Zipper, H., Buta, C., Lämmle, K., Brunner, H., Bernhagen, J., and Frank Vitzthum, F. (2003). Mechanisms underlying the impact of humic acids on DNA quantification by SYBR Green I and consequences for the analysis of soils and aquatic sediments. Nucleic Acids Res. *31*, e39.

Zhou, J. Bruns, M.A., and Tiedje, J. (1996). DNA recovery from soils of diverse composition. Appl. Environ. Microbiol. *62*, 316–322.

Quantification of Microorganisms Using a Functional Gene Approach

6

Lia C.R.S. Teixeira and Etienne Yergeau

Abstract

Quantitative polymerase chain reaction (qPCR) represents an effective method to quantify genes or transcripts within environmental samples. For that reason, qPCR has been widely used to characterize the functional patterns of complex microbial communities. In this chapter we summarize some recent applications of different qPCR approaches targeting functional genes encoding key enzymes in the N-, C- and S-cycles and also functional genes related to antibiotic resistance. We also point out some limitation of qPCR approaches. The ongoing development of new molecular techniques, like metagenomics, will have positive impacts on the specificity and the coverage of qPCR assays, since the availability of more sequence data will help to improve the design of primers targeting functional genes.

Introduction

Quantitative polymerase chain reaction (qPCR) is considered a highly sensitive method for the quantification of microbial communities in environmental samples. Several qPCR assays have been developed to quantify the 16S rRNA gene in order to estimate bacterial population size in the environment (Fierer *et al.*, 2005; Guo *et al.*, 2008). Primers can be designed to target different regions of the 16S gene: highly conserved regions for total bacteria and hypervariable regions for specific taxa, ranging from phylum to species (Smith and Osborn, 2009). However, the number of 16S rRNA genes cannot be easily converted to cell numbers since the copy number of this gene can vary between bacteria (Klappenbach *et al.*, 2000). In addition, even though the overall trends in the results obtained from the 16S rRNA gene analysis and the functional gene analyses could be similar for some key enzymes, it is important to keep in mind that bacteria sharing high 16S rRNA similarity can present very distinct phenotypes (Pereyra *et al.*, 2010). While functional genes are not perfect in every regard, they present a much higher likelihood of corresponding to the function of interest (Pereyra *et al.*, 2010). Therefore, the accurate quantification of functional genes in the environment is key to understand ecosystem functioning. Genes encoding key enzymes that mediate biogeochemical processes can be targeted in order to assess the potential for a particular microbial function in the environment. qPCR is an ideal technique to be used in that context, since not only information about the presence but also the abundance and distribution of these genes in the environment is crucial for a better understanding of microbial functions in the different ecosystems (Smith and Osborn, 2009).

A limitation to the functional gene approach has been the relative lack of characterization of the genes of interest and the subsequent unavailability of target sequences. Because of that, the available primer sets are often more relevant to pure cultures than to complex environmental samples (Pereyra *et al.*, 2010). Another challenge is the high genetic variability of several functional genes generally observed across phylogenetic groups. In order to circumvent this problem, degenerated primers have been used in qPCR (Powell *et al.*, 2006; Steinberg and Regan, 2009; De Long *et al.*, 2010; Pereyra *et al.*, 2010). However, since the specificity of these primers is lower, the reaction conditions should be carefully optimized (Powell *et al.*, 2006). Many other factors – including amount of sample used for DNA extraction, the efficiency of the extractions, and the presence of PCR inhibitors – affect the number of copies of any gene detected in an environmental sample. Despite of all these limitation, the high sensitivity and accuracy of qPCR assays targeting functional genes in environmental samples is well established (Galand *et al.*, 2009; Hai *et al.*, 2009; Pouliot *et al.*, 2009; Yergeau *et al.*, 2009; Bru *et al.*, 2011).

In this chapter we present some selected recent functional genes-based qPCR studies that aimed at characterizing functional microbial communities and at studying biogeochemical cycles (Table 6.1).

Nitrogen cycle

The main steps in the nitrogen cycle are shown in Fig. 6.1. The three processes that have been studied in more details with modern molecular tools are nitrogen fixation, denitrification and nitrification. Anaerobic ammonia oxidation (annamox, combining NO_2 and NH_4 to produce N_2) is also an important step in the N-cycle, but for clarity purposes and since it is not discussed in this text, this step is not shown in Fig. 6.1. Nitrification can only occur under aerobic conditions, while denitrification, in which nitrate is used as a terminal electron acceptor instead of oxygen, can only occur under anaerobic conditions. In complex ecosystems, like soil, these processes can occur simultaneously in different micro-sites submitted to different oxygen conditions. At the ecosystem level, nitrification is generally favoured under soil conditions of ample NH_4 availability, moderate pH and sufficient aeration (except when the soil becomes overly dry), while denitrification is favoured by high availability of labile C (energy) and NO_3 (electron acceptor) in poorly aerated soils and pH close to neutrality (Barnard *et al.*, 2005). For nitrification, ammonia oxidation is the first and usually rate-determining step (Belser, 1979). The major rate-limiting step in the denitrification process is generally recognized to be nitrite reduction, since it is the first committed step that leads to a gaseous intermediate (Zumft, 1997).

Nitrogen fixation

Fixation of N by diazotrophic microorganisms is a crucial step in the N-cycle since it is the only way to bring new nitrogen from the atmosphere to the

Table 6.1 Example of environmental qPCR studies targeting functional genes

Function	Enzyme	Target gene	Applications	Reference
N-fixation	Dinitrogenase	*nifH*	Molecular diversity in the rhizosphere of two sorghum cultivars	Coelho *et al.* (2009)
			Characterization of the diazotrophic communities in a rice paddy field	Martensson *et al.* (2009)
Nitrification	Ammonia monooxygenase	*amoA*	Spatial distribution of ammonia-oxidizing in a region of the central Pacific Ocean	Church *et al.* (2010)
			Distribution of the *amoA* gene in different water mass in the Canadian Arctic	Galand *et al.* (2009)
			Vertical structure and distribution of *amoA* gene variants in two high Arctic lakes	Pouliot *et al.* (2009)
			Comparison between Archaea and Bacteria ammonia-oxidizing activity in soils	Leininger *et al.* (2006), Mertens *et al.* (2009), Gubry-Rangin *et al.* (2010)

Table 6.1 Continued

Function	Enzyme	Target gene	Applications	Reference
Denitrification	Nitrite reductase	*nirK, nirS*	Full nitrogen cycle approach to study agricultural soils	Hai *et al.* (2009)
			Distribution of nitrogen-cycling microbial communities at the landscape scale	Bru *et al.* (2011)
			Effects of silvicultural practices on the abundance of diazotrophs and denitrifiers	Levy-Booth and Winder (2010)
C-degradation	PAH ring hydroxylating dioxygenase	PAH-RHD (Alpha subunit)	Monitoring of bacteria and their functions in polar environments during bioremediation	Yergeau *et al.* (2009)
	Alkane monooxygenase	alkb	Monitoring of bacteria and their functions in polar environments during bioremediation	Powell *et al.* (2006), Yergeau *et al.* (2009)
			Comparison between MNP-PCR vs. Real time PCR for functional genes using alkb as a model.	Schulz *et al.* (2010)
	Chitinase	*chiA*	Functional genetic characterization of polar soils	Yergeau *et al.* (2007, 2010)
	Cellulase	*cel*	Use of functional genes for complex microbial communities characterization	Pereyra *et al.* (2010)
Methanogenesis	Methyl coenzyme M reductase	*mcrA*	Quantification of methanogens communities	Steinberg and Regan (2009)
Methane oxidation	Methane monooxygenase	*pmoA*	Evenness and functioning of methane-oxidizing bacteria in soil after a simulated disturbance-induced mortality	Ho *et al.* (2010)
			Links between methane flux and transcriptional activities of methane oxidizers and methanogens.	Freitag *et al.* (2010)
C-fixation	RubisCO	*ccbL*	Diversity and composition of *ccbL* types in soil bacteria.	Selesi *et al.* (2005)
			Abundance and diversity of CO2-fixing bacteria in soils close to carbon dioxide springs.	Videmsek *et al.* (2009)
Sulfate reduction	Dissimilatory sulfate reductase	*dsrAB*	Quantification of sulfate-reducing microbial communities in a complex environment	Ben-Dov *et al.* (2007)
Antibiotic resistance	Tetracycline resistance	*tet*	Detection of tetracycline resistance genes in samples from swine and bovine manures	Yu *et al.* (2005)
			Effects of UV disinfection on reduction in the amount of detectable *tet* genes on influent and activated sludges	Auerbach *et al.* (2007)

ecosystems. Nitrogenase is the key enzyme in N_2 fixation and has been used as a target to study this pathway. The *nifH* gene encodes the iron protein subunit of the nitrogenase, is highly conserved and produces a phylogeny consistent with the 16S rRNA gene (Coelho *et al.*, 2009). At the DNA level, *nifH* provides evidence for potential nitrogen fixation in complex microbial communities (Coelho *et al.*, 2009). Since this gene has been studied for many years, an extensive collection of *nifH* sequences has been obtained from both cultivated and uncultivated microorganisms isolated from multiple environments. The diversity and abundance of diazotrophic microorganisms,

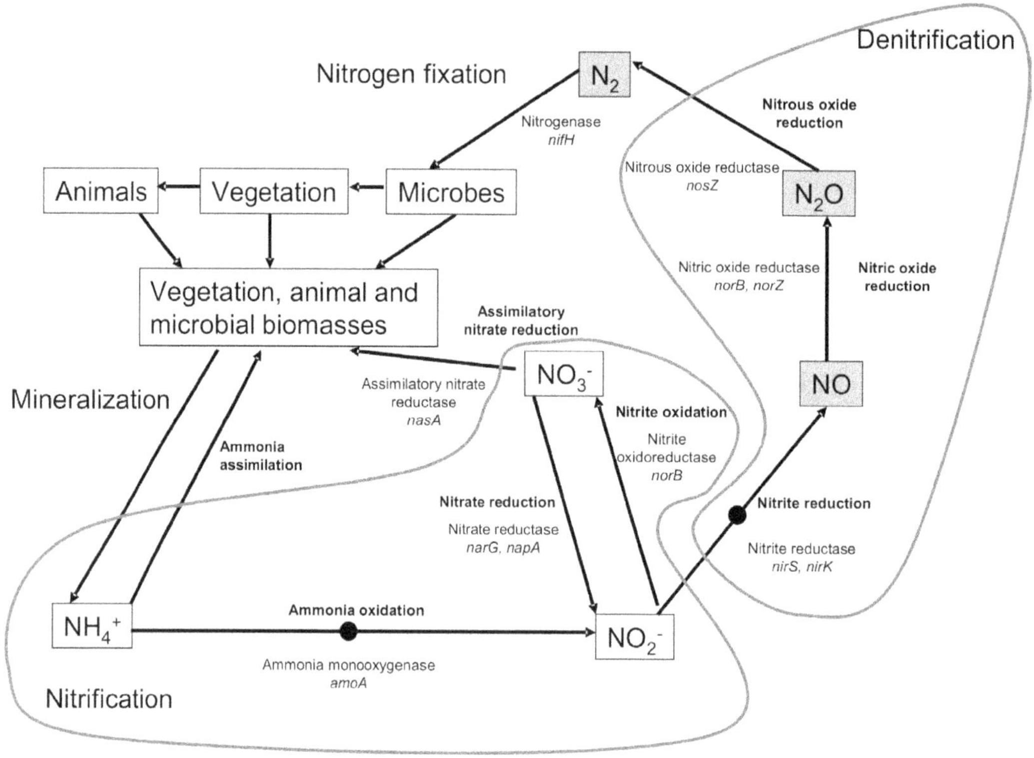

Figure 6.1 The terrestrial nitrogen cycle, including enzymes and associated functional genes. Adapted from Prosser (2007). ● Indicates key/limiting steps and grey-shaded boxes indicate gaseous intermediates.

as determined by *nifH* characterization, was proposed as an indicator of overall soil ecological health (Levy-Booth and Winder, 2010). Real time PCR has been successfully used to determine the abundance of *nifH* genes in a variety of environments. SYBR Green-based qPCR allowed the use of *nifH* primers sets previously designed for other techniques, like T-RFLP and conventional PCR (Poly *et al.*, 2001; Rosch *et al.*, 2002).

Coelho *et al.* (2009) used the primers described by Poly *et al.* (2001) to assess the molecular diversity and the abundance of *nifH* gene sequences by PCR-DGGE and qPCR assays in the rhizosphere of two cultivars of sorghum in soils with contrasting levels of nitrogen fertilizer. DGGE fingerprints showed that the sorghum rhizosphere harboured populations more diverse than those found in the bulk soil. According to the qPCR data, the number of *nifH* genes detected in the rhizosphere soil of both cultivars varied markedly. Overall, diazotroph numbers were higher in the rhizosphere than in the bulk soil, suggesting that the presence of the plant promotes nitrogen fixation. Fertilizer input also influenced diazotroph populations, with a significant decrease in *nifH* abundance being observed in the rhizosphere of both cultivars when high rates of nitrogen were applied. The effect of fertilization interacted with the influence of the rhizosphere, since the nitrogen fertilization did not influence *nifH* abundance in the bulk soil (Coelho *et al.*, 2009).

Using the same set of primers, Martensson *et al.* (2009) combined RT-PCR-DGGE and cloning with RT-qPCR and acetylene reduction activity to identify the active diazotrophic community composition and to quantify *nifH* expression. The study was carried out in two rice paddy fields in Fujian province, south-east China. One of the rice fields was treated with N, K and P (N-fertilized) and the other one was only fertilized with K and P (N-unfertilized). In this study, *nifH* expression together with nitrogenase activity measurements did not reveal any significant differences between the N-fertilized and N-unfertilized soils. The soil analysis performed in the two fields revealed that the differences in the amount of available nitrogen were not large enough to cause any changes (Martensson *et al.*, 2009).

Nitrification

Nitrification is a process involving two different chemolithotrophic prokaryotic groups. In a first step, ammonia is oxidized to nitrite by ammonia-oxidizing bacteria (AOB) or ammonia-oxidizing Archaea (AOA) in a reaction catalysed by the ammonia monooxygenase, for which the gene encoding the alpha subunit (*amoA*) serves as a marker. In the second step, nitrite is oxidized to nitrate by nitrite-oxidizing bacteria (NOB).

The bacterial *amoA* gene was followed by qPCR together with several other N-cycle genes in a microcosm experiment using soil cores from the Antarctic (Yergeau and Kowalchuk, 2008). The abundance of *amoA* was reported to be temperature insensitive when cores were incubated at different constant temperature, but showed some changes when subjected to different frequency of freeze–thaw cycles (FTC). This trend was opposite for vegetated and unvegetated cores, with increasing *amoA* gene abundance with increasing FTC frequency for unvegetated cores and decreasing *amoA* gene abundance with increasing FTC frequency for vegetated cores.

The archaeal *amoA* gene has been detected in a number of marine samples, in both marine and freshwater sediments and in soil (Leininger *et al.*, 2006), indicating a major role for Archaea in the nitrogen cycle (Francis *et al.*, 2007). However, it is still debated whether Archaea or Bacteria are the main ammonia oxidizers. Leininger and colleagues (2006) quantified the *amoA* gene of Crenarcheota and bacteria in 12 pristine and agricultural soils of three climatic zones in Europe in order to evaluate their relative ecological impact. According to the results, archaeal *amoA* genes were more abundant than bacterial *amoA* genes in all analysed soils. To control for methodological bias, related to the fact that different primer sets have different sensitivity and efficiency, five different primer combinations for archaea and bacteria *amoA* and 16S rRNA genes were used in a most probable number (MPN) PCR experiment. These control assays were performed for two of the 12 previously analysed soils and independently of the primer sets used; a predominance of AOA over AOB was detected. This study provided evidence for relatively high abundance of AOA in soils, suggesting that these organisms are adapted to a broad range of growth conditions and might therefore have a more versatile metabolism than AOB (Leininger *et al.*, 2006).

Similarly, high numbers of bacterial and archaeal *amoA* genes were detected in permafrost and active layer soils in the Canadian high Arctic (Yergeau *et al.*, 2010). The archaeal-to-bacterial *amoA* ratio was 31.3 for the active layer and 5.1 for the permafrost, indicating that environmental conditions and soil depth are influencing the relative importance of bacteria and archaea *amoA* genes in Arctic permafrost-affected soils.

In contrast, Mertens and colleagues (2009) reported that AOB, and not AOA, mediate recovery of nitrification after zinc contamination. The number of *amoA* gene copies and transcripts was assessed and compared with patterns of ammonia oxidation activity in an Australian agricultural soil after experimental Zn contamination. In contrast to the general view of archaea being more tolerant to chronic stress conditions than bacteria, this study demonstrated that AOA were more sensitive to Zn contamination than AOB. Interestingly, a significant numbers of AOA *amoA* genes were detected in the Zn-exposed soil, even though no transcripts were detected (Mertens *et al.*, 2009).

The relative contribution of AOA and AOB to ammonia oxidation in two Scottish acid soils (pH 4.5 and 6) was also investigated in microcosms (Gubry-Rangin *et al.*, 2010). Ammonia availability was reduced by the addition of acetylene in the headspace of half of the microcosms and samples were collected at three time points over 1 month. Archaeal *amoA* gene abundance increased significantly in both soils during incubation, while there was no significance increase in the bacterial *amoA* gene. A significant positive correlation was found between nitrification rate and archaeal *amoA* gene abundance, but not with bacterial *amoA* gene abundance. The addition of acetylene did not significantly reduce the abundance of the *amoA* gene for both archaea and bacteria, but it significantly reduced the abundance of archaeal *amoA* gene transcripts. Taken together, these results suggest that AOA, but not AOB, control nitrification in these acidic soils.

Planktonic *Crenarchaeota* are thought to play a key role in ammonia oxidation, a critical step of the marine nitrogen cycle. Church *et al.* (2010) examined the transcriptional activities and distributions of nitrifying marine Crenarchaeota sampled throughout the euphotic and mesopelagic zones (< 1000 m) across a 5200-km region of the Pacific Ocean. Quantitative PCR amplification of crenarchaeal *amoA* and 16S rRNA together with RT-qPCR amplification of *amoA* mRNA transcripts were used. Crenarchaeota abundances showed a strong vertical structure, with greatest abundances observed in the dimly lit waters of the mesopelagic zone. However, the Crenarchaeota *amoA* gene expression was less variable with deep, and generally decreased below the euphotic zone. When these results were normalized to 16S rRNA gene copy numbers, the patterns of transcriptional activity suggest that despite their relatively low abundance, Crenarchaeota may oxidize ammonia very actively throughout the euphotic zone (Church *et al.*, 2010).

Studying three different water masses in the North Water (eastern Canadian Arctic) defined by distinct temperature and salinity characteristics, Galand *et al.* (2009) used qPCR to track the distribution of the Archaeal *amoA* with the primers designed by Wuchter *et al.* (2006). One of the water masses showed very high abundance of archaeal *amoA* gene as compared with other masses, which could result in different biogeochemical capacities between different water masses (Galand *et al.*, 2009).

In another study, the distribution of archaeal *amoA* and 16S rRNA genes was evaluated in two Canadian High Arctic marine-derived lakes characterized by vertical gradients in their physical and chemical properties (Pouliot *et al.*, 2009). The quantitative distribution of *amoA* genes varied with depth for both lakes. A high *amoA* gene copy number was found in the oxycline, while it was less abundant or absent on the surface and in anoxic layers of the lakes. Depths with highest copy numbers of *amoA* genes also had the highest 16S rRNA copies of the Marine Group I (MGI) Crenarcheota. However, the *amoA* to 16S rRNA ratio was always very low and the authors suggested that the *amoA* primers may not be able to amplify all *amoA* genes present in the lakes or that not all MGI Crenarcheota have the *amoA* gene. The primers used in this study were initially designed from marine sequences (Coolen *et al.*, 2007) and may thus contain mismatches to *amoA* genes found in brackish or freshwater environments. The concomitant presence of MGI Crenarcheota and *amoA* genes indicates the potential for crenarcheotal ammonia oxidation in these lakes; however, further experiments and expression measurements are needed to confirm the biogeochemical significance of the presence of archaeal *amoA* genes in these lakes (Pouliot *et al.*, 2009).

Denitrification

Denitrification is the microbial process in the nitrogen cycle, which returns fixed nitrogen to the atmosphere (Fig. 6.1). The reduction of nitrate (NO_3) to nitrite (NO_2) and then to gaseous nitrogen compounds, such as nitric oxide (NO), nitrous oxide (N_2O) or dinitrogen (N_2) is catalysed by various enzymes including nitrate reductases (Nar and Nap), nitrite reductases (NirK and NirS), nitric oxide reductases (cNor and qNor) and nitrous oxide reductase (Nos) (Philippot, 2006). Functional genes related to denitrification for which primers are available include: *narG* and *napA* (nitrate reduction), *nirS* and *nirK* (nitrite reduction), *norB* (nitric oxide reduction) and *nosZ* (nitrous oxide reduction). Denitrifying organisms are taxonomically very diverse, with the majority of them belonging to various subclasses of the Proteobacteria, but also to some Gram-positive phyla and even archaea (Philippot, 2005). Linking density and diversity of denitrifying genes will allow to identify key factors potentially driving nitrogen losses and N_2O and N_2 emissions from the environment (Philippot, 2006). However, it is obvious that studying the diversity of a functional community is of limited value for understanding nitrogen fluxes if the abundance on this functional community is unknown (Philippot, 2006).

Hai *et al.* (2009) investigated nitrogen dynamics on a full-cycle approach analysing the most important N-cycle steps in well-aerated agricultural soils (nitrification, denitrification, and nitrogen fixation). Bacterial genes encoding the nitrogenase reductase (*nifH*), ammonia monooygenase (*amoA*), nitrite reductase (*nirK* and *nirS*), and Archaea ammonia monooxygenase (*amoA*) were quantified by qPCR. According to this study, manure application reduced nitrogen losses by denitrification and enhanced biological nitrogen fixation (Hai *et al.*, 2009).

In a recent study, Bru *et al.* (2011) characterized and detailed the spatial variability in the distribution of microbial community that are involved in nitrogen cycling at the landscape scale. The abundances of *narG*, *napA*, *nirK*, *nirS*, *nosZ* and *amoA* (bacterial and archaeal) genes in soil samples from the Burgundy region were quantified by qPCR. The results showed that the nitrate reducers, nitrite reducers and AOB represented around 5–20%, 1–5% and 0.05–1% of the total bacterial community, respectively. Among the nitrite reducing community (possessing *nirS* or *nirK*), the percentage of bacteria also harbouring the *nosZ* gene was never higher than 56%. This findings support the evidence that a significant fraction of the nitrite-reducing community might lack the genetic ability to perform the reduction of the potent greenhouse gas N_2O into N_2 (Bru *et al.*, 2011). In this study, another interesting result discussed by the authors was that the changes in land use did not strongly influence the abundance of any studied communities other than AOB. This group of bacteria could be used as a biological indicator for monitoring changes in land use. The data also showed that neither the local climate nor the sampling time was a major factor influencing the distribution patterns of functional genes. On the other hand, geographical distance, when separated from the other environmental variables, was a weak but significant predictor of the total bacterial, crenarchaeal, nitrate-reducing and denitrifying communities (Bru *et al.*, 2011).

To compare the effects of silvicultural practices on the abundance of diazotrophs and denitrifiers, Levy-Booth and Winder (2010) used qPCR assays to quantify *nifH*, *nirS*, and *nirK* genes in soils from a long-term experimental installation located on Vancouver Island, British Columbia, Canada. Clear-cutting reduced the abundance of N-cycling genes and also reduced nutrients levels due to the removal of the organic horizons from the soil. Correlations between *nifH* abundance and soil C and N content in the organic horizon, as well as between *nirS* and *nirK* abundance and soil C and N content in the mineral horizons highlighted the relationship between functional gene abundance and soil nutrient status (Levy-Booth and Winder, 2010).

Carbon cycle

The carbon cycle is composed of two major processes, decomposition and carbon fixation (Fig. 6.2). However, the vast array of heterogeneous organic compounds that can be degraded by microorganisms makes this cycle particularly complex and difficult to follow using qPCR-based approaches. The production and oxidation of CH_4 is also of great importance for controlling the levels of this highly potent greenhouse gas.

Carbon degradation

The carbon degradation pathways that are easier to follow by qPCR are the ones involving a clearly defined specific key step involving a single or very few compounds. For this reason, most qPCR work on carbon degradation as been carried out on functional genes encoding key reactions in degradation pathways of environmental pollutants. The accurate quantification of key genes encoding enzymes involved in the catabolic conversions of environmental pollutants can indicate the potential for successful bioremediation and help to follow how microorganisms respond to biostimulation protocols (Smith and Osborn, 2009).

In the presence of oxygen, alkane oxidation is initiated by an activation step where oxygen is introduced to the alkane substrate, after which further catabolic steps can proceed. A number of alkane hydroxylase enzyme systems can catalyse this initial step, including the integral membrane non-haem iron alkan hydroxylases (the alk system). The *alkB* gene is often used as a marker for aerobic alkane degradation. Powell *et al.* (2006) used qPCR to target *alkB*-containing microorganisms in order follow microbial changes in diesel-contaminated Antarctic soil during bioremediation. Results from the qPCR assays

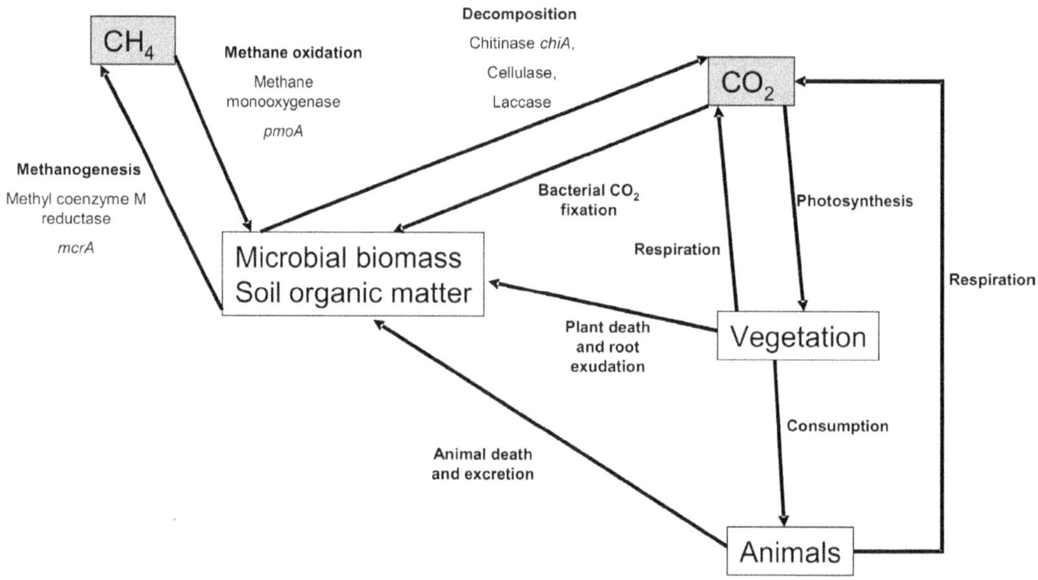

Figure 6.2 The terrestrial carbon cycle, including enzymes and functional genes. Adapted from Prosser (2007). Grey-shaded boxes indicate gaseous intermediates.

were compared with the enumeration of bacteria using the most probable number (MPN) technique. Both analyses suggested that the addition of fertilizer to the contaminated soil increased the total number of alkane-degrading bacteria. Initially, the alkane-degrading bacteria dominated the hydrocarbon-degrading population, but other group of bacteria dominated the bacterial community when the easily degradable alkanes became depleted (Powell *et al.*, 2006).

Polycyclic aromatic hydrocarbons (PAHs) are hydrophobic organic pollutants highly persistent in soil and sediment. The initial step of PAH degradation is carried out by multicomponent aromatic-ring-hydroxylating dioxygenases (RHD) that contain regions conserved among all the different genes encoding PAH-degrading enzymes. Quantification through qPCR, using primers that specifically target the PAH-RHD genes can be used for estimating the PAH biodegradation potential in contaminated environments (Cébron *et al.*, 2008). Cébron and colleagues (2008) designed two sets of PCR primers to target a broad range of Gram-negative and Gram-positive PAH-RHD genes. The qPCR assays subsequently developed were highly specific and allowed accurate quantification of PAH-RHD genes directly from environmental soil and sediment samples (Cébron *et al.*, 2008). Yergeau and colleagues (2009) used these primers together with primers for *alkB*, *nirS* and 16S rRNA genes to analyse the responses of high Arctic soil bacteria to diesel pollution and bioremediation treatments. The expression of degradation genes (*alkB* and PAH-RHD) measured by RT-qPCR showed strong correlation with other more traditional measures of hydrocarbon degradation, like MPN-diesel counts and mineralization potential. This study demonstrated the utility of RT-qPCR as a tool to rapidly monitor bacteria and their functions in polar environments during bioremediation. Using the RT-qPCR strategy, it was possible to observe successions in the expression of functional genes over time, which were most probably related to the successive degradation of different diesel compounds, starting with the most easily degraded components (Yergeau *et al.*, 2009).

Other important pathways related to carbon degradation, like chitin and cellulose degradation, have also been studied using qPCR. The set of primers for *chiA* (chitinase) designed by Williamson *et al.* (2000) was employed in a functional genetic characterization of Antarctic soils

(Yergeau *et al.*, 2007). Combination of a microarray-based functional gene analyses and qPCR were used to target carbon and nitrogen cycle key genes. Bacterial chitinase genes were highly abundant in Antarctic soil with and without vegetation cover as indicated by the microarray assay. However, *chiA* abundances measured by qPCR failed to correlate significantly with microarray results. It should be stressed that qPCR assays relied on different priming sites that the ones used for probing on the array. Also, biases related to both techniques can be responsible for the differences observed (Yergeau *et al.*, 2007). A recent study used the same set of primers to compare high Artic active layer and permafrost soils (Yergeau *et al.*, 2010). Results on the abundance of *chiA* genes, as measured by qPCR, showed a significant difference between the layers, with the active layer harbouring more *chiA* gene copies than the permafrost layer (Yergeau *et al.*, 2010). In another recent study, Pereyra and colleagues (2010) developed qPCR primers to study the cellulose degraders on a mine drainage remediation system. Consensus degenerate primers were designed and validated to amplify both sequenced and unsequenced glycoside hydrolase genes of cellulose-degrading bacteria. As expected, qPCR analyses revealed that the lignocelluloses-based reactor contained higher numbers of this gene than the ethanol fed reactor (Pereyra *et al.*, 2010).

Carbon fixation

CO_2 fixation is a major biogeochemical process in the carbon cycle as it is the central mechanism of primary production in almost all ecosystems. The Calvin–Benson–Bassham cycle is the major and most abundant pathway for CO_2 fixation. This cycle can be found in several organisms (bacteria, algae, plants). The key enzyme of the Calvin cycle is the ribulose-1,5-biphosphate carboxylase/oxygenase (RubisCO) which catalyses the first step in the cycle. RubisCO exists in multiple forms but type I is the most distributed one, occurring in plants, algae and bacteria. The *cbbL* gene encodes the large subunit of type I RubisCO and has been used as a phylogenetic marker in several studies. Based on gene sequences, type I RubisCO can be further subdivided in two major groups: green-like and red-like.

Selesi and colleagues (2005) designed distinct green- and red-like *ccbL*-specific primer sets to determine the diversity and composition of these *ccbL* types in soil bacteria isolated from differently managed agricultural soils in a long-term experiment performed in Germany. In this study, the primers were used to construct a clone library from the different soil types. The clones obtained using the green-like primers had an insufficient number of positives green-like *ccbL* clones, but for the red-like library, a significant diversity of red-like genes were recovered. In a further study, the red-like primers were used to develop a TaqMan quantitative PCR assay to quantify red-like *ccbL* genes in the soils mentioned above (Selesi *et al.*, 2007). Owing to high sequence variation, it was not possible to develop TaqMan probes that match all known red-like *ccbL* sequences. Nonetheless, the developed qPCR assay was able to detect a vast number of bacterial *ccbL* sequences that matched a wide diversity of red-like *cbbL* genes (Selesi *et al.*, 2007). In the three soil types used in the study, 10^6 to 10^7 *ccbL* red-like gene copies per gram of soil were detected. The authors concluded that bacteria with form I Rubisco red-like gene are numerous and widespread in soils and that their importance in soil ecosystems could be underestimated (Selesi *et al.*, 2007).

Videmsek and colleagues (2009) postulated that high CO_2 concentrations in the vicinity of natural carbon dioxide springs could increase the abundance of *ccbL* sequences in soil and change the diversity pattern of this functional group compared with soils that are not under the influence of these springs. To test this hypothesis, the primers mentioned above were used. For bulk and rhizosphere soils, red-like *cbbL* gene copies were significantly less abundant in soils under elevated CO_2. Community structure was also followed using RFLP analysis that showed that red-like *ccbL* genes changed depending on the CO_2 regime. By coupling qPCR and community structure analysis, the authors suggested that only a part of the autotrophic CO_2-fixing microbes could adapt to the very high CO_2 concentrations.

Methane generation

Methanogens are of great importance in carbon cycling, catalysing the production of methane during organic matter degradation in anaerobic environments (Steinberg and Regan, 2009). The key functional enzyme in methane production is the methyl coenzyme-M reductase (Mcr). The *mcrA* gene, coding for a peptide of the Mcr enzyme complex, is a good target to study methanogen communities, since this gene is exclusive to the methanogens (with the exception of the methane-oxidizing archaea), and has a phylogeny that is congruent with the 16S rRNA gene (Steinberg and Regan, 2009). Several studies have used qPCR targeting the *mcrA* gene to quantify methanogens (Freitag and Prosser, 2009; Steinberg and Regan, 2009; Yergeau *et al.*, 2010). Steinberg and Regan (2009) developed a qPCR assay to quantify total numbers of *mcrA* genes and another one using TaqMan probes targeting nine different groups of methanogens. The primers and probes were tested on samples from four anaerobic digesters and on two sample of incubated acidic peat. The number of *mcrA* copies was measured using SYBR Green I method and also by the sum of the methanogen numbers obtained with each TaqMan probe. Total *mcrA* numbers obtained with both the SYBR Green I and the TaqMan probe methods were similar, although not identical. These differences could be explained due to errors in the methods themselves, but also for the inability of the TaqMan technique to detect less dominant community members. The authors concluded that the determination by qPCR of relative numbers of microbes of different clades provides a more in-depth examination of a microbial community than what can be provided by clone libraries alone (Steinberg and Regan, 2009).

Methane oxidation

Methane-utilizing bacteria are a diverse group of Gram-negative bacteria that use methane as a source of energy. Most known aerobic methanotrophs are divided into two major groups (Type I and Type II) based on phylogeny and carbon assimilation pathways (Bowman, 2006). Type I methanotrophs belong to the family Methylococcaceae in the Gammaproteobacteria class, while Type II methanotrophs belong to the family Methylocystaceae in the Alphaproteobacteria class (Bowman, 2006). In addition to these two groups, new methanotrophic bacteria belonging to the phylum Verrucomicrobia were recently described (Dunfield, 2007; Op den Camp *et al.*, 2009). Methane monooxygenase (MMO) is a key enzyme for the methane oxidation process and exists in two forms, the particulate methane monooxygenase (pMMO) and the soluble methane monooxygenase (sMMO). Because of its occurrence in most known methanotrophs, with the exception of *Methylocella* spp., the gene coding for the alpha subunit of the pMMO (*pmoA*) is widely used as a biological marker to study methanotrophic bacteria (Ho *et al.*, 2011).

Although universal primers have been developed to target the *pmoA* gene, the available broad-specificity primers designed to date do not amplify all aerobic methanotrophs. To circumvent this limitation, a number of different probes and primers of narrow specificity have been developed for molecular detection of ecologically and taxonomically relevant clusters (Tuomivirta *et al.*, 2009). The abundance of *pmoA* measured using these primers was positively correlated to methane oxidation potential, suggesting that the primer combination proposed by the authors is reliable and can be used in others studies.

Using methane-oxidizing bacteria as model organisms, a study was conducted on the disturbance caused by a partial die-off of all microbial population, evaluated using a combination of molecular techniques: qPCR, terminal restriction fragment length polymorphism (T-RFLP) and microarrays (Ho *et al.*, 2011). The disturbance was simulated in microcosms mixing native with sterilized paddy soil in two ratios, 1:4 and 1:40, representing moderate and severe die-offs. According to the results, methanotrophs showed a remarkable ability to compensate for die-off. The rapid increase in *pmoA* target numbers revealed by qPCR pointed out a successful re-colonization of the sterile soil by methanotrophs. Analysing the bacterial community along time, the authors observed temporal shifts between type I and type II methanotrophs. Taken together, these results suggest that methanotrophs are not only limited by the availability of methane and oxygen, but also by other nutrients and potentially even by available space (Ho *et al.*, 2011).

Few studies have combined the analysis of both methanogens and methanotrophs communities with measurements of methane fluxes. Therefore, there is a lack of knowledge about the relationships between methanotrophs, methanogens and gaseous emissions. In a recent study, RT-qPCR and qPCR assays targeting *mcrA* and *pmoA* genes were combined to methane measurements in order to check if these genes can be effective predictors of *in situ* methane flux rates (Freitag *et al.*, 2010). The experiments were carried out at two neighbouring sites with contrasting methane fluxes. According to the results, *in situ* flux rates were accurately reflected by the gene/transcript abundance ratio of these key genes involved in methane cycling (Freitag *et al.*, 2010).

Sulfur cycle

Sulfate reduction

Sulfate-reducing bacteria (SRB) are anaerobic microorganisms that use sulfate as a terminal electron acceptor. They are found in anoxic habitats, where they play an important role in both the sulfur and carbon cycles. SRBs belong to different phyla and, when targeting the 16S rRNA gene, several primer sets are needed to follow them in the environment (Daly *et al.*, 2000). Dissimilatory sulfate reductase is the key enzyme in dissimilatory sulfate reduction, and is encoded by the *dsrAB* genes, which can be used as a phylogenetic marker for identification of SRB. In this case, a single set of conserved *dsrAB* primers can be used to target SRBs (Ben-Dov *et al.*, 2007). Other sulfate-reducing prokaryotes possess the adenosine-5-phosphosulfate (APS) reductase encoded by the *apsA* gene.

Ben-Dov *et al.* (2007) developed a rapid method for quantification of SRB communities in complex environments using universal qPCR primers targeting the *dsrA* and *apsA* genes. In their study, samples from five industrial wastewater evaporation pools were used. SRB cell number measured by qPCR was about 100-fold higher than the results obtained using the most probable number (MPN) based on sulfide detection. This suggests that the qPCR method developed can provide valuable information about the abundance of SRB in the environment (Ben-Dov *et al.*, 2007).

In another study, Pereyra *et al.* (2010) developed degenerated primers for the amplification of key functional genes in anaerobic environments, including cellulose degradation, fermentation, sulfate reduction and methanogenesis. Samples were collected from two pilot-scale sulfate-reducing biochemical reactors that received mine drainage: one reactor contained a complex lignocellulose (LC)-based substrate and the second reactor received ethanol (Et) as its primary substrate. The presence of ethanol was expected to select a microbial community rich in SRB. Ethanol is a fermentation product that is not favourably used by most fermenters, but can be directly utilized by SRBs. The qPCR approach developed in the study confirmed this hypothesis, as the Et reactor had a higher *dsrA* gene copy number than the LC reactor (Pereyra *et al.*, 2010).

Antibiotic resistance genes

Antibiotics are widely used to protect the health of human and animals or as a food additive to increase growth rates of animals. The massive indiscriminate use of antibiotics accelerates the development and dispersion of antibiotic resistance genes (ARGs). Hundreds of resistance genes for a broad range of antibiotics have already been found in microorganisms isolated not only from clinical settings, but also from soil and wastewater. There is growing evidence that resistance to antibiotics is increasing both in pathogenic and non-pathogenic bacteria, posing a threat to public and environmental health. Although there is a clinical evidence of increasing antibiotic resistance over the years, almost no quantitative environmental data exist (Knapp *et al.*, 2010). Quantitative measurement of ARGs is required for monitoring the dynamics of antibiotic resistance in the environment.

Tetracycline use is widespread both in the treatment of humans infections and food-animal production. Yu *et al.* (2005) developed a real-time PCR assay to quantify three of the major groups of tetracycline resistance genes: *tet*(A), *tet*(C), and *tet*(G) genes encoding ribosomal protection proteins (RPP). Numerous *tet* gene sequences are available in databases, which facilitated the design of PCR primers. However, it was not possible to design universal primers for all prevalent classes of *tet* genes because of high sequence divergence. The three designed primer pairs were able to target 10 of the major *tet* gene classes. The utility of these qPCR assays was tested on samples of swine and bovine manures, swine wastewater lagoons and composted swine manures. All of the *tet* gene groups targeted in this study were detected in all the samples. The *tet* genes were found to be more abundant in the manures, with the bovine manures containing fewer copies of all three groups of *tet* genes than the swine manures. These differences could be due to the use of tetracycline in swine feeds and also because of the different microbial flora present in the intestinal tract of the two different animal species. The composting of swine manure reduced *tet* gene abundance, while lagoon storage had little effect on *tet* gene abundance (Yu *et al.*, 2005).

The qPCR methodology is not only used for quantitative analysis of ARGs distribution in the environment but also to study the effects of environmental factors or treatment processes on reducing the abundance of some ARGs (Zhang *et al.*, 2009). An investigation on the *tet* genes present in influent water and activated sludge in Germany found that UV disinfection did not reduce the amount of detectable *tet* genes (Auerbach *et al.*, 2007). In another study, DNA was extracted from five Dutch soils from different locations archived from 1940 to 2008, and the abundance of 18 ARG from different major antibiotic classes was quantified. ARGs from all classes tested had significantly increased since 1940, especially the ARGs related to tetracycline resistance. This was somehow surprising, since waste management procedures have largely improved and stricter rules on non-therapeutic antibiotic use in agriculture are being promulgated. These data suggest that basal environmental levels of ARG might still be increasing (Knapp *et al.*, 2010).

Conclusion

qPCR analysis of functional genes has all the limitations related to conventional PCR. The main limitation being the difficulty to design specific primer sets based on the information available in databases. However, with the current interest in environmental sequencing and the ever-increasing sequencing capacity, more and more data are becoming available, helping to design better adapted and more specific primers. One interesting approach is to shotgun-sequence the metagenome of a few samples or pools of samples and then design new primers for functional genes of interest based on the sequences retrieved. These environment-specific primers could then be applied on the full experimental design to follow the functional genes of interest. In such studies, with clearly defined experimental questions, qPCR is the method of choice since it is relatively inexpensive to use, is highly accurate, quantitative and can process many samples rapidly.

Other recent technologies could also be combined with qPCR. For instance, stable isotope probing (SIP) offers the potential to identify microorganisms that actively incorporate specific substrates in their biomass (Dumont and Murrell, 2005). The RNA or the DNA of this labelled active microbial community can be recovered and the expression or the abundance of different functional genes can be monitored by qPCR. In addition, some interesting studies recently started combining open-arrays and real-time PCR to quantify 100s of genes simultaneously (van Doorn *et al.*, 2007, 2009). Adapting such an approach for complex sets of genes, such as those related to carbon degradation for example, would allow quantifying simultaneously all or at least most of the genes involved in a specific function, such as the decomposition of different organic compounds that can be found in the environment.

References

Auerbach, E.A., Seyfried, E.E., and McMahon, K.D. (2007). Tetracycline resistance genes in activated sludge wastewater treatment plants. Water Res. *41*, 1143–1151.

Barnard, R., Leadley, P.W., and Hungate, B.A. (2005). Global change, nitrification, and denitrification: A review. Glob. Biogeochem. Cy. 19, GB1007.

Belser, L.W. (1979). Population ecology of nitrifying bacteria. Annu. Rev. Microbiol. *33*, 309–333.

Ben-Dov, E., Brenner, A., and Kushmaro, A. (2007). Quantification of sulfate-reducing bacteria in industrial wastewater, by real-time polymerase chain reaction (PCR) using *dsrA* and *apsA* genes. Microb. Ecol. *54*, 439–451.

Bowman, J. (2006). The Methanotrophs – The Families *Methylococcaceae* and *Methylocystaceae*. In The

Prokaryotes, Dworkin, M., Falkow, S., Rosenberg, E., Schleifer, K.H., and Stackebrandt, E., eds (Springer, New York), pp. 266–289.

Bru, D., Ramette, A., Saby, N.P., Dequiedt, S., Ranjard, L., Jolivet, C., Arrouays, D., and Philippot, L. (2011). Determinants of the distribution of nitrogen-cycling microbial communities at the landscape scale. ISME J. 5, 532–542.

Cebron, A., Norini, M.P., Beguiristain, T., and Leyval, C. (2008). Real-time PCR quantification of PAH-ring hydroxylating dioxygenase (PAH-RHDalpha) genes from Gram positive and Gram negative bacteria in soil and sediment samples. J. Microbiol. Methods 73, 148–159.

Church, M.J., Wai, B., Karl, D.M., and DeLong, E.F. (2010). Abundances of crenarchaeal *amoA* genes and transcripts in the Pacific Ocean. Environ. Microbiol. *12*, 679–688.

Coelho, M.R., Carneiro, N.P., Marriel, I.E., and Seldin, L. (2009). Molecular detection of *nifH* gene-containing Paenibacillus in the rhizosphere of sorghum (Sorghum bicolor) sown in Cerrado soil. Lett. Appl. Microbiol. *48*, 611–617.

Coolen, M.J.L., Abbas, B., Van Bleijswijk, J., Hopmans, E.C., Kuypers, M.M.M., Wakeham, S.G., and Sinninghe Damsté, J.S. (2007). Putative ammonia-oxidizing Crenarchaeota in suboxic waters of the Black Sea: A basin-wide ecological study using 16S ribosomal and functional genes and membrane lipids. Environ. Microbiol. 9, 1001–1016.

Daly, K., Sharp, R.J., and McCarthy, A.J. (2000). Development of oligonucleotide probes and PCR primers for detecting phylogenetic subgroups of sulfate-reducing bacteria. Microbiology *146*, 1693–1705.

De Long, S.K., Kinney, K.A., and Kirisits, M.J. (2010). qPCR assays to quantify genes and gene expression associated with microbial perchlorate reduction. J. Microbiol. Methods *83*, 270–274.

van Doorn, R., Klerks, M.M., van Gent-Pelzer, M.P., Speksnijder, A.G., Kowalchuk, G.A., and Schoen, C.D. (2009). Accurate quantification of microorganisms in PCR-inhibiting environmental DNA extracts by a novel internal amplification control approach using Biotrove OpenArrays. Appl. Environ. Microbiol. 75, 7253–7260.

van Doorn, R., Szemes, M., Bonants, P., Kowalchuk, G.A., Salles, J.F., Ortenberg, E., and Schoen, C.D. (2007). Quantitative multiplex detection of plant pathogens using a novel ligation probe-based system coupled with universal, high-throughput real-time PCR on OpenArrays. BMC Genomics *8*, 276.

Dumont, M.G., and Murrell, J.C. (2005). Stable isotope probing – linking microbial identity to function. Nat. Rev. Microbiol. *3*, 499–504.

Dunfield, P.F. (2007). The soil methane sink. In Greenhouse gas sinks, Reay, D., Hewitt, C.N., Smith, K., and Grace, J., eds (CABI Publishing, Wallingford, UK,), pp. 152–170.

Fierer, N., Jackson, J.A., Vilgalys, R., and Jackson, R.B. (2005). Assessment of soil microbial community structure by use of taxon-specific quantitative PCR assays. Appl. Environ. Microbiol. *71*, 4117–4120.

Francis, C.A., Beman, J.M., and Kuypers, M.M. (2007). New processes and players in the nitrogen cycle: the microbial ecology of anaerobic and archaeal ammonia oxidation. ISME J. *1*, 19–27.

Freitag, T.E., and Prosser, J.I. (2009). Correlation of methane production and functional gene transcriptional activity in a peat soil. Appl. Environ. Microbiol. 75, 6679–6687.

Freitag, T.E., Toet, S., Ineson, P., and Prosser, J.I. (2010). Links between methane flux and transcriptional activities of methanogens and methane oxidizers in a blanket peat bog. FEMS Microbiol. Ecol. 73, 157–165.

Galand, P.E., Lovejoy, C., Hamilton, A.K., Ingram, R.G., Pedneault, E., and Carmack, E.C. (2009). Archaeal diversity and a gene for ammonia oxidation are coupled to oceanic circulation. Environ. Microbiol. *11*, 971–980.

Gubry-Rangin, C., Nicol, G.W., and Prosser, J.I. (2010). Archaea rather than bacteria control nitrification in two agricultural acidic soils. FEMS Microbiol. Ecol. *74*, 566–574.

Guo, X., Xia, X., Tang, R., Zhou, J., Zhao, H., and Wang, K. (2008). Development of a real-time PCR method for Firmicutes and Bacteroidetes in faeces and its application to quantify intestinal population of obese and lean pigs. Lett. Appl. Microbiol. *47*, 367–373.

Hai, B., Diallo, N.H., Sall, S., Haesler, F., Schauss, K., Bonzi, M., Assigbetse, K., Chotte, J.L., Munch, J.C., and Schloter, M. (2009). Quantification of key genes steering the microbial nitrogen cycle in the rhizosphere of sorghum cultivars in tropical agroecosystems. Appl. Environ. Microbiol. 75, 4993–5000.

Ho, A., Luke, C., and Frenzel, P. (2011). Recovery of methanotrophs from disturbance: population dynamics, evenness and functioning. ISME J. 5, 750–758.

Klappenbach, J.A., Dunbar, J.M., and Schmidt, T.M. (2000). rRNA operon copy number reflects ecological strategies of bacteria. Appl. Environ. Microbiol. *66*, 1328–1333.

Knapp, C.W., Dolfing, J., Ehlert, P.A., and Graham, D.W. (2010). Evidence of increasing antibiotic resistance gene abundances in archived soils since 1940. Environ. Sci. Technol. *44*, 580–587.

Leininger, S., Urich, T., Schloter, M., Schwark, L., Qi, J., Nicol, G.W., Prosser, J.I., Schuster, S.C., and Schleper, C. (2006). Archaea predominate among ammonia-oxidizing prokaryotes in soils. Nature *442*, 806–809.

Levy-Booth, D.J., and Winder, R.S. (2010). Quantification of nitrogen reductase and nitrite reductase genes in soil of thinned and clear-cut Douglas-fir stands by using real-time PCR. Appl. Environ. Microbiol. *76*, 7116–7125.

Martensson, L., Diez, B., Wartiainen, I., Zheng, W., El-Shehawy, R., and Rasmussen, U. (2009). Diazotrophic diversity, *nifH* gene expression and nitrogenase activity in a rice paddy field in Fujian, China. Plant Soil 325, 207–218.

Mertens, J., Broos, K., Wakelin, S.A., Kowalchuk, G.A., Springael, D., and Smolders, E. (2009). Bacteria, not archaea, restore nitrification in a zinc-contaminated soil. ISME J. *3*, 916–923.

Op den Camp, H.J.M., Islam, T., Stott, M.B., Harhangi, H.R., Hynes, A., Schouten, S., Jetten, M.S.M., Birkeland, N.K., Pol, A., and Dunfield, P.F. (2009). Environmental, genomic and taxonomic perspectives on methanotrophic Verrucomicrobia. Environ. Microbiol. Rep. *1*, 293–306.

Pereyra, L.P., Hiibel, S.R., Prieto Riquelme, M.V., Reardon, K.F., and Pruden, A. (2010). Detection and quantification of functional genes of cellulose-degrading, fermentative, and sulfate-reducing bacteria and methanogenic archaea. Appl. Environ. Microbiol. *76*, 2192–2202.

Philippot, L. (2005). Tracking nitrate reducers and denitrifiers in the environment. Biochem. Soc. Trans. *33*, 200–204.

Philippot, L. (2006). Use of functional genes to quantify denitrifiers in the environment. Biochem. Soc. Trans. *34*, 101–103.

Poly, F., Monrozier, L.J., and Bally, R. (2001). Improvement in the RFLP procedure for studying the diversity of *nifH* genes in communities of nitrogen fixers in soil. Res. Microbiol. *152*, 95–103.

Pouliot, J., Galand, P.E., Lovejoy, C., and Vincent, W.F. (2009). Vertical structure of archaeal communities and the distribution of ammonia monooxygenase A gene variants in two meromictic High Arctic lakes. Environ. Microbiol. *11*, 687–699.

Powell, S.M., Ferguson, S.H., Bowman, J.P., and Snape, I. (2006). Using real-time PCR to assess changes in the hydrocarbon-degrading microbial community in Antarctic soil during bioremediation. Microb. Ecol. *52*, 523–532.

Prosser, J.I. (2007). Microorganisms cycling soil nutrients and their diversity. In Modern soil microbiology, van Elsas, J.D., Jansson, J.K., and Trevors, J.T., eds (CRC Press, Boca Raton, FL), pp. 237–261.

Rosch, C., Mergel, A., and Bothe, H. (2002). Biodiversity of denitrifying and dinitrogen-fixing bacteria in an acid forest soil. Appl. Environ. Microbiol. *68*, 3818–3829.

Schulz, S., Perez-de-Mora, A., Engel, M., Munch, J.C., and Schloter, M. (2010). A comparative study of most probable number (MPN)-PCR vs. real-time-PCR for the measurement of abundance and assessment of diversity of *alkB* homologous genes in soil. J. Microbiol. Methods *80*, 295–298.

Selesi, D., Pattis, I., Schmid, M., Kandeler, E., and Hartmann, A. (2007). Quantification of bacterial RubisCO genes in soils by *cbbL* targeted real-time PCR. J. Microbiol. Methods *69*, 497–503.

Selesi, D., Schmid, M., and Hartmann, A. (2005). Diversity of green-like and red-like ribulose-1,5-bisphosphate carboxylase/oxygenase large-subunit genes (*cbbL*) in differently managed agricultural soils. Appl. Environ. Microbiol. *71*, 175–184.

Smith, C.J., and Osborn, A.M. (2009). Advantages and limitations of quantitative PCR (Q-PCR)-based approaches in microbial ecology. FEMS Microbiol. Ecol. *67*, 6–20.

Steinberg, L.M., and Regan, J.M. (2009). *mcrA*-targeted real-time quantitative PCR method to examine methanogen communities. Appl. Environ. Microbiol. *75*, 4435–4442.

Tuomivirta, T.T., Yrjala, K., and Fritze, H. (2009). Quantitative PCR of *pmoA* using a novel reverse primer correlates with potential methane oxidation in Finnish fen. Res. Microbiol. *160*, 751–756.

Videmsek, U., Hagn, A., Suhadolc, M., Radl, V., Knicker, H., Schloter, M., and Vodnik, D. (2009). Abundance and diversity of CO_2-fixing bacteria in grassland soils close to natural carbon dioxide springs. Microb. Ecol. *58*, 1–9.

Williamson, N., Brian, P., and Wellington, E.M. (2000). Molecular detection of bacterial and streptomycete chitinases in the environment. Antonie Van Leeuwenhoek *78*, 315–321.

Wuchter, C., Abbas, B., Coolen, M.J., Herfort, L., van Bleijswijk, J., Timmers, P., Strous, M., Teira, E., Herndl, G.J., Middelburg, J.J., *et al.* (2006). Archaeal nitrification in the ocean. Proc. Natl. Acad. Sci. U.S.A. *103*, 12317–12322.

Yergeau, E., and Kowalchuk, G.A. (2008). Responses of Antarctic soil microbial communities and associated functions to temperature and freeze–thaw cycle frequency. Environ. Microbiol. *10*, 2223–2235.

Yergeau, E., Kang, S., He, Z., Zhou, J., and Kowalchuk, G.A. (2007). Functional microarray analysis of nitrogen and carbon cycling genes across an Antarctic latitudinal transect. ISME J. *1*, 163–179.

Yergeau, E., Arbour, M., Brousseau, R., Juck, D., Lawrence, J.R., Masson, L., Whyte, L.G., and Greer, C.W. (2009). Microarray and real-time PCR analyses of the responses of high-arctic soil bacteria to hydrocarbon pollution and bioremediation treatments. Appl. Environ. Microbiol. *75*, 6258–6267.

Yergeau, E., Hogues, H., Whyte, L.G., and Greer, C.W. (2010). The functional potential of high Arctic permafrost revealed by metagenomic sequencing, qPCR and microarray analyses. ISME J. *4*, 1206–1214.

Yu, Z., Michel, F.C., Jr., Hansen, G., Wittum, T., and Morrison, M. (2005). Development and application of real-time PCR assays for quantification of genes encoding tetracycline resistance. Appl. Environ. Microbiol. *71*, 6926–6933.

Zhang, X.X., Zhang, T., and Fang, H.H. (2009). Antibiotic resistance genes in water environment. Appl. Microbiol. Biotechnol. *82*, 397–414.

Zumft, W.G. (1997). Cell biology and molecular basis of denitrification. Microbiol. Mol. Biol. Rev. *61*, 533–616.

Using Quantitative Polymerase Chain Reaction for Water Microbial Risk Assessments

7

Jorge Santo Domingo, Mary Schoen, Nicholas Ashbolt and Hodon Ryu

Abstract

Microbial risk assessment (MRA) has traditionally utilized microbiological data that was obtained by culture-based techniques that are expensive and time consuming. With the advent of PCR methods there is a realistic opportunity to conduct MRA studies economically, in less time, and simultaneously targeting multiple pathogens and their sources. More importantly, recently developed qPCR assays provide the opportunity to estimate the densities of the reference pathogens and their sources, which is critical to quantitative MRA (QMRA) analyses. In this chapter we discuss the use of qPCR-based methods to identify risks associated with exposure to water, namely drinking and recreational waters. We discuss the advantages associated with the current qPCR approaches used in microbial water quality studies and critically evaluate some of the limitations as they relate to the use of QMRA in the assessment of microbial water quality and public health protection.

Introduction

The Unites States Environmental Protection Agency (EPA) uses risk assessment to characterize the nature and magnitude of health risks to humans (e.g. residents, workers and recreational visitors) and ecological receptors (e.g. birds, fish and wildlife) from microbial and chemical stressors that may be introduced and present in the environment. This information is then used by various agencies to aid in the management of stressors or contaminants. Thus, methods that detect harmful chemicals and pathogens and their surrogates provide a critical role in determining potential exposures to harmful agents and in estimating risks in human and ecosystem health studies. Methods that both detect and quantify the targeted contaminant are preferred in risk assessment studies as exposure outcomes are difficult to predict with binary (presence/absence) data. Methods to quantify chemicals in the environment vary but most can be grouped as spectrophotometric or chromatographic. As the focus of this chapter is on microbial risk assessment, chemical methods will not be discussed in any detail. Additionally, we will focus on environmental risk assessment approach, that is, microbial risks to human. While we will not consider ecological risks in this chapter (i.e. risks associated to non-humans), it is important to note that the introduction of faecal pollution into environmental waters can have an impact on biogeochemical processes and can also be detrimental to macrobiota health.

There are also numerous methods used to quantify microbial agents of diseases (pathogens) in environmental waters. Most classical approaches used to enumerate pathogens rely on growing microorganisms on artificial media or cell lines (http://www.standardmethods.org). While these methods have provided valuable information for a number of pathogens, they have some limitations relevant to risk assessment and pathogen ecology. One limitation has to do with the fact that many pathogens are inherently difficult to grow, while for others no routine media/cell lines have been developed (e.g. human noroviruses). In most cases, culture-based methods are time consuming and therefore not compatible with near real-time measurements that are often required for management decisions. Quantitative methods based on the polymerase chain reaction (PCR) or qPCR can circumvent these problems. For example, qPCR can be used to process hundreds of samples within a matter of hours. Moreover, it is possible to use qPCR-based approaches to quantify multiple pathogens regardless of their culturable status or availability of culturing techniques. Such a multi-targeted approach is useful for pathogens in which numerous genes are required for pathogenicity, as it reduces the number of false-positives in monitoring studies. Nonetheless, false-positives relevant to risk may arise through detecting the genetic targets of pathogens that are no longer infectious (Masago *et al.*, 2008). However, detecting messenger RNA (mRNA) for some pathogens (Garcés-Sanchez *et al.*, 2009; González-Escalona *et al.*, 2009; Deer *et al.*, 2010) may reduce such false-positives.

This chapter focuses on various applications of qPCR for aquatic quantitative microbial risk assessments (QMRA). Herein, qPCR will be used as a synonym to real-time PCR and when data are generated using DNA as the template molecule. When describing quantitative analysis of reverse-transcriptase (RT) products specifically used in the detection of RNA enteric viruses, we will use RT-qPCR. It should be noted that RT-qPCR assays are also used in the detection of mRNA transcripts (from pathogens or cell cultures infected with DNA enteric viruses), which is a particularly useful approach when studying the expression of genes involved in pathogenicity pathways (Béra-Maillet *et al.*, 2009). Technical details associated with PCR-based techniques will only be discussed when relevant to microbial water quality, such as matrix effect on DNA extraction efficiency, PCR inhibition, and detection limits. The main focus of this chapter is on the use of qPCR-based assays to assess pathogen risks in recreational waters; however, the topics addressed are also directly relevant to other environmental sources of pathogens and drinking water. Thus emphasis is placed on the use of qPCR for detecting faecal indicator bacteria and waterborne pathogens that results in gastrointestinal diseases.

Definition of MRA and MRA steps

The term microbial risk assessment (MRA) has been given to any assessment of human health from pathogens, epidemiological or by risk assessment. However, MRA generally and here refers to the evaluation of scientific information on the hazardous properties of pathogens (in hazard identification and characterization), the extent of human exposure to the identified pathogens (in exposure assessment), the dose–response relationship for the selected pathogens (dose–response assessment), and the combination of this information to finally characterize risks (USEPA, 2007). Quantitative microbial risk assessment (QMRA) is applied when further resolution of pathogen risk is required, and as well as including quantitative dose–response information generally attempts to include assessments of parameter variability and uncertainty. For example, MRA has been used internationally in setting water quality guidelines for recreational, drinking and reuse waters by the World Health Organization (www.who.int/water_sanitation_health/en/) and is increasingly been considered to aid in water management through what are often called Water Safety Plans (WHO, 2009). Examples of QMRAs used to manage recreational waters are provided by Ashbolt *et al.* (2010). Overall, MRA fits between risk perception and risk communication/management (Fig. 7.1a).

For waters, MRA represents an alternative or complement to epidemiology studies, which continue to be the primary approach used to assist in setting public health guidance/standards. The field of epidemiology was established more than 150 years ago by John Snow, who was the first to identify that the source of cholera (disease) in London's 1854 outbreak was a drinking water fountain polluted by faeces. Snow's approach was described some 30 years before the agent (*Vibrio choleraee*) was definitively described by the German physician Robert Koch, which cemented 'germ theory' over air (miasma theory) as the cause of the cholera outbreaks. Hence, epidemiology studies were initially used by physicians to identify sources of disease, but have evolved into a science used to investigate outbreaks and design

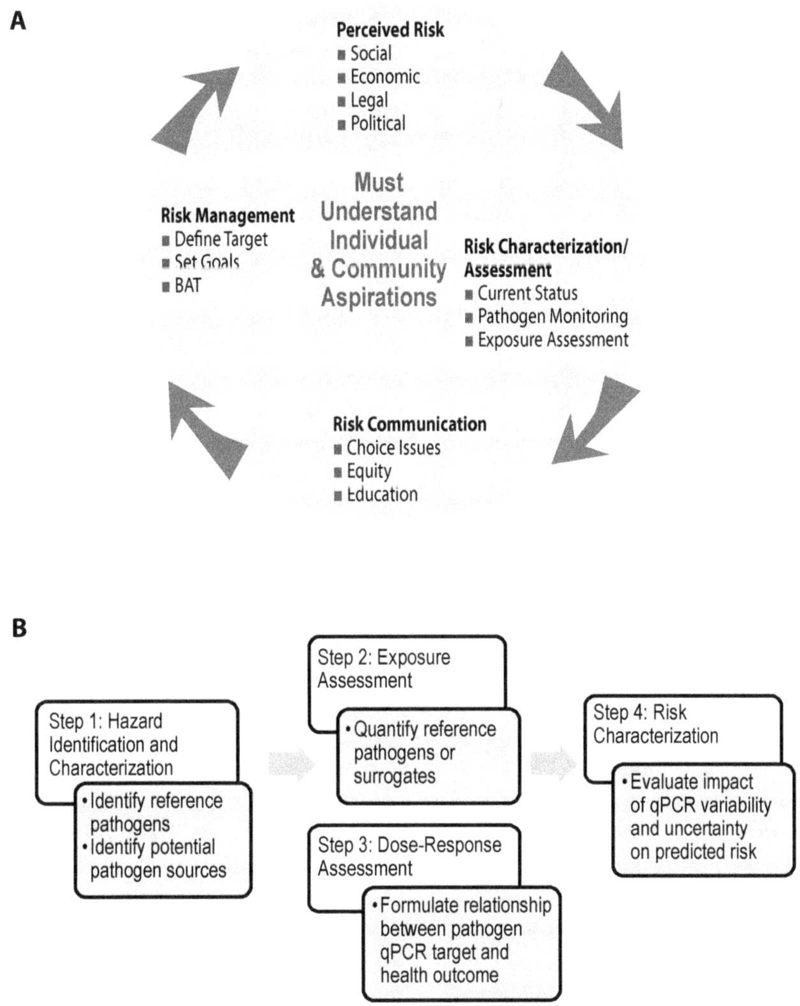

Figure 7.1 (A) The risk cycle: an iterative process (from LeChevallier and Buckley, 2007). (B) Steps in MRA and role of qPCR.

studies to estimate risks associated with exposure to harmful agents. The results from an array of epidemiology studies are typically summarized through what is generally known as a Systematic Review (e.g. http://www.cochrane-handbook.org/), with this information used to establish regulatory thresholds (e.g. EPA IRIS database for chemical contaminants, http://www.epa.gov/IRIS). In contrast to chemical risk assessment, QMRA is a more recent science with less understanding and acceptance in society (Haas *et al.*, 1999). For example, it was first used by the EPA to aid in establishing regulations for waterborne pathogens in the 1990s (Regli *et al.*, 1991, 1999). The state of science used in QMRA is rapidly increasing, particularly through its use to manage pathogens in the food industry (Lammerding and Paoli, 1997; Teunis *et al.*, 2008a), including the role of waterborne pathogens in fresh food production (Mota *et al.*, 2009).

In summary, MRA is a scientifically based process consisting of the following steps: (i) hazard identification and characterization, (ii) exposure assessment, (iii) dose–response assessment, and

(iv) risk characterization (Fig. 7.1b) (US-EPA, 2007). The MRA process starts with the problem formation, an initial step that determines the purpose, scope, players, expected outcomes, analytical approach, schedule, deliverables, QA/QC, resources, and document requirements for the risk assessment. In the problem formation, the route(s) of exposure and target population are identified along with the adverse health outcome (i.e. infection, illness or mortality). Given the pathways and hazardous events that may occur in the system under study, representative (reference) pathogens are identified in the hazard identification component. In is important to consider the range of pathogen types (from viral, bacterial, parasitic protozoan and helminths) that humans may be exposed to during hazard identification and characterization, but this is generally possible with a relatively short list of reference pathogens (e.g. Soller *et al.*, 2010a). The MRA step of exposure assessment refers to the process of estimating the dose or number of reference pathogens the target population is exposed to for each event or scenario modelled. The resulting calculated dose of reference pathogen is used in the dose–response assessment to estimate the selected adverse health outcome. The final step of risk characterization communicates the target population risks for the events identified. Risk characterization combines information on the magnitude of the adverse health outcome as well as other knowledge gained from the MRA process. As such, qPCR methods can play a role in each of the MRA steps. For example, qPCR methods can accelerate the detection of relevant pathogens in the hazard identification step. Most importantly, in the exposure assessment, qPCR may be used to quantify pathogens or their surrogates in environmental samples, and qPCR has been used to derive dose–response data in human volunteer studies. The following sections will discuss the application of qPCR methods for each of the MRA steps: hazard identification, exposure assessment, dose–response assessment and risk characterization. A critical view of some of the advantages and limitations associated with qPCR that may impact on MRA results are also discussed.

Hazard identification and characterization

Detection of pathogens using PCR and qPCR methods

In the hazard identification step, reference pathogens are selected that represent the major pathogens of concern. Common reference pathogens for water exposures include bacterial pathogens like *Escherichia coli* O157:H7, *Campylobacter jejuni* and *Salmonella enterica*, protozoan pathogens like *Cryptosporidium parvum* and *Giardia lamblia*, and enteric viruses such as noroviruses and adenoviruses. Hazard identification requires knowledge of the pathogens present in the environment under investigation, either from direct observation or a sound understanding of the sources of pathogens in the environment. Examples of faecal pathogens found in environmental waters and their sources are presented in Table 7.1. Various qPCR methods can be used for both direct observation and quantification of the pathogens of concern (Table 7.2) or to identify potential sources of pathogens (Table 7.3). Local epidemiology data are also useful in clarifying which pathogens are likely to be present in the study environment, and qPCR is increasingly being used in the examination of clinical samples.

Conventional PCR methods have long been used to identify the pathogens present in the environment. Early on methods focused on targeting bacterial pathogens via 16S rRNA gene (rDNA) sequences. These methods do not always provide the resolution needed to discriminate between pathogenic and non-pathogenic strains from the same bacterial species. On the other hand, the use of 16S rDNA sequence analyses has demonstrated the presence of bacteria relevant to public health. For example, analysis of 954 nearly full-length 16S rRNA gene sequences was used to characterize shower water and biofilms from a stem cell transplant hospital unit over multiple seasons (Perkins *et al.* 2009). Of the bacterial sequences analysed, the most notable potential pathogen identified was *Mycobacterium mucogenicum*. In the same hospital study, clone sequences were identified as Legionella, but none of the sequences were positively identified to a known pathogenic species.

Table 7.1 Examples of faecal pathogens and their faecal sources

Microorganisms	Major diseases	Potential faecal sourcesa	Numbers of indicators and pathogens in faecal sources
Bacteria			
Salmonella enterica Typhi	Typhoid fever	H	Raw sewage (per l)
Salmonella enterica Paratyphi	Paratyphoid fever	H	Total coliform: 108–109 Faecal coliform: 107–108
Salmonella enterica	Salmonellosis	H, A	Faecal streptococci: 106–107
Shigella spp.	Bacillary dysentery	H	Enterococci: 105–106 *Shigella*: present
Vibrio cholerae	Cholera	H	*Salmonella*: 103–105
Enteropathogenic *E. coli*	Gastroenteritis	H, cattle	
Yersinia enterocolitica	Yersiniosis	H, A	
Campylobacter jejuni	Gastroenteritis	H, A	
Enteric viruses			
Polio viruses	Poliomyelitis	H	Raw sewage (per l)
Coxsackie viruses	Aseptic meningitis	H	Enteric viruses: 105–106 Faeces (per gram)
Echo viruses	Aseptic meningitis	H	Enterovirus: 103–108
Other enteroviruses	Encephalities	H	Hepatitis A virus: 108 Rotavirus: 1010–1012
Rotaviruses	Gastroenteritis	H	
Adenoviruses	Upper respiratory and GI illness	H	
Hepatitis A virus	Infectious hepatitis	H	
Hepatitis E virus	Infectious hepatitis	H, A	
Norovirus	Gastroenteritis	H, fomites	
Protozoa			
Acanthamoeba castellani	Amoebic Meningoencephalitis	H	–
Balantidium coli	Balantidosis	H, A	–
Cryptosporidium homonis, C. parvum, C. bailey, C. canis, C. felis, C. meleagridis, C. muris	Cryptosporidiosis (gastroenteritis)	H, A	Raw sewage: 100–103 (per l) Faeces: 106–107 (per gram)
Entamoeba histolytica	Amoebic dysentery	H, A	–
Giardia lamblia	Giardiasis (gastroenteritis)	H, A	Raw sewage: 101–105 (per l) Faeces: 106 (per gram)
Helminths			
Ascaris lumbricoides	Ascariosis	H, A	Raw sewage: 101–104 (ova per l)

Modified from Santo Domingo and Ashbolt (2008).
H, human faeces; A, animal faeces.

While 16S rDNA methods might not always identify the presence of a hazard, they are useful at prioritizing the potential reference pathogens that must be studied further using techniques that can specifically discriminate between hazardous and non-hazardous microbial agents.

Genes implicated in pathogenic pathways have also been used in the development of identification and detection methods for bacterial (Kong *et al.*, 2002), parasitic protozoan (Guy *et al.*, 2003) and viral (Ryu *et al.*, 2010) pathogens. In fact, the use of PCR assays in water and public health microbiology dates over two decades (Bej *et al.*, 1991). PCR-based assays targeting genes

Table 7.2 Examples of qPCR, RT-PCT, and RT-qPCR methods and target genes for waterborne pathogens

Waterborne pathogens	Methods – detection chemistry	Target genes/regions	Reference
Adenovirus	qPCR – TaqMan	Hexon	Heim *et al.* (2003)
Adenovirus	qPCR – TaqMan	Hexon, VA RNA, 1.5 IV-2	Watzinger *et al.* (2004)
Adenovirus	qPCR – TaqMan	Hexon	Ryu *et al.* (2010)
Astrovirus	RT-PCR	ORF1	Guix *et al.* (2002)
Coxsackievirus	RT-qPCR – TaqMan	VP1	Ryu *et al.* (2010)
Echovirus	RT-qPCR – TaqMan	VP1	Ryu *et al.* (2010)
Enteroviruses	RT-PCR	Conserved untranslated region	Schwab *et al.* (1995)
Enteroviruses	RT-qPCR – TaqMan	Conserved untranslated region	Watzinger *et al.* (2004)
Enteroviruses	RT-qPCR – Molecular beacons	Conserved untranslated region	Hwang *et al.* (2007)
Hepatitis A virus	RT-PCR	Interface of the VP1 and VP3 capsid proteins	Schwab *et al.* (1995)
Norovirus	RT-PCR	Genomic region encoding the viral polymerase	Schwab *et al.* (1995)
Norovirus	RT-qPCR – TaqMan	Regions between ORF1 and ORF2	Jothikumar *et al.* (2005)
Rotavirus	RT-PCR	VP6	Villena *et al.* (2003)
Rotavirus	RT-qPCR – TaqMan	Non-structural protein	Pang *et al.* (2004)
Cryptosporidium parvum	RT-PCR	Hsp70	Rochelle *et al.* (1997)
Cryptosporidium parvum	qPCR – TaqMan	Wall protein	Guy *et al.* (2003)
Cryptosporidium parvum	qPCR – TaqMan	Hsp70	Di Giovanni *et al.* (2005)
Cryptosporidium	qPCR – Scorpion Uni-probes (Proligo)	18S rRNA	Stroup *et al.* (2006)
Cryptosporidium	qPCR with TaqMan	18S rRNA	Jothikumar *et al.* (2008)
Giardia duodenalis	RT-PCR	Hsp70	Lee *et al.* (2009)
Giardia	RT-PCR	Hsp70	Abbaszadegan *et al.* (1997)
Giardia lamblia	qPCR – TaqMan	β-giardin	Guy *et al.* (2003)
Toxoplasma gondii	qPCR – TaqMan	B1	Kompalic-Cristo *et al.* (2007)
Escherichia coli O157	qPCR -SYBR Green	*rfbE*	Schets *et al.* (2005)
Escherichia coli O157:H7	qPCR with – Molecular beacons	*rfbE*	Fortin *et al.* (2001)
Escherichia coli	qPCR – Fluorescent hybridization probes	Shiga toxin 1 and 2	Bellin *et al.* (2001)

involved in virulence have been coupled with other techniques such as microarrays to improve pathogen detection (Palka-Santini *et al.*, 2009). As conventional PCR only provides presence/absence data, qPCR methods have become more popular in recent years as they can help estimate the levels of human pathogens in a given sample. Some of the matrices for which qPCR methods have been used to detect pathogens include waters (Liu *et al.*, 2009; Jothikumar *et al.*, 2009), wastewaters (Shannon *et al.*, 2007; Wéry *et al.*, 2007), food (Panicker and Bej, 2005; Huang *et al.*, 2009), faeces (Almeida *et al.*, 2010; Imamovic *et al.*, 2010; Hadfield *et al.*, 2011), and manure (Lebuhn *et al.*, 2003; Klein *et al.*, 2010). Though helminths are pathogens of primary concern via direct contact with manure/faeces and foods grown with wastewaters, there are also environmental reservoirs such as freshwater snails, and PCR methods have been applied to both areas (Kato-Hayashi *et al.*, 2010; Taniuchi *et al.*, 2011). The use of qPCR to identify animal pathogens also notably increased in the last 10 years (Rodríguez-Lázaro *et al.*, 2006; Chaban *et al.*, 2009). This fact is not only relevant to veterinary diagnostics but also to human health as animals play an important role in the dispersal, transmission and evolution of human pathogens (Inglis and Kalischuk, 2004; Smith and Guégan, 2010).

Some care is required when interpreting qPCR results to identify the main pathogens of concern. The pathogen detected most frequently or in greatest magnitude by qPCR may not be viable or infectious; this is further explained using bacterial densities of faecal indicators as surrogates for pathogens. For instance, qPCR has been used to quantify the bacterial density of faecal indicators in wastewater effluent (Varma *et al.*, 2009). In this case, signals that are generated with the qPCR assay may include a fraction of faecal indicator bacteria that are dead or inactivated. Since wastewater treatment is designed to inactivate faecal bacteria as well as pathogens, the detection of non-viable bacteria makes it difficult to identify viable pathogens of concern using qPCR alone, although various pretreatments prior to qPCR show potential (Nocker *et al.*, 2007a; Chen *et al.*, 2010; Kim *et al.*, 2011). Similar problems arise when using qPCR to detect pathogens other than bacterial species. In the case of pathogenic viruses, any nucleic acid fragment is a potential target, although attention is needed to select areas that are not prone to genetic rearrangements or other modifications, including infectious virions following cell culture (Lamberetini *et al.*, 2010). There is also concern that high risk pathogens may be missed due to episodic occurrence or low density. This predicament is not unique to the use of qPCR, but it is also not overcome by the sensitivity of the tool if used with a poor sampling regime. Many sources of pathogens are sporadic in nature, such as resulting from storm events impacting recreational and drinking water sources or release of pathogens from environmental biofilms. Therefore, it remains important to understand the characteristics of the potential sources of pathogens during hazard identification and characterization.

Identify potential sources of contamination using qPCR methods

Various qPCR methods have also been used to help identify potential sources of contamination contributing to environment waters. Collectively, this area of research is known as microbial source tracking (MST) (Simpson *et al.*, 2002). The need to identify faecal sources impacting environmental waters has long been recognized by public health professionals, primarily because different sources may carry different health risks. For example, human faecal sources are presumed to be associated with higher risks than animal faecal sources, particularly due to human pathogens such as *Cryptosporidium hominis*, *Shigella sonnei*, and noroviruses which exhibit a high degree of host-specificity. However, recent increases in the detection of waterborne outbreaks associated with zoonotic agents demonstrate the potential link between animal waste and human illness and ecosystem health (Hrudey *et al.*, 2003; Fayer, 2004; Craun *et al.*, 2010). Non-domesticated animals such as waterfowl are also carriers of human pathogens (Lu *et al.*, 2011a) and can be primary sources of pollution of recreational waters (Edge and Hill, 2007; Lu *et al.*, 2011b). Identifying the sources of pathogen contamination is important from an environmental management standpoint as such information can be used to implement adequate pollution prevention controls, establish remediation strategies, and determine which methods need to be used to assess efficacy of management practices. Additionally, controlling both human and non-human faecal loads in watershed systems is important to the water industry as performance of treatment plants is in part associated with the level of pollution in the influent waters (Hijnen and Medema, 2010).

Molecular assays used in source identification have traditionally been based on bacterial targets, although other microbial groups (i.e. protozoa, enteric viruses and archaea) have also been targeted (Table 7.3). Mitochondrial genes from different animals (Schill and Mathes, 2008) have also been used as targets to develop source tracking markers. Amongst the variety of MST methods proposed in the last 15 years, the use of Bacteroidales PCR-based MST methods seem to be favoured by most laboratories conducting environmental monitoring studies. Various Bacteroidales are among the most numerically dominant bacteria in mammals and are believed to survive poorly outside the gastrointestinal track. More importantly, signature sequences

Table 7.3 Examples of qPCR assays targeting faecal sources of pollution

Assay ID	Target organism	Gene target	Faecal source	Type of assay	Reference
AllBac296F	All Bacteroidales	16S rDNA	Most faeces	TaqMan	Layton *et al.* (2006)
BoBac	*Bacteroides*	16S rDNA	Bovine	TaqMan	Layton *et al.* (2006)
HuBac	*Bacteroides*	16S rDNA	Human	TaqMan	Layton *et al.* (2006)
HF183	Bacteroidales	16S rDNA	Human	SYBR Green	Seurinck *et al.* (2005)
BacR	Bacteroidales	16S rDNA	Ruminant	TaqMan	Reischer *et al.* (2006)
Pig-1-Bac	Bacteroidales	16S rDNA	Swine	TaqMan	Mieszkin *et al.* (2009)
Pig-2-Bac	Bacteroidales	16S rDNA	Swine	TaqMan	Mieszkin *et al.* (2009)
HF183	Bacteroidales	16S rDNA	Human	SYBR Green	Seurinck *et al.* (2005)
Gull-2	*Catellicoccus*	16S rDNA	Gull	SYBR Green	Lu *et al.* (2007)
Gull-2	*Catellicoccus*	16S rDNA	Gull	TaqMan	Shibata *et al.* (2010)
CGOF1- and CGOF2-Bac	Bacteroides	16S rDNA	Goose	TaqMan	Fremaux *et al.* (2010)
Dog-Bact	*Bacteroides*	16S rDNA	Dog	TaqMan	Shibata *et al.* (2010)
HorseBact	*Bacteroidales*	16S rDNA	Horse	TaqMan	Silkie and Nelson (2009)
LA35	*Brevibacterium* sp.	16S rDNA	Chicken	SYBR Green	Weidhaas *et al.* (2011)
BtH-F/R	*B.thetaiotaomicron*	α-1-6, mannanase	Human	TaqMan	Yampara-Iquise *et* al. (2008)
OTU171	*L. sobrius/ amylovorus*	RDA cryptic fragment	Swine	SYBR Green	Marti *et al.* (2010)
BFDFor/BFDRev	*B. thetaiotaomicron* and other *Bacteroides* spp.		Human	TaqMan	Converse *et al.* (2009)
HPyV	*Polyomavirus*	T antigen	Human	TaqMan	McQuaig *et al.* (2009)
Human forward/ reverse	Human mitochondria	Cytochrome b	Human	TaqMan	Schill and Mathes (2008)
Human forward/ reverse	Human mitochondria	NADH dehydrogenase subunit 5	Human	TaqMan	Caldwell *et al.* (2007)

from different *Bacteroidales* species have been shown to exhibit a high level of host specificity (Walters and Field, 2009), explaining why most MST-qPCR assays used in applied studies have employed Bacteroidales-based assays.

To be effective at identifying faecal sources, MST assays must comply with the following: a high degree of host-specificity (ideally the assay only produces a signal when DNA from the targeted host is used as the template); consistent host distribution (i.e. the gene marker should be present in most individual hosts), high detection limits (i.e. the targeted gene is detected at very low concentrations in environmental samples); and a high relative abundance (i.e. the targeted gene must be present in high copies per faecal sample). Fresh samples from various faecal sources impacting different sites need to be collected and used to validate the host specificity, host distribution and abundance of the markers in targeted and non-targeted hosts and environments. While several methods have been used in multiple studies, the performance of each method needs to be determined when starting new projects. Differences in equipment (thermal cycler), nucleic acid extraction kits and enzymes (polymerase) used can all have an impact on the performance of a MST qPCR assay.

For MRA purposes, it is important to ask if a MST approach will identify the important sources of faecal contamination in an environment, noting that minor sources may be the dominant sources of risk (Schoen and Ashbolt, 2010). For example, assume that there are two sources of faecal contamination contributing to pathogens in an environment, in this hypothetical case, raw wastewater (100 ml) and horse faeces (10 g) (Fig. 7.2). If we quantify the total number of bacteria that belong to the Bacteroidales order from each source using a general Bacteroidales assay, the horse faecal source and raw wastewater contributes different levels of the faecal bacteria, specifically 10^{11} and 10^{9} gene copies, respectively (Silkie and Nelson, 2009). When we combine the two sources in our simplified environment, the dominant source of total Bacteroidales is the horse faeces; however, the horse faeces is not the dominant source of human-infectious pathogens (Derlet *et al.*, 2002). If we focus on the presumed dominant reference pathogen in sewage, Norovirus, only the raw wastewater is known to contain human noroviruses, with roughly 10^{2}–10^{5}

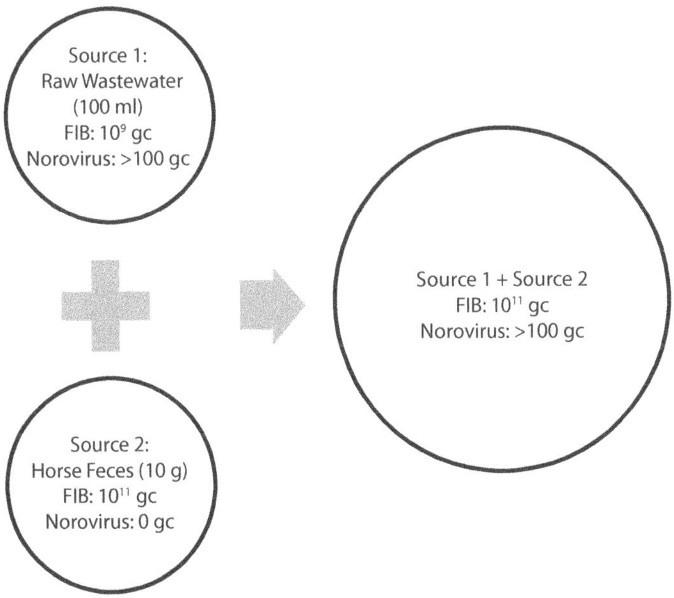

Figure 7.2 Example of dominant pathogen source in a sample containing a source mixture (FIB, faecal indicator bacteria; gc, genome copies).

genome copies per 100 ml (Haramoto *et al.*, 2006; Katayama *et al.*, 2008). As a result, the risk of Norovirus infection to a person exposed to the contaminated environment is driven by the raw wastewater, and not the source contributing the most MST marker. This example illustrates that MST approaches are useful in MRA for identifying potential sources of pathogens, but may not identify the pathogen and the source associated with the highest risk levels. Hence, additional information on pathogen presence in each of the potential sources is ultimately required to assess risk in a quantitative manner (via QMRA).

Exposure assessment

qPCR methods for quantifying reference pathogens of concern

Once hazard assessment and characterization is complete, the reference pathogens and the various scenarios/hazardous events to be studied are described. In exposure assessment, the dose or quantity of each reference pathogen is estimated for each hazardous event or scenario modelled. There are many possible approaches used to estimate the quantity of the reference pathogen dose in QMRA. One approach is to directly quantify the pathogens of interest from environmental samples at the point of exposure. However, pathogens may be too low in density to reliably assay at exposure points, such as in drinking water were we are potentially interested in one virion per million litres (Smeets, 2010). Hence pathogens are often assayed in source materials (sewage, faeces, etc.) by PCR and various surrogates for each reference pathogen are used to estimate loadings, fate and transport from source to human receptor (Petterson *et al.*, 2006, 2009).

Quantifying reference pathogens directly

As mentioned above, there are several genetic targets that have been used to detect pathogens, many of which can be used to quantify their levels directly from environmental samples. Prior to using qPCR assays in QMRA studies there are important technical aspects that must be considered. For example, most pathogens are commonly present in very low densities in environmental waters and therefore biomass concentration from relatively large volumes is needed in order to increase the chances of detection (Smith and Hill, 2009). For some pathogens, purification steps are needed prior to their detection using molecular techniques (Quintero-Betancourt *et al.*, 2002). In cases in which a pre-enrichment step is needed to enhance detection, enumeration is problematic, although relative abundance has been estimated by coupling detection and most probable number approaches (Wright *et al.*, 2007). Additionally, the amplification efficiency might depend on the complexity of the sample and how well the primers used in the assays can target all members of the reference pathogen. All these factors may result in significant underestimation of the risk associated with the environment under study.

The primary strengths of using a qPCR method in the exposure assessment to quantify the reference pathogens (or surrogates) include the potential ability to capture a larger range of microorganisms compared with culture-based methods and the capability to quantify the genotypes of pathogens that infect humans. Decades ago microbiologists discovered that the number of microorganisms determined via microscopical techniques was greater than those estimated by culture-based techniques (Kogure *et al.*, 1979). Differences of several orders of magnitude between microscopic counts and culturable counts are frequently reported for environmental samples, a phenomenon known as the great plate count anomaly (Staley and Konopka, 1985). In most cases, such discrepancies can be explained due to the selective nature of growth media. However, in other cases the differences are attributed to a physiological condition in which the bacterial cells become viable but non-culturable (Oliver, 2005). To be specific, VBNC cells lose their capability to grow on a medium that previously supported growth, yet they remain infectious. Several pathogens have been reported to undergo a VBNC state (Oliver, 2010), of them, *Vibrio* spp. are perhaps the most studied bacteria as their survival seem to be tightly coordinated around environmental cues (such as temperature) that trigger such a response (Constantin de Magney *et al.*, 2008). While the VBNC concept has been challenged (Bogosian *et al.*, 1998; Passerat *et al.*, 2009), there is also evidence suggesting that non-culturable cells can remain active even if they do not show measurable replication. VBNC status can be reversed via pheromone-like substances or resuscitation techniques such as in the case of *Micrococcus luteus* (Mukamolova *et al.*, 2002), *Mycobacterium smegmatis*, (Anuchin *et al.*, 2009) and *S. enterica* (Santo Domingo *et al.*, 2000). Although most MRAs do not address the possibility of VBNC pathogens, there may be environments where they are the dominant fraction within the total infectious pathogen count. For example, *Legionella* spp. are difficult to culture on artificial agar and are known to replicate within drinking water amoebae (Wullings *et al.*, 2010; Thomas and Ashbolt, 2011). Further, Legionella within amoebae may form a VBNC state (Steinert *et al.*, 1997), and a large fraction may simply be obligate intracellular pathogens (Diederen *et al.*, 2007; Wullings *et al.*, 2010). Hence, using qPCR to estimate Legionella could result in more realistic estimates of infectious cells, but would result in significantly higher risks than those based on culturable count data. Similar discrepancies between culture and qPCR method density estimates have been reported for other matrices such as in the detection of *Campylobacter* in chickens (Wolffs *et al.* 2005; Rantsiou *et al.* 2010). While the importance of VBNC in QMRA has not been documented, ignoring VBNC organisms in previous risk assessment studies may have resulted in underestimating the true risk for some environments.

A second advantage of using a qPCR approach to quantify reference pathogens is the ability to discriminating between human-infectious and non-infectious pathogen strains. This is of particular interest when the pathogen sources are non-human (i.e. animals). Pathogen species that infect animals are not all human-infectious. For example, species of *Giardia* that infect animals, but are not thought to be human-infectious, include *Giardia agilis*, *Giardia ardeae* and *Giardia muis*. Whereas, *Giardia lamblia* is human-infectious and assemblages A and B have been found to infect dogs, cats, and livestock (Feng and Xiao, 2011). If qPCR targets all members of the *Giardia* genus (i.e. human and non-human-infectious strains) then the predicted risk from a MRA may be inflated. A QMRA of recreational waters impacted by fresh animal waste estimated the density of human-infectious pathogens in the water by source (Soller *et al.*, 2010b). The risk of gastro-intestinal illness from accidental ingestion of water was calculated two ways. First, the risk was estimated assuming all pathogen species in the water were infectious. The second approach only accounted for the human-infectious pathogens. Using the second approach, the resulting GI illness risk for waters contaminated with chicken and gull faecal matter decreased almost an order of magnitude compared with the risk where all the pathogen species were assumed to be human-infectious.

Frequently, pathogen detection is performed by employing a single assay and then this information is used to estimate the risk associated with exposure to the contaminated medium (water, food, soil or air). Another potential advantage of using qPCR assays for QMRA is the use of multiple gene targets to increase the confidence that an amplification signal is associated with a single pathogen population and not with different strains carrying lesser public health risks. For example, if multiple genes associated with pathogenicity are used to target a pathogen and the qPCR signals are all at similar levels, this could suggest that the signals are coming from the same pathogen population. If not, it is therefore reasonable to assume that there might be populations within the sample carrying many of the genes associated with pathogenicity. This approach might be useful when studying the prevalence and diversity of Shiga toxin-producing *Escherichia coli* in animal faeces (Oporto *et al.*, 2008) and environmental waters (Tanaro *et al.*, 2010). For this approach to work, the genes targeted must be present in the same copy number (ideally as single copies), and the differences in amplification efficiency for each of the assays must be minimal.

Ideally, detection of a signal associated with a pathogen can be used to accurately estimate the relative abundance of that given pathogen. However, there are limitations associated with the qPCR approach that impact on the interpretation of risk. As raised earlier, a limitation of qPCR assays is the quantification of both viable and non-infectious/

dead organisms. Numerous approaches have been proposed to enumerate viable bacteria (Keer and Birch, 2003). Recently, DNA intercalating agents have been proposed to discriminate active from non-active ('dead') bacteria in environmental waters. The general assumption is that propidium monoazide (PMA) and ethidium monoazide (EMA) bind to the DNA of bacteria with compromised cell membranes, inhibiting amplification of genes from dead cells after photoactivation. In particular, PMA and EMA have been used in conjunction with PCR and qPCR assays for enumeration of active faecal indicator bacteria (Varma *et al.*, 2009), faecal source tracking populations (Bae and Wuertz, 2009), and bacterial pathogens (Chang *et al.*, 2009; Wang and Mustapha, 2009; Wang *et al.*, 2010). The value of this general approach was recently shown with wastewater samples. Specifically, differences in Bacteroidales qPCR signals between PMA-treated and non-PMA-treated wastewater samples were significant for both influent and effluent samples (Bae and Wuertz, 2009). The largest differences were observed for effluent samples which is consistent with the exposure of wastewater samples to disinfection treatments. Differences between qPCR signals using these intercalating agents and viable cells can also be due to other reasons. For example, EMA has been reported to inhibit bacterial growth and enter live cells suggesting that this approach might underestimate the number of active bacterial cells in a sample (Flekna *et al.*, 2007, Kobayashi *et al.*, 2009). Similarly, some limitations have been documented when using PMA with UV inactivated cells (Nocker *et al.*, 2007b). Based on these limitations it could be difficult to accurately estimate active cells exposed to sunlight in recreational settings or exposed to UV as part of drinking water or water reuse treatments.

The use of RT-qPCR assays to detect microbial rRNA and mRNA transcripts in environmental samples is becoming an important tool in environmental microbiology (Bodrossy *et al.*, 2006; Smith and Osborn, *et al.*, 2009). By looking at RNA transcripts it is possible to enumerate active populations, although rRNA transcripts may persist for some time in the environment due to the high level of secondary structure. For QMRA, studies targeting RNA instead of DNA could significantly reduce the uncertainty introduced when using qPCR with environmental samples as part of the signals are associated with an unknown amount of nucleic acid from non-active (dead) microbial cells. Thus far, RT-qPCR assays have been used to quantify several bacterial groups in human faecal samples (Matsuda *et al.*, 2009; Kubota *et al.*, 2010) and human blood samples (Sakaguchi *et al.*, 2010). Functional genes have been targeted in studies looking at the expression of genes involved in nitrogen cycle (Levy-Booth *et al.*, 2010), sulfate reduction (Neretin *et al.*, 2003; Villanueva *et al.*, 2008) and biodegradation genes (Lee *et al.*, 2008; Nícolaisen *et al.*, 2008; Kong and Nakatsu, 2010). The use of RNA-targeted approaches to detect pathogens in waters is hampered by the low levels of pathogens and the fact that genes associated with pathogenicity pathways may not be transcribed in cells outside of the host. However, it may be possible to stimulate such genes using culturing techniques to confirm the presence of active pathogens in environmental systems.

The detection limits of qPCR methods are not always lower than culturable and conventional PCR methods, therefore in some instances the advantage of using qPCR methods has more to do with the reduced time it takes to obtain results rather than assay sensitivity. Theoretically, qPCR assays can detect one genetic target per reaction. However, this is not commonly achieved when studying environmental samples. In reality, the specific number will depend on a combination of factors such as nucleic acid extraction and amplification efficiency, detector used, reporter chemistry used, presence of competing hybridization targets, PCR inhibitors, and matrix effects. From a regulatory standpoint, bacterial pathogens present in environmental waters in low numbers are a particular challenge to qPCR assays, particularly when the population in question is present in less than 10 cfu/ml, given that only small aliquots (μl) are actually assayed per PCR reaction. Using US drinking water rules as an example we will illustrate how the detection limits of qPCR might not meet regulatory standards. A similar approach could be used for other water and non-water matrices.

Current international and U.S. regulations

state that drinking water samples should not exceed one CFU of *E. coli* per 100 ml. In other terms, ultimately, this is the detection limit that molecular techniques need to match. This means that if 100 ml are filtered onto a membrane, not only does one cell (or genome equivalent of the targeted population) needs to be part of the DNA extract but also within the template used in a qPCR assay. In practical terms, this detection limit goal is impossible to meet unless there is a very high gene-target copy number per cell and/or a larger volume is filtered. For example, if a commercially available kit is used to extract DNA from a water sample, 100 µl of water/buffer are used in the elution step, 5 µl are used as PCR template, and at least 20 copies of the targeted gene need to be in the final DNA extract to generate a PCR signal. Realistically, several copies are needed per reaction due to factors such as amplification efficiency and presence of PCR inhibitors, factors that may vary significantly depending on the assays (e.g. primers used) and matrix. If anywhere from 5–10 copies are needed to obtain a reliable qPCR signal, a minimum of 100–200 cells need to be present in the original sample if the targeted gene is present as a single copy per genome. This number is smaller for assays targeting multicopy genes such as rRNA genes. The latter genes, however, have limited value in the detection of pathogens as non-pathogenic strains will probably be detected as well.

A notorious limitation of PCR-based approaches is the presence of inhibitory substances that co-precipitate during DNA extraction from environmental samples. Matrixes than contain high amounts of complex organics such as humic acids and iron oxides can interfere with detection limits, plus interfere with the estimation of DNA yields. Amplification efficiency can be improved by increasing salt concentration in the DNA lysis step as it can significantly decrease the amount of humic acid. Using affinity chromatography columns also improves signal detection, but only when A_{260}/A_{280} exceeds 1.5 (LaMontange *et al.*, 2002). The presence of PCR inhibitors is relevant to QMRA particularly due to the low densities at which most pathogens occur in environmental samples. Biomass concentration methods can increase detection of pathogens, although they also tend to increase the amount of inhibitors. Hence, surrogates at higher densities than pathogens are often used to provide data for QMRAs.

Quantify surrogate or indicator concentrations

In exposure assessment, a direct measure of reference pathogen density is not always available. This is often the case for pathogens that are costly or difficult to detect, like viruses. Instead, a surrogate can be measured and used to estimate the dose of the reference pathogen. Surrogates include bacterial spores for parasitic protozoan oocysts and cysts (Hijnen and Medema, 2010) and various bacteriophages for human enteric viruses (Leclerc *et al.*, 2000; Petterson *et al.*, 2001; Ottoson and Stenström, 2003; Storey and Ashbolt 2003). In an MRA of tertiary-treated wastewater effluent, the coliphage that infect male (F+) *Escherichia coli* was selected as the surrogate for enteric viruses (Soller *et al.*, 2003). Coliphage were selected because they were less time-consuming and costly to enumerate by plaque assay than human enteric viruses and exhibited similar resistance to wastewater treatment. Ideally, a surrogate should behave similarly to the reference pathogen in the environment (Noble *et al.*, 2004).

When using qPCR to quantify surrogate density, it is important to understand the persistence of the surrogate and the reference pathogen under the environmental conditions present. Using non-qPCR methods, the persistence of various surrogates has been reported for bacteriophage exposed to sunlight (Sinton *et al.*, 1999; Love *et al.*, 2010); murine norovirus and feline calicivirus exposed to thermal and pH extremes (Cannon *et al.*, 2006); *Clostridium perfringens* subject to chlorine disinfection (Rose *et al.*, 2004); and for bacteriophage exposed to UV treatment (Hijnen *et al.*, 2006). However, the persistence of the qPCR target in the environment may be different from that previously reported. The persistence of the qPCR target of a surrogate should be more persistent than the reference pathogen, or else the pathogen risk will be underestimated. For example, there are mixed views as to the longevity of the Bacteroidales qPCR target in the environment (Harwood *et al.*,

2009; Stapleton *et al.*, 2009) versus the human enteric viruses that are thought to be the major health-related pathogen group in recreational waters (Sinclair *et al.*, 2009). On the other hand, the use of a highly persistent qPCR marker (say *C. perfringens* spores) may overestimate the risk for rapidly inactivated pathogens.

Other MRAs develop predictive models that use the available information about the system to predict pathogen density or dose. The simplest example is a MRA that uses an index organism to estimate pathogen density. This approach is used when a known ratio exists between the index organism and reference pathogen in the environment (Ashbolt *et al.*, 2001). Possible qPCR indexes for water include the qPCR assays for bacterial indicators of faecal pollution that have been developed in recent years (Haugland *et al.*, 2005; Silkie and Nelson, 2009; Chern *et al.*, 2011). Among all the available assays, the qPCR assay targeting the 23S rDNA of enterococci has most frequently been used to assess levels of faecal pollution in a variety of recreational waters (Haugland *et al.*, 2005). The persistence of the qPCR assay targeting the 23S rDNA was estimated in seawaters exposed to sunlight in microcosms and compared with that of infectious enteroviruses (Walters *et al.*, 2009). However, this data may not be applicable to all environmental waters. Microorganisms can respond to environmental changes in a dynamic fashion (i.e. can evolve and reproduce) and data generated for one set of conditions might not be universally applicable for the same pathogens inhabiting a different environment (e.g. low vs high nutrient loads, sunlight intensity or turbidity).

As an alternative, process-based models are built to predict pathogen fate and transport based on pathogen source contributions to the system and information on the processes that affect the pathogen such as removal or inactivation. Process based models for pathogen exposure have been developed for watersheds (Ferguson *et al.*, 2007), drinking water systems (Besner *et al.*, 2011); and drinking water treatment (Summerscales *et al.*, 2010). The same general advice applies to process-based models; all of these approaches require knowledge about the behaviour of the reference pathogen over a range of conditions.

Dose–response assessment

Compatibility of qPCR methods

Each reference pathogen requires some description of its dose–response relationship that translates the pathogen dose into the health outcome. Dose-response data are most often reported using culture-based methods (Table 7.4). Three approaches have been used to address the discrepancy between available dose–response relationships expressed using non-qPCR methods and reference pathogen dose estimates expressed using qPCR.

The least time demanding approach is to assume that the qPCR estimate of dose is a conservative estimate. Such a conservative approach could result in overestimating the dose associated with the presence of a particular pathogen, surrogate or faecal source identifier. For example, Ahmed *et al.* (2010) tested 214 rainwater samples from roof-tanks used as potable or non-potable water for the presence of *S. enterica*, *L. pneumophila*, *C. jejuni*, *G. lamblia* and *C. parvum* using qPCR assays targeting pathogen-specific genes. With the exception of *C. parvum*, all reference pathogens were detected anywhere from 0.4% to 10.7% in the water samples tested. While the risks of infection were low for most pathogens, the estimated risks of salmonellosis and giardiasis using a QMRA approach were higher than expected from the reported incidences of gastroenteritis. Such a discrepancy could have been the result of not being able to discriminate viable from non-viable cells, individuals acquiring immunity due to frequent exposure to pathogens (changing the dose–response relationship), or errors due to the sampling regime. To be on the conservative side, the authors recommended the treatment of roof water, particularly if used for drinking purposes, resulting in added costs to each household. Clearly, overestimation could also lead to other economic losses, such as beaches being closed even when pathogen risks are minimal or in the food industry, food stocks being unnecessarily labelled as contaminated and causing costly losses and potential recalls.

A second approach is to convert the qPCR results into cell equivalents. Cell equivalents are estimated by converting Ct values to the number

Table 7.4 Dose–response relationships for waterborne pathogens

Reference pathogen	Published dose–response model	Dose unit	Health endpoint	Reference
Norovirus	Hypergeometric	Genome copies	Infection	Teunis *et al.* (2008a)
Adenovirus 4	Exponential	MPN	Infection	Crabtree *et al.* (1997)
Cryptosporidium spp.	Exponential	oocysts	Infection	Haas *et al.* (1999); USEPA (2005)
	Generalized beta-Poisson	oocysts	Illness	Englehardt *et al.* (2006)
Giardia lamblia	Exponential	Cysts	Infection	Rose *et al.* (1991) ; Haas *et al.* (1999)
Campylobacter jejuni	Beta-Poisson	cfu	Infection	Medema *et al.* (1996); Teunis *et al.* (1996, 2005)
Salmonella enterica	Beta-Poisson	cfu	Illness	Haas *et al.* (1999)
	Gompertz log	cfu	Illness	Coleman *et al.* (1998, 2000, 2004); Soller *et al.* (2007)
E. coli O157:H7	Beta-Poisson	cfu	Infection	Teunis *et al.* (2008b)
Legionella pneumophila	Exponential	cfu	Infection	Armstrong *et al.* (2008)

of target copies present in the reaction. Plasmids within cells containing the targeted DNA fragments have been used as the reference material for estimating the number of target copies. The genome size of a type species is then used as the reference to convert number of copies to genome copies. Genes present as single copies are more amenable to estimate cell equivalents using the rationale that one gene copy equals one cell. Converting gene copies directly to cell equivalents may be inaccurate when cells are actively growing and as a result harbouring multiple partial chromosomes (Bremer and Dennis, 1987). Moreover, genome size of targeted populations may vary and therefore cell equivalents may vary from one population to another. As genomes of multiple strains are sequenced, it will be possible to determine the level of variation between different strains and species.

In the case of 16S rRNA gene-based assays, the number of operons can vary within targeted populations. The level of variation among a population has not been taken into consideration. This is an important omission as it has been shown that the average number of rRNA operon copies changes during different successional stages (Shrestha *et al.* 2007) and in fact may increase due to selection processes such as resource availability (Klappenbach *et al.* 2000). Very little is known on the survival capabilities of populations used in MST and how the environment can select for populations that have a certain number of rrn operons. Moreover, since the number of replication forks per cell vary depending on growth rate (Bremer and Dennis, 1987), it is also possible that the number of operon copies will vary depending on the *in situ* physiological status of the organism, a particularly good survival strategy when nutrients are plentiful. The position of the operons relative to the origin of replication can also have an impact on estimating the number of cell equivalents. Specifically, genes that are closer to the origin of replication may be at times more numerous than those that are farther away. It should be noted that rrn operons are often close to the origin, and therefore estimating cell equivalents targeting rrn genes is prone to such caveats. Use of non-multicopy genes is an option for estimating cells densities or cell equivalents, however, detection limits of such assays tend to be lower than assays targeting rrn genes (Chern *et al.*, 2011). In order to convert qPCR values to cell equivalents, a number of conditions should be met. Ideally, the PCR assay should be 100% efficient under a wide variety of conditions or else well estimated. This is a very difficult proposition as the kinetics of amplification not only depend on the G+C content, length of the primers, and amplification product size, but also on the presence of inhibitors and the complexity of the template used in the assay.

A third solution is to develop a new dose–response relationship using the qPCR method with reference pathogens in a form very similar to what the target human population is exposed to. Members of the genus Norovirus provide a good case example, being a major cause of gastrointestinal illness and hospitalizations, they have escaped routine cultivation thus far. Hence the dose–response relationship commonly used in risk assessments for human noroviruses was based on a TaqMan RT-qPCR enumeration of virions (Teunis *et al.*, 2008a). This approach has a major limitation in that the actual count of infectious virions in the inocula used to parameterize the dose–response relationship remains unknown. The noroviruses were purified from both fresh stool samples and stored stock suspensions. To use the dose–response for a dose estimated from an environmental setting, the risk assessor must assume that the ratio of total to infectious numbers of virions is the same as that of the dose–response inocula, as too the nature of virion clumping, which may be unrealistic.

Risk characterization

Risk characterization is arguably the most important step of the MRA process when the knowledge about risk is summarized and communicated. This step may include quantitative analysis, such as calculating total risk from multiple exposures; however, risk characterization is more descriptive in nature. Risk characterization should communicate the scope and limitations of the MRA approach. For MRAs that incorporate qPCR, we have previously identified many potential limitations including variable performance and uncertainty about the infectivity of the detected organisms. In risk characterization, the risk assessor communicates how these limitations effect the interpretation of results.

The potential limitations of the qPCR approach for MRA can generally be categorized into two groups. The first group of limitations introduces variability into the estimate of pathogen dose. Sources of variability may include DNA extraction efficiency, PCR amplification efficiency, estimation of viable cells, and estimation of cell equivalents among others, and subsets that are infections to humans. The variability from each step of the qPCR method can be carried through to the estimated risk using probabilistic techniques, assuming the variability can be characterized. The variability that cannot be easily characterized falls into the second group, the limitations that introduce uncertainty into the estimation of pathogen dose. Uncertain inputs are those that could be more precisely characterized with more data collection or analysis. Typically, in the risk characterization step, the risk assessor evaluates if the limitations contributing uncertainty are causing large uncertainties in the predicted risk.

The most simplistic way to identify important sources of uncertainty is through parametric sensitivity analysis. Parametric sensitivity analysis estimates the predicted outcome's sensitivity to possible changes in each input variable separately, holding all other variables at the best estimate. In QMRA, the change in the best estimate risk can be quantified when each uncertain parameter is held at its maximum and minimum limits. Some sources of uncertainty may not result in large changes in risk while others may drastically change the interpretation. Consider the MRA for the assessment of pathogen risk associated with potable use of roof collected rainwater in south-east Queensland, Australia (Ahmed *et al.*, 2010). The aim was to determine if ingestion of rainwater caused more than one extra infection per 10,000 persons per exposure, a commonly used risk benchmark. One qPCR related parameter in the exposure assessment with uncertainty is the number of genome copies per *G. lamblia* cyst. The best estimate cyst equivalent was estimated at 16 gene copies, but the value in reality is uncertain with a potential range of values. Does this uncertainty have an impact on the interpretation of risk? The range of detected genome copies in samples of rainwater collected was roughly $9.0–5.7 \times 10^1$ genome copies per 1000 ml, accounting for recovery. Using the best estimate cyst equivalent, the number of cysts per 1000 ml was estimated as 5.6×10^{-1} to 3.6 cysts. Assuming an ingestion rate per day of 1000 ml per person, the infection risk was estimated as 110–690 infections per 10,000 persons assuming all cysts were infectious (Ahmed *et al.*, 2010). This is well above the threshold of one extra infection per 10,000 persons per exposure.

Here, we will assume the cyst equivalent based on gene target copy numbers is between a minimum of 10 and a maximum of 20 gene copies. Using the low and high cyst equivalent limits, the infection risk is recalculated and results in 180 to 1100 and 89 to 550 infections per 10,000 persons. Again, even when considering potential uncertainty in the cyst equivalent, the predicted risk is well above the identified threshold of one per 10,000 persons per exposure. Therefore, the uncertainty in the cyst equivalent appears not to have an important impact on the final interpretation of risk in this example. However, other assumptions may have significant impact.

In reality, there are multiple sources of uncertainty in an MRA that potentially influence the estimation of risk. When the number of uncertain input variables is large, Monte Carlo approaches can be used to estimate the resulting uncertainty in the predicted risk due to uncertain input parameters (Fig. 7.3) (Haas *et al.*, 1999). Using rank order correlation, the sources of uncertainty that contribute most to the spread in the predicted risk can be identified and ranked (Fig. 7.3). Similar to parametric sensitivity analysis, Monte Carlo approaches also require assumptions about the possible range of the uncertain variable. It may be challenging to set limits for or characterize the sources of uncertainty associated with qPCR and as a result challenging to assess the impact on risk. In QMRA, perhaps the most significant source of uncertainty when using molecular assays such as qPCR is the fact that we are working with samples of unknown molecular diversity and physiological status. For example, assuming uniformity among subsamples, it is possible to estimate recovery yields for different DNA extraction methods. Variation in extraction efficiency and yields are common when using different methods (Ariefdjohan *et al.*, 2010; Lear *et al.*, 2010; Nylund *et al.*, 2010). More troublesome is the fact that studies have shown that using the same methods and soil samples inter- and intra-laboratory variation can introduce some significant biases (Pan *et al.*, 2010). However, these yields cannot be used to

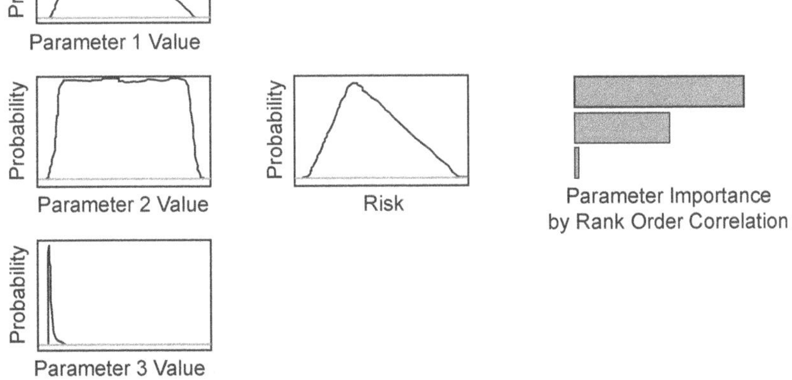

Figure 7.3 Monte Carlo approach to estimating risk from uncertain input parameters (random samples, typically some 10,000) from each parameter distribution are individually selected based on their frequency of occurrence, so giving the final risk estimate as a probability distribution).

calculate DNA extraction efficiency if the level of PCR inhibition is not determined accurately and the structure of environmental microbial communities is unknown. Even if membership or community structure can be elucidated using molecular methods (Cabrol *et al.*, 2010; Pan *et al.*, 2010), in most cases it is difficult to accurately estimate *in situ* abundances, and in turn, the extraction efficiency of that particular group is difficult to estimate. Moreover, differences in cell lysis among different cell types can vary significantly, a reason why different extraction protocols have been proposed for different matrices (Zhou *et al.*, 1996; Johnson *et al.*, 2005; Bey *et al.*, 2010; Lear *et al.*, 2010). Furthermore, lysis efficiency may vary depending on the physiological status of a given bacterial group (e.g. vegetative cells vs. spores), rendering calculations subject to considerable variation. In spite of their importance, some of these issues are rarely addressed in significant detail in microbial water quality studies (Stoeckel *et al.*, 2009). In fact, only few efforts have been made to standardize nucleic extraction procedures for environmental samples (i.e. soils; Petric *et al.*, 2011). One important goal in risk characterization is to understand the effect of these potential uncertainties in the final estimate of risk and identify areas where additional information should be collected to reduce the uncertainty.

Conclusions

The use of qPCR assays is becoming common practice in environmental microbiology and public health studies. The number of QMRA studies using qPCR to quantify risk associated with environmental waters is small, yet growing, as qPCR methods are becoming mainstream. The information acquired by qPCR can be used throughout the risk assessment process to identify potential reference pathogens of concern, quantify pathogen dose and characterize risk. There are a number of advantages to using qPCR to identify/quantify targeted microorganisms over more traditional culture-based methods such as rapid results, detection of viable but non-culturable pathogens, and discrimination of human-infectious pathogens. However, there remain a number of limitations including the overestimation of risk due to detection of non-infectious/viable organisms, low recoveries due to inhibition, and the lack of compatibility with culture-derived dose–response relationships. Like other methods used for detecting pathogens, qPCR has a number of performance issues that can be accounted for, if quantifiable, in the QMRA process, like recovery, cross-amplification, and sensitivity.

Future trends

Like other molecular based techniques, the development of qPCR in public health will, to a large extent, depend on the availability of sequence data, identification of genetic networks associated with pathogenicity, and availability of dose–response relationships. As new microbial pathogens emerge there will also be a need to generate genetic data associated with them and in turn the need to develop methods to detect such novel pathogens. In this regard, high throughput sequencing technologies will accelerate genetic database development for genomes of different pathogens (Harris *et al.*, 2010) and non-pathogenic strains. Pathogens do not exist in isolation and in many cases depend on other organisms to become infectious and to survive within and outside of the host (Lau and Ashbolt, 2009). Techniques such as phylogenetic microarrays (Brodie *et al.*, 2007; DeSantis *et al.*, 2007; Rajilić-Stojanović *et al.*, 2009) and barcoding sequencing strategies (Andersson *et al.*, 2008; Caporaso *et al.*, 2010) are beginning to create a more robust picture of the diversity of bacteria within a variety of different environments. These approaches will probably be applied to drinking water systems showing a high incidence of outbreaks and to polluted recreational waters impacted with different sources of pollution to better understand which organisms are associated with high levels of risk, and how the presence of other microorganisms might impact on pathogen fate and transport.

The availability of metagenomic data will help in the discovery of pathogens from complex samples (Nakamura *et al.*, 2009). Moreover, metagenomic data from human microbiome projects will accelerate our understanding of how microorganisms may work synergistically with others to prevent disease and how pathogenic strains overcome such antagonistic barriers within complex microbial communities to inflict disease to a given host (NIH HMP Working Group *et al.*, 2009). Metagenomic data will also be relevant to understand the complexity of genetic transfer that is relevant to public health (Qu *et al.*, 2008; Jones *et al.*, 2010). Similarly, transcriptomics research will enhance our understanding on the genes that are responsible for pathogenicity in individual populations while metatranscriptomics will be useful at elucidating how microbial communities respond to the presence of environmental insults that are relevant to the fate of pathogens. For these techniques to be successful, the field of bioinformatics will need to become an integral part to research in public health sciences, applied environmental microbiology, and QMRA.

The use of QMRA to predict risk for scenarios or conditions that are difficult to observe is increasing. This may include predicting the pathogen dose and risks associated with source water under extreme events, climate change, or land-use change. The models are becoming increasingly more complex, accounting for fate and transport of pathogens. This use of QMRA requires knowledge on the removal, fate, and infectivity of qPCR targets throughout the environment under various conditions. Metadata on physiochemical parameters encountered by the pathogens and their surrogates as well as hydrological and geographical landscape data are critical to further strengthen fate and transport predictive models. The need to look at pathogens as ecological units is necessary to have a better handle of how microbial agents evolve novel pathogenic pathways that may be the result of changes in urban landscape and climatological conditions. As a result, future QMRA models may need to incorporate more sophisticated representation of ecosystem frameworks to predict risks and to help in assessing which and when ecological factors drive the evolutionary fate of pathogenic strains in the environment and how the human microbiome may mediate pathogen infectivity/effect. How this information can be integrated into managing risks will be an important challenge that public health officials will be facing in the not so distant future.

Additional web resources

Microbial genomes

http://mbgd.genome.ad.jp/
http://cmr.jcvi.org/tigr-scripts/CMR/CmrHomePage.cgi

QMRA

http://camra.msu.edu/qmra.htmland

Public health and fecal pollution

http://water.epa.gov/type/rsl/monitoring/vms511.cfm
http://water.epa.gov/drink/contaminants/basicinformation/pathogens.cfm
http://water.epa.gov/scitech/swguidance/waterquality/standards/criteria/health/recreation/index.cfm
http://www.epa.gov/nrmrl/pubs/600r05064/600r05064.htm

References

Abbaszadegan, M., Huber, M.S., Gerba, C.P., and Pepper, I.L. (1997). Detection of viable *Giardia* cysts by amplification of heat shock-induced mRNA. Appl. Environ. Microbiol. *63*, 324–328.

Ahmed, W., Vieritz, A., Goonetilleke, A., and Gardner, T. (2010). Health Risk from the use of roof-harvested rainwater in southeast Queensland, Australia, as potable or nonpotable water, determined using quantitative microbial risk assessment. Appl. Environ. Microbiol. *76*, 7382–7391.

Almeida, A., Pozio, E., and Cacciò, S.M. (2010). Genotyping of *Giardia duodenalis* cysts by new real-time PCR assays for detection of mixed infections in human samples. Appl. Environ. Microbiol. *76*, 1895–1901.

Andersson, A.F., Lindberg, M., Jakobsson, H., Bäckhed, F., Nyrén, P, and Engstrand, L. (2008). Comparative analysis of human gut microbiota by barcoded pyrosequencing. PLoS ONE *3*, e2836.

Anuchin, A.M., Mulyukin, A.L., Suzina, N.E., Duda, V.I., El-Registan, G.I., and Kaprelyants, A.S. (2009). Dormant forms of *Mycobacterium smegmatis* with distinct morphology. Microbiology *155*, 1071–1079.

Ariefdjohan, M.W., Savaiano, D.A., and Nakatsu, C.H. (2010). Comparison of DNA extraction kits for PCR-DGGE analysis of human intestinal microbial communities from fecal specimens. Nutr. J. *9*, 23.

Armstrong, T.W., and Haas, C.N. (2008). Legionnaires' disease: evaluation of a quantitative microbial risk assessment model. J. Water Health *6*, 149–166.

Ashbolt, N.J., Grabow, W.O.K., and Snozzi, M. (2001). Indicators of microbial water quality. In Water Quality: Guidelines, Standards and Health. Risk Assessment and Management for Water-related Infectious Disease,

Fewtrell, L., and Bartram, J. eds (IWA Publishing, London), pp. 289–316.

Ashbolt, N.J., Schoen, M.E. Soller, J.A., and Roser, D.J. (2010). Predicting pathogen risks to aid beach management: The real value of quantitative microbial risk assessment (QMRA). Water Res. *44*, 4692–4703.

Bae, S., and Wuertz, S. (2009). Discrimination of viable and dead fecal *Bacteroidales* bacteria by quantitative PCR with propidium monoazide. Appl. Environ. Microbiol. *75*, 2940–2944.

Bej, A.K., Mahbubani, M.H., and Atlas, R.M. (1991). Detection of viable *Legionella* pneumophila in water by polymerase chain reaction and gene probe methods. Appl. Environ. Microbiol. *57*, 597–600.

Bellin, T., Pulz, M., Matussek, A., Hempen, H., and Gunzer, F. (2001) Rapid detection of enterohemorrhagic *Escherichia coli* by real-Time PCR with fluorescent hybridisation probes. J. Clin. Microbiol. *39*, 370–374.

Béra-Maillet, C., Mosoni, P., Kwasiborski, A., Suau, F., Ribot, Y., and Forano. E. (2009). Development of a RT-qPCR method for the quantification of Fibrobacter succinogenes S85 glycoside hydrolase transcripts in the rumen content of gnotobiotic and conventional sheep. J. Microbiol. Methods 77, 8–16.

Besner, M.C., Prévost, M., and Regli, S. (2011). Assessing the public health risk of microbial intrusion events in distribution systems: conceptual model, available data, and challenges. Water Res. *45*, 961–979.

Bey, B.S., Fichot, E.B., Dayama, G., Decho, A.W., and Norman, R.S. (2010). Extraction of high molecular weight DNA from microbial mats. Biotechniques *49*, 631–640.

Bogosian, G., Morris, P.J., and O'Neil, J.P. (1998) A mixed culture recovery method indicates that enteric bacteria do not enter the viable but nonculturable state. Appl. Environ. Microbiol. *64*, 1736–1742.

Bremer, H., and Dennis, P.P. (1987) Modulation of chemical composition and other parameters of the cell by growth rate. In *Escherichia coli* and *Salmonella* typhimurium: cellular and molecular biology, Neidhardt, F.C., Ingraham, J.L., Low, K.B., Magasanik, B., Schaechter, M., and Umbarger H.E., eds (American Society for Microbiology, Washington, DC), pp. 1527–1542

Brodie, E.L., DeSantis, T.Z., Parker, J.P., Zubietta, I.X., Piceno, Y.M., and Andersen, G.L. (2007). Urban aerosols harbor diverse and dynamic bacterial populations. Proc. Natl. Acad. Sci. U.S.A. *104*, 299–304.

Bodrossy, L., Stralis-Pavese, N., Konrad-Köszler, M., Weilharter, A., Reichenauer, T.G., Schöfer, D., and Sessitsch, A. (2006). mRNA-based parallel detection of active methanotroph populations by use of a diagnostic microarray. Appl. Environ. Microbiol. *72*, 1672–1676.

Cabrol, L., Malhautier, L., Poly, F., Lepeuple, A.S., and Fanlo, J.L. (2010). Assessing the bias linked to DNA recovery from biofiltration woodchips for microbial community investigation by fingerprinting. Appl. Microbiol. Biotechnol. *85*, 779–790.

Caldwell, J.M., Raley, M.E., and Levine, J.F. (2007). Mitochondrial multiplex real-time PCR as a source tracking method in fecal-contaminated effluents. Environ. Sci. Technol. *41*, 3277–3283.

Cannon, J.L., Papafragkou, E., Park, G.W., Osborne, J., Jaykus, L.A., and Vinje, J. (2006). Surrogates for the study of norovirus stability and inactivation in the environment: a comparison of murine norovirus and feline calicivirus. J. Food Prot. *69*, 2761–2765.

Caporaso, J.G., Lauber, C.L., Walters, W.A., Berg-Lyons, D., Lozupone, C.A., Turnbaugh, P.J., Fierer, N., and Knight, R. (2010). Microbes and Health Sackler Colloquium: Global patterns of 16S rRNA diversity at a depth of millions of sequences per sample. Proc. Natl. Acad. Sci. U.S.A. *108*, 4516–4522.

Chaban, B., Musil, K.M., Himsworth, C.G., and Hill, J.E. (2009). Development of cpn60-based real-time quantitative PCR assays for the detection of 14 *Campylobacter* species and application to screening of canine fecal samples. Appl. Environ. Microbiol. *75*, 3055–3061.

Chang, B., Sugiyama, K., Taguri, T., Amemura-Maekawa, J., Kura, F., and Watanabe, H. (2009). Specific detection of viable Legionella cells by combined use of photoactivated ethidium monoazide and PCR/real-time PCR. Appl. Environ. Microbiol. *75*, 147–153.

Chen, N.T., and Chang, C.W. (2010). Rapid quantification of viable legionellae in water and biofilm using ethidium monoazide coupled with real-time quantitative PCR. J. Appl. Microbiol. *109*, 623–634.

Chern, E., Siefring, S., Paar, J., Doolittle, M., and Haugland, R. (2011). Comparison of quantitative PCR assays for *Escherichia coli* targeting ribosomal RNA and single copy genes. Lett. Appl. Microbiol. *52*, 298–306.

Coleman, M., and Marks, H. (1998). Topics in dose–response modeling. J. Food Prot. *61*, 1550–1559.

Coleman, M., Marks, H., Golden, N., and Latimer, H. (2004). Discerning strain effects in microbial dose–response data. J. Toxicol. Environ. Health, Part A *67*, 667–685.

Coleman, M.E., and Marks, H.M. (2000). Mechanistic modeling of salmonellosis. Quant. Microbiol. *2*, 227–247.

Constantin de Magny, G., Murtugudde, R., Sapiano, M.R., Nizam, A., Brown, C.W., Busalacchi, A.J., Yunus, M., Nair, G.B., Gil, A.I., Lanata, C.F., *et al.* (2008). Environmental signatures associated with cholera epidemics. Proc. Natl. Acad. Sci. U.S.A. *105*, 17676–17681.

Converse, R.R., Blackwood, A.D., Kirs, M., Griffith, J.F., and Noble, R.T. (2009). Rapid QPCR-based assay for fecal *Bacteroides* spp. as a tool for assessing fecal contamination in recreational waters. Water Res. *43*, 4828–4837.

Crabtree, K.D., Gerba, C.P., Rose, J.B., and Haas, C.N. (1997). Waterborne adenovirus: A risk assessment. Water Sci. Technol. *35*, 1–6.

Craun, G.F., Brunkard, J.M., Yoder, J.S., Roberts, V.A., Carpenter, J., Wade, T., Calderon, R.L., Roberts, J.M., Beach, M.J., and Roy, S.L. (2010). Causes of outbreaks associated with drinking water in the United States from 1971 to 2006. Clin. Microbiol. Rev. *23*, 507–528.

Deer, D.M., Lampel, K.A., and González-Escalona, N. (2010). A versatile internal control for use as DNA in real-time PCR and as RNA in real-time reverse transcription PCR assays. Lett. Appl. Microbiol. *50*, 366–372.

Derlet, R.W., and Carlson, J.R. (2002). An analysis of human pathogens found in horse/mule manure along the John Muir Trail in Kings Canyon and Sequoia and Yosemite National Parks. Wilderness Environ. Med. *13*, 113–118.

DeSantis, T.Z., Brodie, E.L., Moberg, J.P. Zubieta, I.X., Piceno, Y.M., and Andersen, G.L. (2007). High-density universal 16S rRNA microarray analysis reveals broader diversity than typical clone library when sampling the environment. Microb. Ecol. *53*, 371–383.

Di Giovanni, G., and LeChevallier, M.W. (2005). Quantitative-PCR assessment of *Cryptosporidium parvum* cell culture infection. Appl. Environ. Microbiol. *71*, 1495–1500.

Diederen, B.M., de Jong, C.M.A., Aarts, I., Peeters, M.F., and van der Zee, A. (2007). Molecular evidence for the ubiquitous presence of *Legionella* species in Dutch tap water installations. J. Water Health *5*, 375–383.

Edge T.A., and Hill. S. (2007). Multiple lines of evidence to identify the sources of fecal pollution at a freshwater beach in Hamilton Harbour, Lake Ontario. Water Res. *41*, 3585–3594.

Englehardt, J.D., and Swartout, J. (2006). Predictive Bayesian microbial dose–response assessment based on suggested self-organization in primary illness response: *C. parvum*. Risk Anal. *26*, 543–554.

Fayer, R. (2004). *Cryptosporidium*: a water-borne zoonotic parasite. Vet. Parasitol. *126*, 37–56.

Feng, Y., and Xiao, L. (2011). Zoonotic potential and molecular epidemiology of *Giardia* species and giardiasis. Clin. Microbiol. Rev. *24*, 110–140.

Ferguson, C.M., Croke, B.F.W., Beatson, P.J., Ashbolt, N.J., and Deere, D.A. (2007) Development of a process-based model to predict pathogen budgets for the Sydney drinking water catchment. J. Water Health *5*, 187–208.

Flekna, G., Štefanič, P., Wagner, M., Smulders, F.J.M., Možina, S.S., and Hein, I. (2007). Insufficient differentiation of live and dead *Campylobacter jejuni* and *Listeria monocytogenes* cells by ethidium monoazide (EMA) compromises EMA/real-time PCR. Res. Microbiol. *158*, 405–412.

Fortin, N.Y., Mulchandani, A., and Chen, W. (2001). Use of real-time polymerase chain reaction and Molecular beacons for the detection of *Escherichia coli* O157:H7. Anal. Biochem. *289*, 281–288.

Fremaux, B., Boa, T., and Yost, C.K. (2010). Quantitative real-time PCR assays for sensitive detection of Canada goose-specific fecal pollution in water sources. Appl. Environ. Microbiol. *76*, 4886–4889.

Garcés-Sanchez, G., Wilderer, P.A., Munch, J.C., Horn, H., and Lebuhn, M. (2009). Evaluation of two methods for quantification of hsp70 mRNA from the waterborne pathogen *Cryptosporidium parvum* by reverse transcription real-time PCR in environmental samples. Water Res. *43*, 2669–2678.

González-Escalona, N., Hammack, T.S., Russell, M., Jacobson, A.P., De Jesús, A.J., Brown, E.W., and Lampel, K.A. (2009). Detection of live *Salmonella* sp. cells in produce by a TaqMan-based quantitative reverse transcriptase real-time PCR targeting *invA* mRNA. Appl. Environ. Microbiol. *75*, 3714–3720.

Guix, S., Caballero, S., Villena, C., Bartolomé, R., Latorre, C., Rabella, N., Simó, M., Bosch, A., and Pintó, R.M. (2002). Molecular epidemiology of astrovirus infection in Barcelona, Spain. J. Clin. Microbiol. *40*, 133–139.

Guy, R.A., Payment, P., Krull, U.J., and Horgen, P.A. (2003). Real-time PCR for quantification of *Giardia* and *Cryptosporidium* in environmental water samples and sewage. Appl. Environ. Microbiol. *69*, 5178–5185.

Haas, C., Rose, J., and Gerba, C. (1999). Quantitative Microbial Risk Assessment (John Wiley and Sons, Inc., New York).

Hadfield, S.J., Robinson, G., Elwin, K., Chalmers, R.M. (2011). Detection and differentiation of *Cryptosporidium* spp. in human clinical samples by use of real-time PCR. J. Clin. Microbiol. *49*, 918–924.

Haramoto, E., Katayama, H., Oguma, K., Yamashita, H., Tajima, A., Nakajima, H., and Ohgaki, S. (2006). Seasonal profiles of human Noroviruses and indicator bacteria in a wastewater treatment plant in Tokyo, Japan. Water Sci. Technol. *54*, 301–308.

Harris, S.R., Feil, E.J., Holden, M.T., Quail, M.A., Nickerson, E.K., Chantratita, N., Gardete, S., Tavares, A., Day, N., Lindsay, J.A., *et al.* (2010). Evolution of MRSA during hospital transmission and intercontinental spread. Science *327*, 469–474.

Harwood, V.J., Brownell, M., Wang, S., Lepo, J., Ellender, R.D., Ajidahun, A., Hellein, K.N., Kennedy, E., Ye, X., and Flood, C. (2009). Validation and field testing of library-independent microbial source tracking methods in the Gulf of Mexico. Water Res. *43*, 4812–4819.

Haugland, R.A., Siefring, S.C., Wymer, L.J., Brenner, K.P., and Dufour, A.P. (2005). Comparison of Enterococcus measurements in freshwater at two recreational beaches by quantitative polymerase chain reaction and membrane filter culture analysis. Water Res. *39*, 559–568.

Heim, A., Ebnet, C., Harste, G., and Pring-Akerblom, P. (2003). Rapid and quantitative detection of human adenovirus DNA by real-time PCR. J. Med. Virol. *70*, 228–239.

Hijnen, W.A.M., Beerendonk, E.F., and Medema, G.J. (2006). Inactivation credit of UV radiation for viruses, bacteria and protozoan (oo)cysts in water: a review. Water Res. *40*, 3–22.

Hijnen, W.A.M., and Medema, G.J. (2010). Elimination of micro-organisms by drinking water treatment orocesses. A review – 3rd Edition. (IWA Publishing, London).

Hrudey, S.E., Payment, P., Huck, P.M., Gillham, R.W., and Hrudey, E.J. (2003). A fatal waterborne disease epidemic in Walkerton, Ontario: comparison with other waterborne outbreaks in the developed world. Water Sci. Technol. 47, 7–14.

Huang, J., Zhu, Y., Wen, H., Zhang, J., Huang, S., Niu, J., and Li, Q. (2009). Quadruplex real-time PCR assay for detection and identification of *Vibrio cholerae* O1 and O139 strains and determination of their toxigenic potential. Appl. Environ. Microbiol. 75, 6981–6985.

Hwang, Y.C., Leong, O.M., Chen, W., and Yates, M.V. (2007). Comparison of a reporter assay and immunomagnetic separation real-time reverse transcription-PCR for the detection of enteroviruses in seeded environmental water samples. Appl. Environ. Microbiol. 73, 2338–2340.

Imamovic, L., Ballesté, E., Jofre, J., and Muniesa, M. (2010). Quantification of Shiga toxin-converting bacteriophages in wastewater and in fecal samples by real-time quantitative PCR. Appl. Environ. Microbiol. 76, 5693–5701.

Inglis, G.D., and Kalischuk, L.D. (2004). Direct quantification of *Campylobacter jejuni* and *Campylobacter lanienae* in feces of cattle by real-time quantitative PCR. Appl. Environ. Microbiol. 70, 2296–2306.

Jones, B.V., Sun, F., and Marchesi, J.R. (2010). Comparative metagenomic analysis of plasmid encoded functions in the human gut microbiome. BMC Genomics 11, 46.

Johnson, D.J., Martin, L.R., and Roberts, K.A. (2005). STR-typing of human DNA from human fecal matter using the QIAGEN QIAamp stool mini kit. J. Forensic Sci. 50, 802–808.

Jothikumar, N., Lowther, J.A., Henshilwood, K., Lees, D.N., Hill, V.R., and Vinje, J. (2005). Rapid and sensitive detection of norovirus by using TaqMan-based one-step reverse transcription-PCR assays and application to naturally contaminated shellfish samples. Appl. Environ. Microbiol. 71, 1870–1875.

Jothikumar, N., da Silva, A.J., Moura, I., Qvarnstrom, Y., and Hill, V.R. (2008). Detection and differentiation of *Cryptosporidium hominis* and *Cryptosporidium parvum* by dual TaqMan assays. J. Med. Microbiol. 57, 1099–1105.

Jothikumar, N., Kang, G., and Hill, V.R. (2009). Broadly reactive TaqMan® assay for real-time RT-PCR detection of rotavirus in clinical and environmental samples. J. Virol. Methods 155, 126–131.

Katayama, H., Haramoto, E., Oguma, K., Yamashita, H., Tajima, A., Nakajima, H., and Ohgaki, S. (2008). One-year monthly quantitative survey of noroviruses, enteroviruses, and adenoviruses in wastewater collected from six plants in Japan. Water Res. 42, 1441–1448.

Kato-Hayashi, N., M. Kirinoki, Y. Iwamura, T. Kanazawa, V. Kitikoon, H. Matsuda, and Y. Chigusa. (2010). Identification and differentiation of human schistosomes by polymerase chain reaction. Exp. Parasitol. 124, 325–329.

Keer, J.T., and Birch, L. (2003). Molecular methods for the assessment of bacterial viability. J. Microbiol. Methods 53, 175–183.

Klappenbach, J., Dunbar, J., and Schmidt, T. (2000). rRNA operon copy number reflects ecological strategies of bacteria. Appl. Environ. Microbiol. 66, 1328–1333.

Kim, K., Katayama, H., Kitajima, M., Tohya, Y., and Ohgaki, S. (2011). Development of a real-time RT-PCR assay combined with ethidium monoazide treatment for RNA viruses and its application to detect viral RNA after heat exposure. Water Sci. Technol. 63, 502–507.

Klein, M., Brown, L., Tucker, R.W., Ashbolt, N.J., Stuetz, R.M., and Roser, D.J. (2010). Diversity and abundance of zoonotic pathogens and indicators in manures of feedlot cattle in Australia. Appl. Environ. Microbiol. 76, 6947–6950.

Kobayashi, H., Oethinger, M., Tuohy, M.J., Hall, G.S., and Bauer, T.W. (2009). Unsuitable distinction between viable and dead *Staphylococcus aureus* and *Staphylococcus epidermidis* by ethidium bromide monoazide. Lett. Appl. Microbiol. 48, 633–638.

Kogure, K., Simidu, U., and Taga, N. (1979). A tentative direct microscopic method for counting living marine bacteria. Can. J. Microbiol. 25, 415–420.

Kompalic-Cristo, A., Frotta, C., Suarez-Mutis, M., Fernandes, O., and Britto, C. (2007). Evaluation of a real-time PCR assay based on the repetitive B1 gene for the detection of *Toxoplasma gondii* in human peripheral blood. Parasitol. Res. 101, 619–625.

Kong, R.Y.C., Lee, S.K.Y., Law, T.W.F., Law, S.H.W., and Wu, R.S.S. (2002). Rapid detection of six types of bacterial pathogens in marine waters by multiplex PCR. Water Res. 36, 2802–2812.

Kong, W., and Nakatsu, C.H. (2010). Optimization of RNA extraction for PCR quantification of aromatic compound degradation genes. Appl. Environ. Microbiol. 76, 1282–1284.

Kubota, H., Tsuji, H., Matsuda, K., Kurakawa, T., Asahara,T., and Nomoto, K. (2010). Detection of human intestinal catalase-negative, Gram-positive cocci by rRNA-targeted reverse transcription-PCR. Appl. Environ. Microbiol. 76, 5440–5451.

Lambertini, E., Spencer, S.K. Bertz, P.D. Loge, F.J., and Borchardt, M.A. (2010). New mathematical approaches to quantify human infectious viruses from environmental media using integrated cell culture-qPCR. J. Virol. Methods 163, 244–252.

Lammerding, A.M., and Paoli, G.M. (1997). Quantitative risk assessment: an emerging tool for emerging foodborne pathogens. Emerg. Infect. Dis. 3, 483–487.

LaMontagne, M.G., Michel, F.C. Jr., Holden, P.A., and Reddy, C.A. (2002). Evaluation of extraction and purification methods for obtaining PCR-amplifiable DNA from compost for microbial community analysis. J. Microbiol. Methods 49, 255–264.

Lau, H.Y., and Ashbolt, N.J. (2009). The use of *Acanthamoeba* as a tool for understanding Legionella pathogenesis: implications for drinking water. J. Appl. Microbiol. 107, 368–378.

Layton, A., McKay, L., Williams, D., Garrett, V., Gentry, R., and Sayler, G. (2006). Development of Bacteroides 16S rRNA gene TaqMan-based real-time PCR assays for estimation of total, human, and bovine fecal pollution in water. Appl. Environ. Microbiol. 72, 4214–4224.

Lear, G., Dong, Y., and Lewis, G. (2010). Comparison of methods for the extraction of DNA from stream epilithic biofilms. Antonie Van Leeuwenhoek *98*, 567–571.

Lebuhn, M., Effenberger, M., Gronauer, A., Wilderer, P.A, and Wuertz, S. (2003). Using quantitative real-time PCR to determine the hygienic status of cattle manure. Water Sci. Technol. *48*, 97–103.

LeChevallier, M., and Buckley, M. (2007) Clean Water. What is Acceptable Microbial Risk? A Report from the American Academy of Microbiology (American Society for Microbiology, Washington DC).

Leclerc, H., Edberg, S., Pierzo, V., and Delattre, J.M. (2000). Bacteriophages as indicators of enteric viruses and public health risk in groundwaters. J. Appl. Microbiol. *88*, 5–21.

Lee, G., Nam, S., Chae, J., and Lee., C. (2009). *Giardia duodenalis*: improved detection of viable cysts by reverse transcription-PCR of heat shock-inducible hsp70 gene. Exp. Parasitol. *123*, 377–380.

Levy-Booth, D.J., and Winder, R.S. (2010). Quantification of nitrogen reductase and nitrite reductase genes in soil of thinned and clear-cut Douglas-fir stands by using real-time PCR. Appl. Environ. Microbiol. *76*, 7116–7125.

Liu, Y.J., Zhang, C.M., and Wang, X.C. (2009). Simultaneous detection of enteric bacteria from surface waters by QPCR in comparison with conventional bacterial indicators. Environ. Monit. Assess. *158*, 535–544

Love, D.C., Silverman, A., and Nelson, K.L. (2010). Human virus and bacteriophage inactivation in clear water by simulated sunlight compared with bacteriophage inactivation at a southern California beach. Environ. Sci. Technol. *44*, 6965–6970.

Lu, J., Santo Domingo, J.W., Lamendella, R., Edge, T., and Hill, S. (2008). Phylogenetic diversity and molecular detection of bacteria in gull feces. Appl. Environ. Microbiol. *74*, 3969–3976.

Lu, J., Ryu, H., Hill, S., Schoen, M., Ashbolt, N., Edge, T., and Santo Domingo, J.W. (2011b). Distribution and potential significance of a gull faecal marker in urban coastal and riverine areas of southern Ontario, Canada. Water Res. *45*, 3960–3968.

Lu, J., Ryu, H., Santo Domingo, J.W., Griffith, J.F., and Ashbolt, N. (2011a). Molecular detection of *Campylobacter* spp. in California Gull (Larus californicus) excreta. Appl. Environ. Microbiol. *77*, 5034–5039.

McQuaig, S.M., Scott,T.M., Lukasik, J.O., Paul, J.H., and Harwood, V.J. (2009). Quantification of human polyomaviruses JC Virus and BK Virus by TaqMan quantitative PCR and comparison to other water quality indicators in water and fecal samples. Appl. Environ. Microbiol. 75, 3379–3388.

Marti, R., Dabert, P., Ziebal, C., and Pourcher, A.M. (2010). Evaluation of *Lactobacillus sobrius/L. amylovorus* as a new microbial marker of pig manure. Appl. Environ. Microbiol. *76*, 1456–1461.

Masago, Y., Shibata, T., and Rose, J.B. (2008). Bacteriophage P22 and *Staphylococcus aureus* attenuation on nonporous fomites as determined by plate assay and quantitative PCR. Appl. Environ. Microbiol. *74*, 5838–5840.

Matsuda, K., Tsuji, H., Asahara,T., Matsumoto, K., Takada, T., and Nomoto, K. (2009). Establishment of an analytical system for the human fecal microbiota, based on reverse transcription-quantitative PCR targeting of multicopy rRNA molecules. Appl. Environ. Microbiol. *75*, 1961–1969.

Medema, G.J., Teunis, P.F.M., Havelaar, A.H., and Haas, C.N. (1996). Assessment of the dose–response relationship of *Campylobacter jejuni*. Int. J. Food Micro. *30*, 101–111.

Mieszkin, S., Furet, J.P., Corthier, G., and Gourmelon, M. (2009). Estimation of pig fecal contamination in a river catchment by real-time PCR using two pig-specific Bacteroidales 16S rRNA genetic markers. Appl. Environ. Microbiol. *75*, 3045–3054.

Mota, A., Mena, K.D., Soto-Beltran, M., Tarwater, P.M., and Chaidez, C. (2009). Risk assessment of *Cryptosporidium* and *Giardia* in water irrigating fresh produce in Mexico. J. Food Prot. *72*, 2184–2188.

Mukamolova, G.V., Turapov, O.A., Kazarian, K., Telkov, M., Kaprelyants, A.S., Kell, D.B., and Young, M. (2002). The rpf gene of *Micrococcus luteus* encodes an essential secreted growth factor. Mol. Microbiol. *46*, 611–621.

Nakamura, S., Yang, C.-S., Sakon, N., Ueda, M., Tougan, T., Yamashita, A., Goto, N., Takahashi, K., Yasunaga, T., Ikuta, K., *et al.* (2009). Direct metagenomic detection of viral pathogens in nasal and fecal specimens using an unbiased high-throughput sequencing approach. PLoS ONE *4*, e4219.

Neretin, L.N., Schippers, A., Pernthaler, A., Hamann, K., Amann, R., and Jørgensen, B.B. (2003). Quantification of dissimilatory (bi)sulphite reductase gene expression in *Desulfobacterium autotrophicum* using real-time RT-PCR. Environ. Microbiol. *5*, 660–671.

Nícolaisen, M.H., Bælum, J., Jacobsen, C.S., and Sørensen, J. (2008). Transcription dynamics of the functional *tfdA* gene during MCPA herbicide degradation by *Cupriavidus necator* AEO106 (pRO101) in agricultural soil. Environ. Microbiol. *10*, 571–579.

NIH HMP Working Group, Peterson, J., Garges, S., Giovanni, M., McInnes, P., Wang, L., Schloss, J.A., Bonazzi, V., McEwen, J.E., Wetterstrand, K.A. (2009). The NIH Human Microbiome Project. Genome Res. *19*, 2317–2323.

Noble, R., Lee, I., and Schiff, K. (2004) Inactivation of indicator micro-organisms from various sources of faecal contamination in seawater and freshwater. J. Appl. Microbiol. *96*, 464–472.

Nocker, A., Sossa, P., Burr, M., and Camper, A.K. (2007a). Use of propidium monoazide for live–dead distinction in microbial ecology. Appl. Environ. Microb. *73*, 5111–5117.

Nocker, A., Sossa, K.E., and Camper, A.K. (2007b). Molecular monitoring of disinfection efficacy using propidium monoazide in combination with quantitative PCR. J. Microbiol. Methods *70*, 252–260.

Nylund, L., Heilig, H.G., Salminen, S., de Vos, W.M., and Satokari, R. (2010). Semi-automated extraction of microbial DNA from feces for qPCR and phylogenetic microarray analysis. J. Microbiol. Methods. *83*, 231–235.

Oliver, J.D. (2005). The viable but nonculturable state in bacteria. J. Microbiol. *43*, 93–100.

Oliver, J.D. (2010). Recent findings on the viable but nonculturable state in pathogenic bacteria. FEMS Microbiol. Rev. *34*, 415–425.

Oporto, B., Esteban, J.I., Aduriz, G., Juste, R.A., and Hurtado, A. (2008). *Escherichia coli* O157:H7 and non-O157 Shiga toxin-producing *E. coli* in healthy cattle, sheep and swine herds in Northern Spain. Zoonoses Public Health *55*, 73–81.

Ottoson, J., and Stenström, T.A. (2003). Faecal contamination of greywater and associated microbial risks. Water Res. *37*, 645–655.

Palka-Santini, M., Cleven, B.E., Eichinger, L., Krönke, M., and Krut, O. (2009). Large scale multiplex PCR improves pathogen detection by DNA microarrays BMC Microbiol. *9*, 1.

Pan, Y., Bodrossy, L., Frenzel, P., Hestnes, A., Krause, S., Lüke, C., Meima-Franke, M., Siljanen, H., Svenning, M., and Bodelier, P. (2010). Impacts of inter- and intralaboratory variations on the reproducibility of microbial community analyses. Appl. Environ. Microbiol. *76*, 7451–7458.

Pang, X.L., Lee, B., Boroumand, N., Leblanc, B., Preiksaitis, J.K., and Ip, C.C.Y. (2004). Increased detection of rotavirus using a real time reverse transcription-polymerase chain reaction (RT-PCR) assay in stool specimens from children with diarrhea. J. Med. Virol *72*, 496–501.

Panicker, G., and Bej, A.K. (2005). Real-time PCR detection of *Vibrio vulnificus* in oysters: comparison of oligonucleotide primers and probes targeting *vvhA*. Appl. Environ. Microbiol. *71*, 5702–5709.

Passerat, J., Got, P., Dukan, S., and Monfort, P. (2009). Respective roles of culturable and viable-but-nonculturable cells in the heterogeneity of *Salmonella enterica* serovar Typhimurium invasiveness. Appl. Environ. Microbiol. *75*, 5179–5185.

Perkins, S.D., Mayfield, J., Fraser, V., and Angenent, L.T. (2009). Potentially pathogenic bacteria in shower water and air of a stem cell transplant unit. Appl. Environ. Microbiol. *75*, 5363–5372.

Petric, I., Philippot, L., Abbate, C., Bispo, A., Chesnot, T., Hallin, S., Laval, K., Lebeau, T., Lemanceau, P., Leyval, C., *et al.* (2011). Inter-laboratory evaluation of the ISO standard 11063 "Soil quality – Method to directly extract DNA from soil samples". J. Microbiol. Methods *84*, 454–460.

Petterson, S.R., Dumoutier, N., Loret, J.F., and Ashbolt, N.J. (2009). Quantitative Bayesian predictions of source water concentration for QMRA from presence/absence data for *E. coli* O157:H7. Water Sci. Technol. *59*, 2245–2252.

Petterson, S., Signor, R. Ashbolt, N., and Roser, D. (2006). QMRA Methodology. In Quantitative Microbial Risk Assessment in the Water Safety Plan. In Final Report on the EU MicroRisk Project, Medema, G., Loret, J.-C., Stenström, T.A., and Ashbolt, N., eds (European Commission, Brussels), pp. 3–64. Available at: http://www.microrisk.com/uploads/microrisk_qmra_methodology.pdf

Petterson, S.R., Teunis, P.F.M., and Ashbolt, N. (2001). Modeling virus inactivation on salad crops using microbial count data. Risk Anal. *21*, 1097–1107.

Qu, A., Brulc, J.M., Wilson, M.K., Law, B.F., Theoret, J.R., Joens, L.A., Konkel, M.E., Angly, F., Dinsdale, E.A., Edwards, R.A., *et al.* (2008). Comparative metagenomics reveals host specific metavirulomes and horizontal gene transfer elements in the chicken cecum microbiome. PLoS One *3*, e2945.

Quintero-Betancourt, W., Peele, E.R., and Rose, J.B. (2002). *Cryptosporidium parvum* and Cyclospora cayetanensis: a review of laboratory methods for detection of these waterborne parasites. J. Microbiol. Methods *49*, 209–224.

Rajilić-Stojanović, M., Heilig, H.G., Molenaar, D., Kajander, K., Surakka, A., Smidt, H., de Vos, W.M. (2009). Development and application of the human intestinal tract chip, a phylogenetic microarray: analysis of universally conserved phylotypes in the abundant microbiota of young and elderly adults. Environ. Microbiol. *11*, 1736–1751.

Rantsiou, K., Lamberti, C., and Cocolin, L. (2010). Survey of *Campylobacter jejuni* in retail chicken meat products by application of a quantitative PCR protocol. Int. J. Food Microbiol. *141*, S75-S79.

Regli, S., Odom, R. Cromwell, J. Lustic, M., and Blank, V. (1999). Benefits and costs of the IESWTR. J. Am. Water Works Assoc. *91*, 148–158.

Regli, S., Rose, J.B. Haas, C.N., and Gerba, C.P. (1991). Modeling the risk from *Giardia* and viruses in drinking water. J. Am. Water Works Ass. *83*, 76–84.

Reischer, G.H., Kasper, D.C., Steinborn, R., Mach, R.L., and Farnleitner, A.H. (2006). Quantitative PCR method for sensitive detection of ruminant fecal pollution in freshwater and evaluation of this method in alpine karstic regions. Appl. Environ. Microbiol. 72, 5610–5614.

Rochelle, P.A., Ferguson, D.M., Handojo, T.J., Leon, R.D., Stewart, M.H., and Wolfe, R.L. (1997). An assay combining cell culture with reverse transcriptase PCR to detect and determine the infectivity of waterborne *Cryptosporidium parvum*. Appl. Environ. Microbiol 63, 2029–2037.

Rodríguez-Lázaro, D., Lewis, D.A., Ocampo-Sosa, A.A., Fogarty, U., Makrai, L., Navas, J., Scortti, M., Hernández, M., and Vázquez-Boland, J.A. (2006). Internally controlled real-time PCR method for quantitative species-specific detection and *vapA* genotyping of *Rhodococcus equi*. Appl. Environ. Microbiol. 72, 4256–4263.

Rose, J.B., Haas, C.N., and Regli, S. (1991). Risk assessment and control of waterborne giardiasis. Am. J. Public Health *81*, 709–713.

Rose, J.B., Nowlin, H., Farrah, S.R., Harwood, V., Levine, A., Lukasik, J., Mendendez, P., and Scott, T.M. (2004). Reduction of pathogens, indicator bacteria, and alternative indicators by wastewater treatment and reclamation processes. Water Environment Research Foundation Report 00-PUM-2T (IWA Publishing, London, UK).

Ryu, H., Mayer, B., and Abbaszadegan, M. (2010). Applicability of quantitative PCR for determination of removal efficacy of enteric viruses and *Cryptosporidium* by water treatment processes. J. Water Health *8*, 101–108.

Sakaguchi, S., Saito, M., Tsuji, H., Asahara, T., Takata, O., Fujimura, J., Nagata, S., Nomoto, K., and Shimizu, T. (2010). Bacterial rRNA-targeted reverse transcription-PCR used to identify pathogens responsible for fever with neutropenia. J. Clin. Microbiol. *48*, 1624–1628.

Santo Domingo, J.W., Harmon, S., and Bennett, J. (2000). Survival of *Salmonella* species in river water. Curr. Microbiol. *40*, 409–417.

Santo Domingo, J.W., and Ashbolt, N.J. (2008). Fecal pollution of water. In Encyclopedia of Earth, Cleveland, C.J., ed (Environmental Information Coalition, National Council for Science and the Environment Washington, DC). Available at: http://www.eoearth.org/article/Fecal_pollution_of_water

Schets, F.M., During, M., Italiaander, R., Heijnen, L., Rutjes, S.A., Zwaluw, W.K.v.d., and Husman, A.M.d.R. (2005). *Escherichia coli* O157:H7 in drinking water from private water supplies in the Netherlands. Water Res. *39*, 4485–4493.

Schill, W.B., and Mathes, M.V. (2008). Real-time PCR detection and quantification of nine potential sources of fecal contamination by analysis of mitochondrial cytochrome b targets. Environ. Sci. Technol. *42*, 5229–5234

Schoen, M.E., and Ashbolt, N.J. (2010). Assessing pathogen risk to swimmers at non-sewage impacted recreational beaches. Environ. Sci. Technol. *44*, 2286–2291.

Schwab, K.J., Leon, R.d., and Sobsey, M.D. (1995). Concentration and purification of beef extract mock eluted from water samples for the detection of enteroviruses, hepatits A virus, and Norwalk virus by reverse transcriptase-PCR. Appl. Environ. Microbiol. *61*, 531–537.

Seurinck, S., Defoirdt, T., Verstraete, W., and Siciliano, S.D. (2005). Detection and quantification of the human-specific HF183 Bacteroides 16S rRNA genetic marker with real-time PCR for assessment of human faecal pollution in freshwater. Environ. Microbiol. *7*, 249–259.

Shannon, K.E., Lee, D.-Y., Trevors, J.T., and Beaudette, L.A. (2007). Application of real-time quantitative PCR for the detection of selected bacterial pathogens during municipal wastewater treatment. Sci. Total Environ. *382*, 121–129.

Shibata, T., Solo-Gabriele, H.M., Sinigalliano, C.D., Gidley, M.L., Plano, L.R., Fleisher, J.M., Wang, J.D., Elmir, S.M., He, G., Wright, M.E., *et al.* (2010). Evaluation of conventional and alternative monitoring methods for a recreational marine beach with nonpoint source of fecal contamination. Environ. Sci. Technol. *44*, 8175–8181.

Shrestha, P., Noll, M., and Liesack, W. (2007). Phylogenetic identity, growth-response time and rRNA operon copy number of soil bacteria indicate different stages of community succession. Environ. Microbiol. *9*, 2464–2474.

Silkie, S.S., and Nelson, K.L. (2009). Concentrations of host-specific and generic fecal markers measured by quantitative PCR in raw sewage and fresh animal feces. Water Res. *43*, 4860–4871.

Simpson, J.M., Santo Domingo, J.W., and Reasoner, D.J. (2002). Microbial source tracking: state of the science. Environ. Sci. Technol. *36*, 5279–5288

Sinclair, R.G., Jones, E.L., and Gerba, C.P. (2009). Viruses in recreational water-borne disease outbreaks: a review. J. Appl. Microbiol. *107*, 1769–1780.

Sinton, L.W., Finlay, R.K., and Lynch, P.A. (1999). Sunlight inactivation of fecal bacteriophages and bacteria in sewage-polluted seawater. Appl. Environ. Microbiol. *65*, 3605–3613.

Smith, C.J., and Osborn, A.M. (2009). Advantages and limitations of quantitative PCR (Q-PCR)-based approaches in microbial ecology. FEMS Microbiol. Ecol. *67*, 6–20.

Smith, C.M., and Hill, V.R. (2009). Dead-end hollow-fiber ultrafiltration for recovery of diverse microbes from water. Appl. Environ. Microbiol. *75*, 5284–5289.

Smith, K.F., and Guégan, J.-F. (2010). Changing geographic distributions of human pathogens. Ann. Rev. Ecol. Evol. Syst. *41*, 231–250.

Smeets, P.W.H. (2010). Stochastic Modelling of Drinking Water Treatment in Quantitative Microbial Risk Assessment (IWA Publishing, London).

Soller, J.A., Olivieri, A.W., Crook, J., Cooper, R.C., Tchobanoglous, G., Parkin, R.T., Spear, R.C., and Eisenberg, J.N.S. (2003). Risk-based approach to evaluate the public health benefit of additional wastewater treatment. Environ. Sci. Technol. *37*, 1882–1891.

Soller, J.A., Seto, E.Y., and Oliveri, A.W. (2007). Application of microbial risk assessment techniques to estimate risk due to exposure to reclaimed waters. WaterReuse Foundation (Alexandria, VA).

Soller, J.A., Bartrand, T., Ashbolt, N.J., Ravenscroft, J., and Wade, T.J. (2010a). Estimating the primary etiologic agents in recreational freshwaters impacted by human sources of faecal contamination. Wat Res. *44*, 4736–4747.

Soller, J.A., Schoen, M.E., Bartrand, T., Ravenscroft, J.E., and Ashbolt, N.J. (2010b). Estimated human health risks from exposure to recreational waters impacted by human and non-human sources of faecal contamination. Water Res. *44*, 4674–4691.

Staley, J.T., and Konopka, A. (1985). Measurements of *in situ* activities of nonphotosynthetic microorganisms in aquatic and terrestrial habitats. Annu. Rev. Microbiol. *39*, 321–346.

Stapleton, C.M., Kay, D., Wyer, M.D., Davies, C., Watkins, J., Kay, C., McDonald, A.T., Porter, J., and Gawler,

A. (2009). Evaluating the operational utility of a Bacteroidales quantitative PCR-based MST approach in determining the source of faecal indicator organisms at a UK bathing water. Water Res. *43*, 4888–4899.

Steinert, M., Emody, L. Amann, R., and Hacker, J. (1997). Resuscitation of viable but nonculturable Legionella pneumophila Philadelphia JR32 by Acanthamoeba castellanii. Appl. Environ. Microbiol. *63*, 2047–2053.

Stoeckel, D.M., Stelzer, E.A., and Dick, L.K. (2009). Evaluation of two spike-and-recovery controls for assessment of extraction efficiency in microbial source tracking studies. Water Res. *43*, 4820–4827.

Storey, M.V., and Ashbolt, N.J. (2003). Enteric virions and microbial biofilms – a secondary source of public health concern. Water Sci. Technol. *48*, 97–104.

Stroup, S.E., Roy, S., McHele, J., Maro, V., Ntabaguzi, S., Siddique, A., Kang, G., Guerrant, R.L., Kirkpatrick, B.D., Fayer, R., *et al.* (2006). Real-time PCR detection and speciation of *Cryptosporidium* infection using Scorpion probes. J. Med. Microbiol. *55*, 1217–1222.

Summerscales, I.M., and McBean, E.A. (2010). Evaluation of Risk Assessment Tools to Predict Canadian Waterborne Disease Outbreaks. Water Qual. Res. J. Canada *45*, 1–11.

Tanaro, J.D., Leotta, G.A., Lound, L.H., Galli, L., Piaggio, M.C., Carbonari, C.C., Araujo, S., and Rivas, M. (2010). *Escherichia coli* O157 in bovine feces and surface water streams in a beef cattle farm of Argentina. Foodborne Pathog. Dis. *7*, 475–477.

Taniuchi, M., Verweij, J.J., Noor, Z., Sobuz, S.U., Lieshout, L., Petri, W.A., Jr., Haque, R., and Houpt, E.R. (2011). High throughput multiplex PCR and probe-based detection with Luminex beads for seven intestinal parasites. Am. J. Trop. Med. Hyg. *84*, 332–337.

Teunis, P.F.M., van der Hiejden, O.G., van der Giessen, J.W.B., and Havelaar, A.H. (1996). The dose–response relation in human volunteers for gastro-intestinal pathogens. Rijksinstituut voor Volksgezondheid en Milieu (RIVM, Bilthoven, the Netherlands).

Teunis, P., Van den Brandhof, W., Nauta, M., Wagenaar, J., Van den Kerkhof, H., and Van Pelt, W. (2005). A reconsideration of the *Campylobacter* dose–response relation. Epidemiol. Infect. *133*, 583–592.

Teunis, P.F., Moe, C.L., Liu, P., S, E.M., Lindesmith, L., Baric, R.S., Le Pendu, J., and Calderon, R.L (2008a). Norwalk virus: How infectious is it? J. Med. Virol. *80*, 1468–1476.

Teunis, P.F., Ogden, I.D., and Strachan, N.J. (2008b). Hierarchical dose response of *E. coli* O157:H7 from human outbreaks incorporating heterogeneity in exposure. Epidemiol. Infect. *136*, 767–770.

Thomas, J.M., and Ashbolt, N.J. (2011). Do free-living amoebae in treated drinking water systems present an emerging health risk? Environ. Sci. Technol. *45*, 860–869.

USEPA (U.S. Environmental Protection Agency). (2005). Economic analysis for the final long term 2 enhanced surface water treatment rule. EPA 815-R-06-001 (US EPA, Washington DC).

USEPA (U.S. Environmental Protection Agency). (2007). Thesaurus of terms used in microbiological risk assessment. Office of Water (US EPA, Washington, DC).

Varma, M., Field, R., Stinson, M., Rukovets, B.,Wymer L., and Haugland, R. (2009). Quantitative real-time PCR analysis of total and propidium monoazide-resistant fecal indicator bacteria in wastewater. Water Res. *43*, 4790–4801.

Villanueva, L., Haveman, S.A., Summers, Z.M., and Lovley, D.R. (2008). Quantification of Desulfovibrio vulgaris dissimilatory sulfite reductase gene expression during electron donor- and electron acceptor-limited growth. Appl. Environ. Microbiol. *74*, 5850–5853.

Villena, C., El-Senousy, W.M., Abad, F.X., Pinto, R.M., and Bosch, A. (2003). Group A rotavirus in sewage samples from Barcelona and Cairo: emergence of unusual genotypes. Appl. Environ. Microbiol. *69*, 3919–3923.

Walters, S.P., and Field, K.G. (2009). Survival and persistence of human and ruminant-specific faecal Bacteroidales in freshwater microcosms. Environ. Microbiol. *11*, 1410–1421.

Walters, S.P., Yamahara, K.M., and Boehm, A.B. (2009). Persistence of nucleic acid markers of health-relevant organisms in seawater microcosms: Implications for their use in assessing risk in recreational waters. Water Res. *43*, 4929–4939.

Wang, L., Li, Y., and Mustapha, A. (2009). Detection of viable *Escherichia coli* O157:H7 by ethidium monoazide real-time PCR. J. Appl. Microbiol. *107*, 1719–1728.

Wang, L., and Mustapha, A. (2010). EMA-real-time PCR as a reliable method for detection of viable *Salmonella* in chicken and eggs. J. Food Sci. *75*, M134–139.

Watzinger, F., Suda, M., Preuner, S., Baumgartinger, R., Ebner, K., Baskova, L., Niesters, H.G., Lawitschka, A., and Lion, T. (2004). Real-time quantitative PCR assays for detection and monitoring of pathogenic human viruses in immunosuppressed pediatric patients. J. Clin. Microbiol. *42*, 5189–5198.

Weidhaas, J.L., Macbeth, T.W., Olsen, R.L., and Harwood, V.J. (2011). Correlation of quantitative PCR for a poultry-specific brevibacterium marker gene with bacterial and chemical indicators of water pollution in a watershed impacted by land application of poultry litter. Appl. Environ. Microbiol. *77*, 2094–2102.

Wéry, N., Lhoutellier, C., Ducray, F., Delgenès, J.P., and Godon, J.J. (2008). Behaviour of pathogenic and indicator bacteria during urban wastewater treatment and sludge composting, as revealed by quantitative PCR. Water Res. *42*, 53–62.

WHO. (2009). Water Safety Plan Manual: Step-by-step Risk Management for Drinking-water Suppliers (World Health Organization, Geneva).

Wolffs, P., Norling, B., and Radstrom, P. (2005). Risk assessment of false-positive quantitative real-time PCR results in food, due to detection of DNA originating from dead cells. J. Microbiol. Methods *60*, 315–323.

Wong, M., Kumar, L., Jenkins, T.M., Xagoraraki, I., Phanikumar, M.S., and Rose, J.B. (2009). Evaluation of public health risks at recreational beaches in Lake Michigan via detection of enteric viruses and a human-specific bacteriological marker. Water Res. *43*, 1137–1149.

Wright, A.C., Garrido, V, Debuex, G, Farrell-Evans, M, Mudbidri, A.A., and Otwell, W.S. (2007). Evaluation of postharvest-processed oysters by using PCR-based most-probable-number enumeration of *Vibrio vulnificus* bacteria. Appl. Environ. Microbiol. *73*, 7477–7481.

Wullings, B.A., Bakker, G., and van der Kooij, D. (2011). Concentration and diversity of uncultured *Legionella* spp. in two unchlorinated drinking water supplies with different concentrations of natural organic matter. Appl. Environ. Microbiol. *77*, 634–641.

Yamara-Iquise, H., Zheng, G., Jones, J.E., and Carson, C.A. (2008). Use of a *Bacteroides* thetaiotaomicron-specific alpha-1–6, mannanase quantitative PCR to detect human faecal pollution in water. J. Appl. Microbiol. *105*, 1686–1693.

Zhou, J., Bruns, M.A., and Tiedje, J.M. (1996). DNA recovery from soils of diverse composition. Appl. Environ. Microbiol. *62*, 316–322.

Quantitative Polymerase Chain Reaction in Food Microbiology

8

Luca Cocolin and Kalliopi Rantsiou

Abstract

Since its first application in food microbiology in the late 1990s, quantitative PCR (qPCR) has attracted the interest of researchers, working mainly in the field of food safety, but lately also of microbiologists studying spoilage and fermentation processes. In addition to the advantages that conventional PCR offers in microbiological testing, i.e. specificity, reduced time of analysis and detection of viable but not culturable cells, qPCR allows quantification of target populations. This aspect is particularly relevant for foodborne pathogens, for which specific microbiological criteria exist, but also for spoilage and technological important microorganisms, in order to follow their population kinetics in foods. Although advancements in food microbiology have been made from its application, qPCR has not yet been utilized to its full potential: the quantification step is only rarely carried out and qPCR is often used as an alternative of conventional PCR.

In this chapter we will critically describe the application of qPCR in food microbiology based on the available literature, taking into account the specific problems and suggesting some possible solutions.

Introduction

Quantitative PCR (qPCR), or alternatively real-time PCR, has opened a wide range of possibilities in the field of food microbiology. Traditionally, microbes in foods were quantified by colony plate count, a method that not only requires several days from analysis to result, but also entails the growth of microorganisms from single cells to colonies, thereby enumeration is totally dependent on the capability that microorganisms have to grow on synthetic media. With the increasing application of molecular biology tools, it has been shown that classical techniques do not properly address the quantification step in the microbiological examination of food products. This refers, for example, to cells that are in a viable but not culturable state, possessing metabolic activity (thereby able to cause diseases in the case of pathogens or to spoil foods for spoilers) (Rowan, 2004), but that are not able to form a colony on the plate. Moreover, populations that are numerically limited or microorganisms that are stressed or in a sublethal state cannot be recovered, thereby they are eliminated from consideration. One last aspect to consider, related to traditional enumeration in foods, is the limited selectivity of synthetic microbiological media. Basically, the quantification of a specific microbial population classically relies on the capability of such population to grow in defined conditions, in terms of composition of the growth media, temperature, pH and oxygen. For instance, for the enumeration of lactic acid bacteria (LAB), the use of the MRS (for rods) and M17 (for cocci) media, is scientifically accepted. Studies that have applied molecular biology approaches to identify populations grown on such media have underlined the lack of consistence in terms of specificity: lactobacilli could be easily recovered from M17 and lactococci from MRS (Dolci *et al.*, 2008; Rantsiou *et al.*, 2008). This example is highlighting an important pitfall in the classical enumeration of microorganisms in food products, which can potentially be circumvented, considering the specificity of PCR. By designing specific primers and probes, it becomes possible to specifically enumerate a microbial population without the need for its isolation, thereby avoiding the biases related to the culturing.

Pioneering studies in the field of food microbiology to quantify microorganisms by molecular methods date back to the 90s. The approaches used were mainly based on the quantification of PCR products by enzyme linked immunosorbent assay (Gutierrez *et al.*, 1998; Manzano *et al.*, 1998; Wang and Hong, 1999) and electrochemiluminescence (Venkitanarayanan *et al.*, 1997). Only at the beginning of the 2000s the first papers using real-time monitoring of PCR amplification in food microbiology, by the means of fluorescence probes, became available (McKillip and Drake, 2000). Since then, an increasing interest in the application of this method has been manifested by the scientists, witnessed by the numbers of papers published since 1995 and their citations recorded in the last decade. Remarkably, on a total of 567 papers available at the time of writing this chapter, about half of them were published in the last 3 years (Fig. 8.1). Analysing the papers available on the application of qPCR in food microbiology, it becomes obvious that the majority of the efforts were dedicated to study the prevalence of foodborne pathogens, while only a few examples exist on quantification of spoilage and technologically important microorganisms. A similar trend was observed when conventional PCR was introduced from the clinical microbiology to the food microbiology field. The first applications referred to the detection of pathogenic microorganisms in food products, using PCR as a rapid and selective method for their detection. Only later, the potential of PCR was also used to study the other categories of microorganisms relevant to food microbiology. Foodborne pathogens represent an important concern for food business operators (FBOs), who have to guarantee that food products that enter in the market are safe for the consumers. The food safety aspect, in the past considered as part of the quality, is now a pre-requisite that foods must possess in order to be commercialized. Moreover, since 1 January 2006, a new food legislation entered into force in Europe (EU Regulation 2073/2005 amended by 1441/2007), imposing for the first time specific microbiological criteria that have to be respected for foods intended for human consumption. For example, *Listeria monocytogenes* has to be absent in foods for consumers at risk (i.e. elderly, newborns and pregnant women), thereby confirming the zero tolerance criterion used in the USA (Shank *et al.*, 1996), while a limit of 100 colony-forming units per g or ml at the end of shelf life for foods that

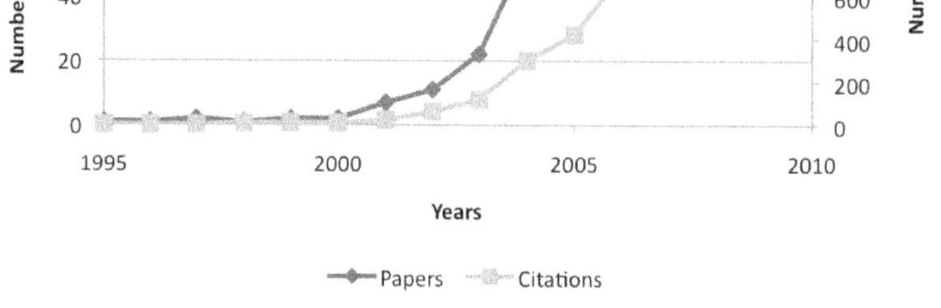

Figure 8.1 Number of qPCR papers in food microbiology published in the last 15 years and their correspondent citations (source: Scopus, www.scopus.com, accessed December 2010).

are supporting the growth of *L. monocytogenes*, or at the moment the foods exit the production plant if they do not support its growth, is admitted. As a consequence, the availability of methods that could respond appropriately to the needs of FBOs in terms of pathogenic microorganisms detection and quantification in foods is deemed necessary. To this regard, qPCR represents a tool that can potentially be used in order to quantify rapidly foodborne pathogens, in order to assess compliance of the food products to the current legislation. In this chapter, the application of qPCR in food microbiology will be described, critically evaluating its potentials and problems.

Application of qPCR in food microbiology: pros and cons

Food microbiological testing is experiencing an exciting momentum due to the introduction of new analytical approaches that do not necessarily involve the growth of a microorganism for its detection. Particularly relevant in the analysis of food-borne pathogens are the methods based on molecular biology. More specifically, at the end of the 1980s, Kary Mullis invented the polymerase chain reaction (PCR) (Mullis *et al.*, 1986), a method that revolutionized the traditional microbiological analysis by allowing a culture-independent detection directly from a sample such as food, without the necessity for isolation and identification of the microorganism of interest. Technological advancements then allowed conventional PCR, which is a qualitative approach, to become quantitative. In the late 1990s, instruments able to detect PCR products while they are synthesized became available, making quantitative analysis of PCR amplicons possible. Quantitative PCR or qPCR revolutionized the molecular approaches used in food microbiology. Not only it was now possible to quantify one specific microorganism in food, but it was also possible to study its behaviour as a consequence of the influence of the environment (i.e. food composition, temperature, pH, oxygen, etc.). Moreover, since the process is monitored in real time, there is no need for post-amplification treatment of the samples, such as gel electrophoresis, saving time and efforts. qPCR, undoubtedly, possesses some features that make it extremely useful in the field of food microbiology:

- The target microorganism is detected and quantified independently of its physiological state, thereby stressed or injured cells are picked up by the method, avoiding the false-negative results described above for the traditional microbiological methods;
- The time required for the analysis is much shorter if compared with the cultural methods. Responses on the presence of specific microorganisms can be obtained in a minimum of 3 hours to a maximum of 24 hours, if an enrichment step is carried out, a step that becomes necessary for the detection of small numbers of pathogenic microorganisms;
- The method is highly specific for the target microorganism and in this way false positive results, due to misidentifications, can be avoided.

On the other hand, qPCR methods have some disadvantages that are still playing against its wide application in the food sector:

- Since the target for the amplification is mainly represented by DNA, a molecule characterized by high stability also after cell death, a positive result by PCR does not necessarily indicate the presence of living microorganisms. This aspect is extremely important for processed foods, i.e. thermally processed, where the production includes a step in which food-borne pathogens are killed. This pitfall of the PCR has been addressed in different ways by researchers (i.e. RNA analysis to quantify living populations (Rantsiou *et al.*, 2008) and use of ethidium monoazide (EMA) for the selective quantification of living populations (Josefsen *et al.*, 2010). Furthermore, it is generally accepted that if positive results are obtained after an enrichment step, this implies the presence of living populations that were able to multiply and their DNA was subsequently amplified by PCR (Rossen *et al.*, 1992);
- There is a need for a significant investment in terms of equipment and trained personnel, even if, in the last years, the market has seen a

significant decrease in the prices of the thermal cyclers, dedicated to the amplification of the DNA.

The correct implementation of qPCR in food microbiology must take into account key steps that have to be carefully evaluated and examined. These are represented by: the sample preparation, through which nucleic acids are extracted from the food matrix; the quantification procedure; and an eventual necessity to enrich the food samples, especially in cases where the numbers of target microorganisms is below the quantification limit of the method.

Sample preparation

It is well accepted that foods contain a number of inhibitors that may impair PCR performance. In the 1990s, studies have highlighted that fats, proteins, culture media components, specific classes of compounds related to the origin of the food (polyphenols in vegetables or blood in meat and meat products) can dramatically influence the efficiency of the amplification reaction, thereby increasing the risk of false-negative results (Rossen *et al.*, 1992). Moreover, the inhibitory activity can also be exerted by nucleic acids extraction reagents (Rodriguez-Lazaro *et al.*, 2007). The problem related to the inhibitors in foods becomes even more relevant if we take into consideration qPCR. The presence of compounds able to interfere with DNA polymerase activity can result in lack of amplification or underestimation of the target microorganism. Thus, there is a need for a dedicated (often labour-intensive) sample preparation to recover cells from foods to acquire good quality DNA for amplification. Furthermore, additional process controls must be taken into consideration to evaluate the recovery of target bacteria during the extraction procedures.

In recent years, sample preparation rather than the PCR technique has been shown to be the bottleneck for quantification of microorganisms using qPCR. It has been shown that the use of complex and laborious methods do not always yield a better recovery (Baert *et al.*, 2008). Separation and concentration methods are possible strategies in order to remove inhibitory components and to concentrate the target microorganisms, and they have been reviewed in more details elsewhere (see Brehm-Stecher *et al.*, 2009). A potential risk of using concentration methods, to increase the sensitivity of qPCR, is the concomitant concentration of inhibitory factors from the food matrix. This can jeopardize the efficient functioning of the qPCR reaction, which is prone to inhibition, and in turn might lead to false negative results. Protocols for concentrating bacteria or DNA from large volumes [e.g. by the use of bacteriophage-based capture (Bennett *et al.*, 1997), the use of nanoparticles (Yang *et al.*, 2008), or the use of devices enabling upscaling of present available capture (Warren *et al.*, 2007)] have been described. Fukushima *et al.* (2007) reported a sample preparation method combining filtration, low- and high-speed centrifugation and buoyant density centrifugation to concentrate food pathogens up to 250-fold without enrichment and detected as low as 10 to 10^3 cfu/g. Negative qPCR results were observed that may have occurred due to inhibition by food constituents that were difficult to eliminate when applying the described protocol. Clearly, choosing an appropriate sample preparation approach is a critical issue prior to qPCR detection. Sample preparation methods should be selected with care depending upon the food matrices under analysis and the sensitivity of the test required. Choosing the best protocol is always a compromise between cost, rapidity and universal application with regard to the food matrix.

Quantification of the target microorganism in food

As mentioned above, development of qPCR has enabled a quantitative approach. However, in order to keep the relation between the number of bacteria present in food and the enumeration by PCR, a prior enrichment step is prohibited. Nevertheless, a short pre-enrichment step, followed by qPCR, has been suggested in order to obtain quantitative data of *Salmonella* in meat (Malorny *et al.*, 2009). A prerequisite for the application of qPCR to quantify foodborne pathogens in foods is the generation of calibration curves. Calibration curves associate the parameter of colony forming units (cfu)/g or ml, commonly used in traditional microbiological analysis, with the Ct or Cq value, the threshold cycle in qPCR. In this way, results of qPCR can be translated to a value that relates with microbiological criteria.

For application in food diagnostics, it should be stressed out that calibration curves should be created in food matrices, because this way all steps involved in sample preparation and amplification are taken into account. One should avoid creating standard curves for the quantification of a microorganism in a specific food, by diluting a pure culture, extracting the DNA from each dilution and then performing the qPCR. In this way, the influence that the food matrices have when extracting the DNA, but also during the amplification step, would not be considered, thereby resulting in an underestimation of the microbial load. Moreover, in order to have a direct correlation of quantification with internationally accepted ways to report microbial counts, standard curves cannot be constructed from DNA dilutions, as is the usual approach in molecular biology applications, and the results should not be reported as genome equivalents. This approach does not take into account the extraction efficiency from food. In addition, extraction efficiency may differ from one food product to another. Thus, it is recommended to include with each extraction step a process control (cells inoculated in the food item/suspension taken up in parallel during the extraction of the food under investigation). Process controls can provide information on the efficiency of the extraction and of the PCR procedure.

As already mentioned above, the majority of papers applying qPCR to food microbiology aimed at quantifying foodborne pathogens. Analysing the available data, a quantification limit of approximately 10^3–10^4 cells per gram is often reported, too high for those pathogens for which the maximum tolerable level is set at 100/g. The explanation for this relatively low performance of the qPCR is not related to the PCR technique itself (which is intrinsically sensitive), but rather with the problems of sample preparation and DNA extraction from food samples. As mentioned above, this relates to PCR inhibitors present in food but also to the use of small input volumes for PCR versus large test portions to be analysed. The quantification limits of qPCR as reported above are thus too high for most practical application, since for most samples taken along the food chain the contamination is low (usually less than 100/g). For this reason, in the field of food safety, the main application of qPCR is for qualitative, presence/absence testing, performed after an enrichment step. Only in few cases (if compared with the total number of papers published in the last 5 years) authors put efforts in developing qPCR protocols that could be used for quantification purposes without enrichment.

Even in the era of qPCR using a closed tube format, the threat of contamination (and thus false positive results) is still to be taken into account. Optimization of qPCR has made it an intrinsic extremely sensitive technique and DNA is a very persistent molecule. Aerosols created during operation of PCR may transfer small DNA fragments between wells resulting in false positive results. Therefore, qPCR positive results with Ct values >40 (overall indication of very low levels of DNA near the theoretical detection limit) should be interpreted with care (Stals *et al.*, 2009). In case of positive PCR results with high Ct values, an attempt should be made to confirm the positive result by means of a culture-based method. Such confirmation is important since it also allows isolation of the bacterial cells for further characterization and/or epidemiological studies.

The enrichment step

The terminology 'rapid' and 'sensitive' in the description of the qPCR technique as a quantification method is relative. PCR, as such, is taking only approx. 90–120 minutes. For detection of low numbers, i.e. 1 to 10 cells per 25 g (as in the case of pathogenic microorganisms in foods), a prior cultural enrichment procedure is normally applied. Overall, DNA extraction starts from approximately one ml enriched food suspension and concentrates DNA into 100 µl of which a maximum of 10 µl can be used in the PCR reaction. At its best, detection of 1–10 DNA copies per PCR tube corresponds to approximately 1–10 cells in the 10 µl DNA extract, thus 10–100 cells in the 1 ml primary suspension. For reliable qPCR detection, low numbers of cells (for example 1 in 10 g of food sample) need to grow during prior enrichment to levels of at least 10–100 cells per ml of the primary suspension. This corresponds to approximately 10 multiplications. The minimum time needed for the prior enrichment to enable a positive PCR detection might be calculated from the mean generation time at optimum temperature in the enrichment medium, for example 20–30 minutes, and shows, for this example, that 4–5 h of prior enrichment is needed to obtain a positive test result, if starting from low numbers. However, it should also be mentioned that microorganisms in foods are often stressed and/or sub-lethally injured, due to processing and preservation techniques (salting, acidification, freezing, packaging, etc.), requiring an increased lag phase (2–3 h) prior to the exponential growth phase, prolonging the minimum time to detection (Jasson *et al.*, 2007).

Therefore, a 6–8 h enrichment is considered as a prerequisite for reliable detection of low numbers of microorganisms in a variety of foods. The enrichment time may be further prolonged (commonly used overnight enrichment) for slower growing organisms or if selective (less growth supporting) broths are used for enrichment (Jasson *et al.*, 2009). Shortened enrichment procedures, overlooking the need to repair sublethally injured cells, may result in a true risk of not reaching the detection limit of qPCR and can lead to false-negative results.

Analysis of the available literature on qPCR in food microbiology

As stated above, the number of papers published on qPCR in the last 5 years has tremendously increased. The vast majority of them focus on the development of protocols for quantification of foodborne pathogens in food, thereby the area of food safety is at the moment the richest in qPCR applications. Several bacterial pathogens, as well as viruses have been investigated in many food matrices, providing an important mass of data that can be usefully used for risk assessment analysis. Although the number of papers dealing with foodborne pathogens is important, studies on spoilage and technologically important microorganisms are generally lacking. There are only few examples of qPCR methodologies applied to investigate the presence and dynamics of spoilage organisms in foods, and even less on microorganisms responsible for food transformations, such as fermentations. It should be pointed out that the expectations that scientists have from qPCR are totally different based on the type of microorganism that is targeted. While for foodborne pathogens a low limit of quantification is required, for the other two categories (spoilage and technological microorganisms), the necessity to count low quantities is less important. Spoilage processes, as well as fermentations, take place when a critical number of microorganisms is reached. This is generally above 10^6 cfu/ml or g, thereby there are no concerns associated with the lack of sensitivity of qPCR due to food matrix inhibition, although this parameter should not be neglected in order to get a correct quantification of the target organism.

Table 8.1 is reporting an extensive, but not complete list of studies using qPCR as a quantification tool in food microbiology. The target organism, the food matrix and the limit of quantification are also included.

Table 8.1 Examples of studies in which quantification of the target microorganism(s) was done by the mean of qPCR without enrichment

Microorganism	Food matrix	Limit of quantification	Reference
Foodborne pathogens			
Bacillus cereus	Liquid egg, infant formula	60 cfu/ml	Martinez-Blach *et al.* (2009)
Campylobacter jejuni	Chicken meat	10 cfu/g	Rantsiou *et al.* (2010)
Campylobacter spp.	Chicken carcass rinse	100 cfu/ml	Josefsen *et al.* (2010)
Campylobacter spp.	Chicken carcass rinse	2×10^3 cfu/g	Botteldoorn *et al.* (2008)

Table 8.1 Continued

Microorganism	Food matrix	Limit of quantification	Reference
Clostridium botulinum	Rice	100 cells/g	Takahashi *et al*. (2010)
Escherichia coli O157:H7	Milk Dairy foods	10^3 cfu/g	Singh *et al*. (2009)
Escherichia coli O157:H7	Vegetable leaves	2.3×10^3 cfu/g	Ibekwe *et al*. (2006)
Escherichia coli O157:H7	Ground beef	10^4 cfu/g	Wang *et al*. (2009)
Hepatitis A	Raspberries	50 pfu/25 g fresh raspberry 100 pfu/25 g frozen raspberries	Blaise-Boisseau *et al*. (2010)
Klebsiella pneumoniae	Infant formula	2×10^3 cfu/25 g	Sun *et al*. (2010)
Listeria monocytogenes	Meat and meat products Milk and milk products Ready to eat salads	10^3 to 10^4 cfu/ml or g	Rantsiou *et al*. (2008)
Listeria monocytogenes	Salmon	10^3 cfu/g	Rodriguez-Lazaro *et al*. (2005)
Listeria monocytogenes	Meat	10^3 cfu/g	Rodriguez-Lazaro *et al*. (2004)
Listeria monocytogenes	Salads	10 cfu/g	Berrada *et al*. (2006)
Mycobacterium avium subsp. *paratuberculosis*	Milk	5–6 cells/ml	Slana *et al*. (2008)
Salmonella enterica *Listeria monocytogenes*	Yogurt	10 cells/10 g	D'Urso *et al*. (2009)
Salmonella enterica	Pork meat (pig carcass swab)	4.4×10^2 cfu/ml	Lofstrom *et al*. (2011)
Salmonella enterica	Seafood	20 cfu/g	Kumar *et al*. (2010)
Salmonella typhimurium	Meat products	60 cfu/g	Chen *et al*. (2009)
Staphylococcus aureus	Milk Meat	10^3 cfu/ml or g	Chiang *et al*. (2007)
Vibrio alginolyticus	Seafood	100 cells/1 ml tissue homogenate	Zhou *et al*. (2007)
Vibrio vulnificus	Shellfish Drinkwater	10^2 to 10^3 cells/g or ml	Gordon *et al*. (2008)
Yersinia pestis	Milk Ground beef	10 to 10^3 cfu/ml 10^2 to 10^5 cfu/g	Amoako *et al*. (2010)
Spoilage organisms			
Brettanomyces bruxellensis	Wine	100 cfu/ml	Campolongo *et al*. (2010)
Brettanomyces bruxellensis *Pediococcus damnosus*	Wine	10^4 cfu/ml 40 cfu/ml	Delaherche *et al*. (2004)
Brettanomyces bruxellensis	Wine	1 cell/ml	Phister and Mills (2003)
Brochothrix thermosphacta	Fresh and spoiled raw meat	2.2×10^2 cfu/g	Pennacchia *et al*. (2009)
Clostridium tyrobutiricum	Milk	25 spores/25 ml	López-Enríquez *et al*. (2007)
Hanseniaspora uvarum	Wine	100 cfu/ml	Phister *et al*. (2007)
Technologically important organisms			
Lactic acid bacteria	Wine	10^3 cells/ml	Neeley *et al*. (2005)
Lactobacillus sakei	Meat Fermented sausages	3×10^3 cfu/g	Martin *et al*. (2006)
Saccharomyces cerevisiae	Wine	78 cfu/ml	Salinas *et al*. (2009)

Apart from being used for the quantification of microorganisms, qPCR has recently shown good applicability in the environmental control of foodborne pathogens. Cross contaminations during or post production represent one of the most important factors of contamination by foodborne pathogens. Especially for specific microorganisms, such as *L. monocytogenes, Salmonella, Staphylococcus aureus* and *Escherichia coli*, contaminations occurring after the production step, often become the most important aspect responsible for food contamination. An important role is played by the operator; however the environment, and more specifically the equipment, represents a key factor to control. In this context, qPCR is a relevant tool to exploit for its capability to quantify, in a culture-independent way, foodborne pathogens. In 2010, two papers, one dealing with *L. monocytogenes* in dairy processing plants (Alessandria *et al.*, 2010) and one with Shiga-like producing *Escherichia coli* (STEC) in the lamb production food chain (Oses *et al.*, 2010), underlined how qPCR can be used to quantify loads on pathogens in the plants. The results of these studies highlighted that it is vital for food producing companies to have control of the contamination routes of foodborne pathogens in their plants. This can be achieved with the implementation of correct sampling schemes and adequate cleaning and disinfection procedures. In addition, the choice of the analytical procedure should also be considered as a relevant decision. As reported in both studies, traditional microbiological procedures only partially monitored the presence of *L. monocytogenes* and STEC in the processing plants. When environmental conditions influence the fitness and behaviour of the pathogen, such as in brines (with 15–20% salt) or in equipment swabs (where cells have to respond to several stresses such as disinfectants, starvation and dried conditions), molecular methods should be used to properly quantify pathogens. The application of molecular methods can result in a better comprehension of the spread of a specific pathogen in a processing plant, thereby allowing the implementation of corrective actions to eliminate or decrease the risk associated with its presence in the final product.

Conclusions and future perspectives

Recently, we have been experiencing important improvements in food microbiology analyses, thanks to the availability of novel techniques that provide significant information regarding the quality and safety of food products. The invention of conventional PCR, initially characterized by an extreme rapidity when compared with traditional microbiological methods, but not adequate for the enumeration of the cells of the microorganisms, became, with technical improvement, a quantitative technique (qPCR) allowing to monitor the amplification of the PCR product in real time.

PCR-based assays offer versatility in reducing the incidence of false negative in the quality assurance screening measures, sensitivity in processing heterogeneous food matrices, specificity in differentiating among closely related bacterial strains, and speed, due to the use of real time fluorescent chemistries currently on the market. Although these assays will probably not completely replace conventional culture, biochemical, or antibody-based testing methods, they offer significant potential for high throughput and reliability, advantages that offset the cost of initial equipment needs and ongoing maintenance or reagent expenses. Molecular diagnostics also offer a reliable initial screening tool so that follow-up, using conventional methods, is often only necessary if a sample tests positive for the pathogen of interest.

The new challenge for food microbiologists that emerges in the meta-genomic era is the understanding of the behaviour of microorganisms *in vitro* and *in situ*. How are molecular mechanisms of virulence influenced by food matrices? Previous knowledge, coupled with the data from the recently completed and currently undergoing sequencing projects, has identified several potential virulence factors. Also, pathogenicity characters that are directly related with disease symptoms (e.g. the Shiga-like toxin produced by pathogenic *E. coli*) as well as attributes of foodborne pathogens that allow them to survive and respond to environmental stresses encountered during food processing, distribution and most importantly, after ingestion and within the human body, should be considered. These factors interact to ultimately determine if a bacterium will eventually be able to cause a disease. Based on DNA sequences and genome mining, several genes, encoding potential virulence and pathogenicity factors, can be identified. The goal now is to determine if these genes are expressed in any of the stages 'from farm to fork', and, if yes, whether the synthesized proteins are acting in the predicted way and contribute to diseases.

Global approaches are now available, that allow scientists to look at the total RNA synthesized in a cell (transcriptomics) (Chatterjee *et al.*, 2006) and the total proteins produced by a cell (proteomics), at any given moment or condition. These are powerful techniques that allow pinpointing target genes and gene products that may require further investigation. In this context, reverse transcriptase (RT)-qPCR offers the possibility to quantify the expression of target genes in an accurate manner (Sharkey *et al.*, 2004; Kubista *et al.*, 2006). The expression of target genes is measured by RT-qPCR, under control and also under conditions that resemble the food environment or the human intestinal environment. The up-regulation of a target gene under conditions that resemble the food environment can be a strong indication that this gene is important for the adaptation of the microorganism in this particular environment. On the contrary, a gene that is down-regulated is considered less essential under these conditions. This strategy has been used to study virulence and stress responses of foodborne pathogens in synthetic media (Critzer *et al.*, 2008; Olesen *et al.*, 2009), as well as in real food matrices. Although expression studies in foods are scarce, the first evidence gathered strongly suggests the need to perform gene expression experiments is real food samples instead of artificial systems (Duquenne *et al.*, 2010; Olesen *et al.*, 2010; Rieu *et al.*, 2010). The general trend emerging from these studies is that relative transcription of certain virulence or pathogenicity genes is higher in laboratory broths compared with real food matrices. Furthermore, previous studies suggest that a food matrix, in particular a meat-based food matrix, may influence virulence potential of *L. monocytogenes*, possibly through down-regulation of virulence genes (Mahoney and Henriksson, 2003). The possibility to study the behaviour of foodborne pathogens *in situ* will allow a better comprehension of the risks associated with a specific category of food, in terms of composition (pH, Aw, salt), storage temperature and atmosphere. New information can be generated that in the future will substitute the classical risk assessment based on the numbers of pathogenic microorganisms. Application of these approaches has highlighted differences in virulence expression between strains of the same species, thereby introducing new parameters that must be evaluated rather than simply determining cfu/g or ml. Availability of such data will elucidate the molecular mechanism by which foodborne bacteria elicit disease and, furthermore, will allow us to (i) design tools in order to predict the risk related with the presence of foodborne pathogens in food and (ii) to develop intervention strategies to reduce such risk.

References

Alessandria, V., Rantsiou, K., Dolci, P., and Cocolin, L. (2010). Molecular methods to assess *Listeria monocytogenes* route of contamination in a dairy processing plant. Int. J. Food Microbiol. *141*, S156–S162.

Amoako, K.K., Goji, N., Macmillan, T., Said, K.B., Druhan, S., Tanaka, E., and Thomas, E.G. (2010). Development of multitarget real-time PCR for the rapid, specific, and sensitive detection of *Yersinia pestis* in milk and ground beef. J. Food Prot. *73*, 18–25.

Anonymous. (2005). Commission Regulation (EC) No. 2073/2005 of 15 November 2005 on microbiological criteria for foodstuffs.

Anonymous. (2007). Commission Regulation (EC) No. 1441/2007 of 5 December 2007 amending Regulation (EC) No 2073/2005 on microbiological criteria for foodstuffs.

Baert, L., Uyttendaele, M., and Debevere, J. (2008). Evaluation of viral extraction methods on a broad range of ready-to-eat foods with conventional and real-time RT-PCR for Norovirus GII detection. Int. J. Food Microbiol. *123*, 101–108.

Bennett, A.R., Davids, F.G.C., Vlahodimou, S., Banks, J.G., and Betts, R.P. (1997). The use of bacteriophage-based systems for the separation and concentration of *Salmonella*. J. Appl. Microbiol. *83*, 259–265.

Berrada, H., Soriano, J.M., Pico, Y., and Manes, J. (2006). Quantification of *Listeria monocytogenes* in salads by real time quantitative PCR. Int. J. Food Microbiol. *107*, 202–206.

Blaise-Boisseau, S., Hennechart-Collette, C., Guillier, L., and Perelle, S. (2010). Duplex real-time qRT-PCR for the detection of hepatitis A virus in water and raspberries using the MS2 bacteriophage as a process control. J. Virol. Methods *166*, 48–53.

Botteldoorn, N., Van Coillie, E., Piessens, V., Rasschaert, G., Debruyne, L., Heyndrickx, M., Herman, L., and Messens, W. (2008). Quantification of *Campylobacter* spp. in chicken carcass rinse by real-time PCR. J. Appl. Microbiol. *105*, 1909–1918.

Brehm-Stecher, B., Young, C., Jaykus, L.A., and Tortorello, M.L. (2009). Sample preparation: the forgotten beginning. J. Food Prot. 72, 1774–1789.

Campolongo, S., Rantsiou, K., Giordano, M., Gerbi, V., and Cocolin L. (2010). Prevalence and biodiversity of *Brettanomyces bruxellensis* in wine from the north-west of Italy. Am. J. Enol. Vitic. *61*, 486–491.

Chatterjee, S.S., Hossain, H., Otten, S., Kuenne, C., Kuchmina K., Machata, S., Domann, E., Chakraborty, T., and Hain, T. (2006). Intracellular gene expression profile of *Listeria monocytogenes*. Infect. Immun. *74*, 1323–1338.

Cheng, C.Y., Chi, J.R., Lin, S.R., Chou, C.C., and Huang, C.C. (2009). Rapid quantification of *Salmonella* Typhimurium inoculated to meat products by real-time PCR. Acta Vet. Hung. *57*, 25–38.

Chiang, Y.C., Fan, C.M., Liao, W.W., Lin, C.K., and Tsen, H.Y. (2007). Real-time PCR detection of *Staphylococcus aureus* in milk and meat using new primers designed from the heat shock protein gene *htrA* sequence. J. Food Prot. *70*, 2855–2859.

Critzer, F.J., Dsouza, D.H., and Golden, D.A. (2008). Transcription analysis of *stx1*, *marA*, and *eaeA* genes in *Escherichia coli* O157:H7 treated with sodium benzoate. J. Food Prot. *71*, 1469–1474.

D'Urso, O.F., Poltronieri, P., Marsigliante, S., Storelli, C., Hernandez, M., and Rodriguez-Lazaro, D. (2009). A filtration-based real-time PCR method for the quantitative detection of viable *Salmonella enterica* and *Listeria monocytogenes* in food samples. Food Microbiol. *26*, 311–316.

Delaherche, A., Claisse, O., and Lonvaud-Funel, A. (2004). Detection and quantification of *Brettanomyces bruxellensis* and 'ropy' *Pediococcus damnosus* strains in wine by real-time polymerase chain reaction. J. Appl. Microbiol. *97*, 910–915.

Dolci, P., Alessandria, V., Rantsiou, K., Rolle, L., Zeppa, G., and Cocolin, L. (2008). Microbial dynamics of Castelmagno PDO, a traditional Italian cheese, with a focus on lactic acid bacteria ecology. Int. J. Food Microbiol. *122*, 302–311.

Duquenne, M., Fleurot, I., Aigle, M., Darrigo, C., Borezée-Durant, E., Derzelle, S., Bouix, M., Deperrois-Lafarge, V., and Delacroix-Buchet, A. (2010). Tool for quantification of staphylococcal enterotoxin gene expression in cheese. Appl. Environ. Microbiol. *76*, 1367–1374.

Fukushima, H., Katsube, K., Hata, Y., Kishi, R., and Fujiwara, S. (2007). Rapid separation and concentration of food-borne pathogens in food samples prior to quantification by viable-cell counting and real-time PCR. Appl. Environ. Microbiol. *73*, 92–100.

Gordon, K.V., Vickery, M.C., DePaola, A., Staley, C., and Harwood, V.J. (2008). Real-time PCR assays for quantification and differentiation of *Vibrio vulnificus* strains in oysters and water. Appl. Environ. Microbiol. *74*, 1704–1709.

Gutiérrez, R., Garća, T., González, I., Sanz, B., Hernández, P.E., and Martín, R. (1998). Quantitative detection of meat spoilage bacteria by using the polymerase chain reaction (PCR) and an enzyme linked immunosorbent assay (ELISA). Lett. Appl. Microbiol. *26*, 372–376.

Ibekwe, A.M., Shouse, P.J., and Grieve, C.M. (2006). Quantification of survival of *Escherichia* coli O157:H7 on plants affected by contaminated irrigation water. Eng. Life Sci. *6*, 566–572.

Jasson, V., Rajkovic, A., Debevere, J., and Uyttendaele, M. (2009). Kinetics of resuscitation and growth of *L. monocytogenes* as a tool to select appropriate enrichment conditions as a prior step to rapid detection methods. Food Microbiol. *26*, 88–93.

Jasson, V., Uyttendaele, M., Rajkovic, A., and Debevere, J. (2007). Establishment of procedures provoking sub-lethal injury of *Listeria monocytogenes, Campylobacter jejuni* and *Escherichia coli* O157 to serve method performance testing. Int. J. Food Microbiol. *118*, 241–249.

Josefsen, M.H., Jacobsen, N.R., and Hoorfar, J. (2004). Enrichment followed by quantitative PCR both for rapid detection and as a tool for quantitative risk assessment of food-borne thermotolerant campylobacters. Appl. Environ. Microbiol. *70*, 3588–3592.

Kubista, M., Andrade, J.M., Bengtsson, M., Forootan, A., Jonák, J., Lind, K., Sindelka, R., Sjöback, R., Sjögreen, B., Strömbom, L., *et al.* (2006). The real-time polymerase chain reaction. Mol. Aspects Med. *27*, 95–125.

Kumar, R., Surendran, P.K., and Thampuran, N. (2010). Rapid quantification of salmonella in seafood using real-time PCR assay. J. Microbiol. Biotechnol. *20*, 569–573.

Lofstrom, C., Schelin, J., Norling, B., Vigre, H., Hoorfar, J., and Radstrom, P. (2011). Culture-independent quantification of *Salmonella enterica* in carcass gauze swabs by flotation prior to real-time PCR. Int. J. Food Microbiol. 145, S103–S109.

López-Enríquez, L., Rodríguez-Lázaro, D., and Hernández, M. (2007). Quantitative detection of *Clostridium tyrobutyricum* in milk by real-Time PCR. Appl. Environ. Microbiol. 73, 3747–3751.

McKillip, J.L., and Drake, M. (2000). Molecular beacon polymerase chain reaction detection of *Escherichia coli* O157:H7 in milk. J. Food Prot. *63*, 855–859.

Mahoney, M., and Henriksson, A. (2003). The effect of processed meat and meat starter cultures on gastrointestinal colonization and virulence of *Listeria monocytogenes* in mice. Int. J. Food Microbiol. *84*, 255–261.

Malorny, B., Huehn, S., Dieckmann, R., Kramer, N., and Helmuth, R. (2009). Polymerase Chain Reaction for the rapid detection and serovar identification of *Salmonella* in food and feeding stuff. Food Anal. Method. 2, 81–95.

Manzano, M., Cocolin, L., Cantoni, C., and Comi, G. (1998). Detection and identification of *Listeria monocytogenes* in food by PCR and oligonucleotide-specific capture plate hybridization. Food Microbiol. *15*, 651–657.

Martín, B., Jofré, A., Garriga, M., Pla, M., and Aymerich, T. (2006). Rapid quantitative detection of Lactobacillus sakei in meat and fermented sausages by real-time PCR. Appl. Environ. Microbiol. *72*, 6040–6048.

Martínez-Blanch, J.F., Sánchez, G., Garay, E., and Aznar, R. (2009). Development of a real-time PCR assay for detection and quantification of enterotoxigenic members of *Bacillus cereus* group in food samples. Int. J. Food Microbiol. *135*, 15–21.

Mullis, K., Faloona, F., Scharf, S., Saiki, R., Horn, G., and Erlich, H. (1986). Specific enzymatic amplification of DNA *in vitro*: the polymerase chain reaction. Cold Spring Harb. Symp. Quant. Biol. *51*, 263–373.

Neeley, E., Phister, T., and Mills, D.A. (2005). A differential real-time PCR assay for enumeration of lactic acid bacteria in wine. Appl. Environ. Microbiol. *71*, 8954–8957.

Olesen, I., Vogensen, F.K., and Jespersen, L. (2009). Gene transcription and virulence potential of *Listeria monocytogenes* strains after exposure to acidic and NaCl stress. Foodborne Pathog Dis. *6*, 669–680.

Olesen, I., Thorsen, L., and Jespersen, L. (2010). Relative transcription of *Listeria monocytogenes* virulence genes in liver pâtés with varying NaCl content. Int. J. Food Microbiol. *141*, S60-S68.

Osés, S.M., Rantsiou, K., Cocolin, L., Jaime, I., and Rovira, J. (2010). Prevalence and quantification of shiga-toxin producing *Escherichia coli* along the lamb food chain by quantitative PCR. Int. J. Food Microbiol. *141*, S163-S169.

Pennacchia, C., Ercolini, D., and Villani, F. (2009). Development of a real-time PCR assay for the specific detection of *Brochothrix thermosphacta* in fresh and spoiled raw meat. Int. J. Food Microbiol. *134*, 230–236.

Phister, T.G., and Mills, D.A. (2003). Real-time PCR assay for detection and enumeration of *Dekkera bruxellensis* in wine. Appl. Environ. Microbiol. *69*, 7430–7434.

Phister, T.G, Rawsthorne, H., Lucy Joseph, C.M., and Mills, D.A. (2007). Real-time PCR assay for detection and enumeration of *Hanseniaspora* species from wine and juice. Am. J. Enol. Vitic. *58*, 229–233.

Rantsiou, K., Alessandria, V., Urso, R., Dolci, P., and Cocolin, L. (2008). Detection, quantification and vitality of *Listeria monocytogenes* in food as determined by quantitative PCR. Int. J. Food Microbiol. *121*, 99–105.

Rantsiou, K., Lamberti, C., and Cocolin, L. (2010). Survey of *Campylobacter jejuni* in retail chicken meat products by application of a quantitative PCR protocol. Int. J. Food Microbiol. *141*, S75-S79.

Rantsiou, K., Urso, R., Dolci, P., Comi, G., and Cocolin, L. (2008). Microflora of Feta cheese from four Greek manufacturers. Int. J. Food Microbiol. *126*, 36–42.

Rieu, A., Guzzo, J., and Piveteau, P. (2009). Sensitivity to acetic acid, ability to colonize abiotic surfaces and virulence potential of *Listeria monocytogenes* EGD-e after incubation on parsley leaves. J. Appl. Microbiol. *108*, 560–570.

Rodriguez-Lazaro, D., Jofré, A., Aymerich, T., Hugas, M., and Pla, M. (2004). Rapid quantitative detection of *Listeria monocytogenes* in meat products by realtime PCR. Appl. Environ. Microbiol. *70*, 6299–6301.

Rodriguez-Lazaro, D., Jofré, A., Aymerich, T., Garriga, M., and Pla, M. (2005). Rapid quantitative detection of *Listeria monocytogenes* in salmon products: evaluation of pre-real-time PCR strategies. J. Food Prot. *68*, 1467–1471.

Rodriguez-Lazaro, D., Lombard, B., Smith, H., Rzezutka, A., D'Agostino, M., Helmuth, R., Schroeter, A., Malorny, B., Miko, A., Guerra, B., *et al.* (2007). Trends in analytical methodology in food safety and quality: monitoring microorganisms and genetically modified organisms. Trends Food Sci. Tech. *18*, 306–319.

Rossen, L., Norskov, P., Holmstrom, K., and Rasmussen, O.F. (1992). Inhibition of PCR by components of food samples, microbial diagnostic assays and DNA-extraction solutions. Int. J. Food Microbiol. *17*, 37–45.

Rowan, N.J. (2004). Viable but non culturable forms of food and waterborne bacteria: Quo Vadis? Trends Food Sci. Tech. *15*, 462–467.

Salinas, F., Garrido, D., Ganga, A., Veliz, G., and Martínez, C. (2009). TaqMan real-time PCR for the detection and enumeration of *Saccharomyces cerevisiae* in wine. Food Microbiol. *26*, 328–332.

Shank F.R., Elliot, E.L., Kaye Wachsmuth, I., and Losikoff, M.E. (1996). US position on *Listeria monocytogenes* in foods. Food Control 7, 229–234.

Sharkey, F.H., Banat, I.M., and Marchant, R. (2004). Detection and quantification of gene expression in environmental bacteriology. Appl. Environ. Microbiol. *70*, 3795–3806.

Singh, J., Batish, V.K., and Grover, S. (2009). A scorpion probe-based real-time PCR assay for detection of *E. coli* O157:H7 in dairy products. Foodborne Pathog. Dis. *6*, 395–400.

Slana, I., Kralik, P., Kralova, A., and Pavlik, I. (2008). On-farm spread of *Mycobacterium avium* subsp. paratuberculosis in raw milk studied by IS900 and F57 competitive real time quantitative PCR and culture examination. Int. J. Food Microbiol. *128*, 250–257.

Stals, A., Baert, L., Botteldoorn, N., Werbrouck, H., Herman, L., Uyttendaele, M., and Van Coillie, E. (2009). Multiplex real-time RT-PCR for simultaneous detection of GI/GII noroviruses and murine norovirus 1. J. Virol. Methods *161*, 247–253.

Sun, F., Wu, D., Qiu, Z., Jin, M., Wang, X., and Li, J. (2010). Development of real-time PCR systems based on SYBR Green for the specific detection and quantification of *Klebsiella pneumoniae* in infant formula. Food Control *21*, 487–491.

Takahashi, H., Takakura, C., and Kimura, B.O.N. (2010). A quantitative real-time PCR method for monitoring *Clostridium botulinum* type A in rice samples. J. Food Prot. 73, 688–694.

Venkitanarayanan, K.S., Faustman, C., Crivello, J.F., Khan, M.I., Hoagland, T.A., and Berry, B.W. (1997). Rapid estimation of spoilage bacterial load in aerobically stored meat by a quantitative polymerase chain reaction. J. Appl. Microbiol. *82*, 359–364.

Wang, C., and Hong, C. (1999). Quantitative PCR for *Listeria monocytogenes* with colorimetric detection. J. Food Prot. *62*, 35–39.

Wang, L., Li, Y., and Mustapha, A. (2009). Detection of viable *Escherichia coli* O157:H7 by ethidium monoazide real-time PCR. J. Appl. Microbiol. *107*, 1719–1728.

Warren, B.R., Yuk, H.G., and Schneider, K.R. (2007). Detection of *Salmonella* by flow-through immunocapture real-time PCR in selected foods within 8 hours. J. Food Prot. *70*, 1002–1006.

Yang, W.C., Mi, L.J., Cao, X.Y., Zhang, X.D., Fan, C.H., and Hu, J. (2008). Evaluation of gold nanoparticles as the additive in real-time polymerase chain reaction with SYBR Green I dye. Nanotechnology. 19, 255101-1–255101-9.

Zhou, S., Hou, Z., Li, N., and Qin, Q. (2007). Development of a SYBR Green I real-time PCR for quantitative detection of *Vibrio alginolyticus* in seawater and seafood. J. Appl. Microbiol. *103*, 1897–1906.

Part III

RT-qPCR for Studying Microbial Gene Expression

Studying Microbial Gene Expression in Complex Environmental Matrices Using Quantitative Real-time Polymerase Chain Reaction

9

Vijay J. Gadkar and Martin Filion

Abstract

Environmental matrices are highly diverse in their composition and range from simple (e.g. water) to highly complex (e.g. organic soils/biosolids). Analysis of microbial gene expression from such substrates is done for variety of purposes which could range from bio-surveillance to elucidation of biological function of a target microbe. Quantitative real-time PCR (RT-qPCR) has become a technique of choice for studying such bio-processes, due to its unique ability to both detect and quantify a target transcript in real-time. Challenges in extracting inhibitor-free, structurally intact RNA, amenable for a sensitive technique like RT-qPCR, has however proved to be a major impediment in our ability to rigorously implement this highly versatile technology. Despite these 'substrate defined' limitations, many attempts have been made to implement the RT-qPCR technology. Efforts like these have given us invaluable insight into the expression status of a particular transcript and hence, the biological functioning of the microbe, specifically under natural *in situ* conditions. As a result, it has enhanced our understanding of the role and diversity of many microbial populations which, previously was not possible using conventional molecular approaches. In this chapter, we have sought to summarize such technical problems faced by molecular environmental microbiologist and solutions developed to mitigate those challenges.

Introduction

The key to studying gene expression quantitatively in environmental samples lies in one's ability to effectively and most importantly, specifically detect messenger RNA (mRNA), corresponding to the activity of interest. In this regard, the use of reverse transcription (RT) in conjunction with quantitative real-time polymerase chain reaction (qPCR) has become a method of choice for quantifying specific mRNA's.

The unique ability of the RT-qPCR based method to analyse large number of samples rapidly, with high degree of sensitivity and specificity, makes it a logical choice over other forms of RNA detection methodologies for example Northern blotting, *in situ* hybridization, RNase protection assays and semi-quantitative RT-PCR analyses amongst others. Some of the aforementioned non-qPCR methods are highly laborious and time consuming to perform, making them unsuitable for high-throughput experimental designs. Moreover, some of them, for example RNase protection assay and Northern blotting, have a lower threshold for detection compared with RT-qPCR (Sharkey *et al.*, 2004). In summary, RT-qPCR is a technique of choice to detect target mRNA, if high degree of specificity, sensitivity and rapidity is desired. This chapter focuses on the methods, variants and technical limitation in quantifying mRNA's from complex environmental matrices.

Microbial gene expression in environmental matrices

Detection and monitoring of microbial gene expression from an environmental sample has become an integral part of microbial ecology, bioremediation, biothreat analysis and diversity monitoring, amongst others (Sharkey *et al.*, 2004). While there are several molecular tools available to analyse microbial gene expression *per se* (Fleming and Sayler, 1995), the introduction of the RT-qPCR technology adds a very different dimension, not afforded by any of the earlier techniques (Sharkey *et al.*, 2004). With RT-qPCR technology, we have the unique ability to perform two specific functions simultaneously – monitor and quantify transcripts from target microbes, all under real-time conditions. The ease by which these two functions can be implemented by RT-qPCR, has significantly enhanced our ability to understand the microbial processes occurring under natural conditions at such fine scale and resolution, not previously afforded by any of the known conventional gene targeting techniques.

Analysing microbial gene expression in an environmental matrix involves multifaceted handling and processing steps. The complexity of these initial steps is heavily dependent on the type of matrix in question. While there are simple matrices generated through 'controlled' experimental conditions, there are myriad of matrices one has to contend with, especially naturally sourced, starting from simple (e.g. water) to highly diverse and complex types (e.g. soil, raw sewage/biosolids and food). Since each of the aforementioned matrices possess different levels of compositional complexity which, in most cases is poorly defined, one's ability to reliably implement a sensitive technology like RT-qPCR, has hinged on one major factor – 'pre-processing' of target matrices. This, in other words, implies extraction of target components from the matrix in such forms that, it becomes amenable for RT-qPCR analysis. In practical terms, this involves the use of specific RNA extraction protocols which, in principle are not only optimized for an efficient recovery of nucleic acids (RNA in the present context), but are also capable of mitigating co-extraction of endogenous confounding substances. These 'confounding' substances or inhibitors could be broadly defined as chemical moieties which, directly/indirectly interact with the components of the RT-qPCR reaction, leading to dampening of the final fluorescence signal. Such undesirable interactions invariably lead to an aberrant, and in many instances, an irreproducible gene expression quantification pattern (Suslov and Steindler, 2005).

Based on the wide diversity and hence the compositional complexity of environmental matrices *per se*, the array of such chemical inhibitors, documented till date, have been extremely vast (Wilson, 1997). These inhibitors, whose list is continually increasing, has an amazing range of diversity, from simple forms of heavy metals, commonly found in food/water-based matrices (Kreader, 1996; Ijzerman *et al.*, 1997), to the highly amorphous 'brown contaminating' substances, endemic to organic/sedimentary soils (Watson and Blackwell, 2000). Developing a viable RNA extraction protocol, capable of mitigating the effects of these endogenous inhibitory substances from each type of environmental matrix documented till date, is one major challenge faced by microbiologists working in this area of research. This is in sharp contrast to the wide scale application of RT-qPCR technology in other areas of microbial research, where inhibitor mitigation is easily achieved through the use of highly controlled/defined microbial cultivation substrates (e.g. artificial growth media) and a well-optimized RNA extraction protocol. Since it is impossible to 'dictate' the composition of an environmental matrix, especially those sourced naturally, the only option till date has been to manipulate/optimize the nucleic acid extraction protocols, so that a high quality, inhibitor-free RNA, compatible to the sensitive requirements of RT-qPCR, is obtained. This, however, has proven to be a major challenge, since developing highly optimized protocols for each type of matrix, or class of matrices, is not only technically impossible, but also logistically impractical. In summary, this 'substrate defined' technical limitation, is a major bottleneck in our ability to rigorously implement the highly versatile RT-qPCR technology in environmental microbiology. Achieving a minimal 'quality benchmark' for this first and highly critical pre-processing step, has been a major and a continual challenge for environmental microbiologist.

In the subsequent sections, we aim to describe the technical challenges faced by many molecular environmental microbiologists interested in analysing microbial gene expression from environmental matrices, especially complex ones. Some of the areas covered include different type of RNA extraction methodologies, the invariant problem of endogenous inhibitors (which are co-extracted with RNA), common strategies employed to mitigate the effects of the same, finally followed by representative examples of specific environmental matrices namely soil, food/water and sludge/biosolids where RT-qPCR has successfully been implemented to quantify microbial gene expression.

Problems in extracting RNA from environmental matrices and quality issues

The first step towards analysing gene expression in any system is the invariant process of extracting RNA using an extraction protocol, optimized to yield high-quality, intact RNA. Low-quality RNA may compromise the derived expression results as previously observed (Raeymarkers, 1993; Imbeaud *et al.*, 2005). Developing an optimized RNA extraction protocol, suitable for a wide-range or even a subset of environmental matrices with a common compositional makeup, has been a major challenge for researchers. Notwithstanding the availability of 'off the shelf' commercial kits capable of extracting high quality total RNA from environmental samples, the end user still has to empirically test a particular kit's suitability on his/her target matrix. This in many cases is extremely cumbersome and especially difficult if the sample amounts are exceedingly limiting. Moreover, commercial kits are only customized to work optimally for certain matrix type(s) and thus, may not be suitable for a particular user-defined matrix. Also, if the experimental set-up demands a high throughput RNA extraction, the overall cost of nucleic acid preparation can be a major deciding factor for a user to take into consideration.

The inclusion of a robust internal/external extraction control(s), is one way to circumvent the issue of variable extraction efficiencies between different extraction protocols, as previously reported (Hoorfar *et al.*, 2003, 2004; Nolte 2004; van Doorn *et al.*, 2009). However, the unavailability of an universal external control(s), specifically designed for environmental matrices, at least until recently (González-Escalona *et al.*, 2009; Deer *et al.*, 2010; DeCoste *et al.*, 2011), and lack of unanimity amongst researchers on the choice of an optimal internal control, are some of the technical bottlenecks which have prevented researchers to implement testing RNA extraction protocols, either user developed or commercially available kits, in order to test their suitability for a particular matrix or class of matrices.

RNA extraction protocols

Although many protocols for extracting genomic DNA from environmental samples have been established till date (Bakken and Lindahl, 1995; Bakken and Frostegård, 2006), developing protocols capable of yielding high-quality, intact mRNA from the same substrate, remains a major obstacle. This is primarily due to the extreme labiality of the mRNA *per se*, especially of the prokaryotic origin which has a very short (in minutes) half-life (Takayama and Kjelleberg, 2000). Of the limited number of RNA extraction protocols developed till date for environmental matrices, one is able to always infer a well defined logical work flow. This work flow could be at best described into two steps: (a) cell lysis and release of RNA into the aqueous solution and (b) selective elution/precipitation of RNA. Any improvements to the current extraction protocols or development of new ones have always focused on modifying sub-steps within these two main steps.

The first step in extracting any RNA target is its release, preferably complete from the living cells. An inefficient or even an incomplete lysis could result in low or unrepresentative RNA amounts which, if submitted to RT-qPCR analysis would give either an aberrant or irreproducible gene expression pattern (Smith *et al.*, 2006). Cell lysis is easily achieved in simple matrices (e.g. water) however; this has proven to be extremely difficult for highly complex environmental matrices like sludge, natural soil and unprocessed food. This difficulty is most probably attributed to the components of the complex matrix which bind/adsorb the target components hence, making it difficult to access the cells for a complete or at least a representative lysis (Hahn *et al.*, 1990). Cell lysis is easily achieved through the use of chaotropic agents capable of permeabilizing the cells in simple matrices such as food (e.g. yoghurt) and water. Some of the commonly used agents include detergents such as sodium lauryl sulfate (SLS), lauryl sarcosyl and Triton X-100, EDTA and lysozyme, amongst others (reviewed by Bakken and Frostegård, 2006). Since the major goal is to isolate RNA, these disrupting agents are used in combination with chemicals which inhibit the activity of the ubiquitously present RNase's which are released post cell lysis into the extraction solution. Some of the most commonly used agents include: lithium chloride (LiCl), cetyl-trimethyl ammonium bromide (CTAB), guanidinium thiocyanate (GdnHCl) and guanidine isothocyanate (GTC) (Sambrook and Russel, 2001). Complex matrices like soil and biosolids require a harsh physical disruption to release the microbial RNA contents into the extraction media (reviewed by Bakken and Frostegård, 2006). Some of the commonly employed methods to achieve this include, but are not limited to, mortar grinding with/without liquid nitrogen (Zhou *et al.*, 1996; Frostegård *et al.*, 1999; Hurt *et al.*, 2001) and bead beating (Bürgmann *et al.*, 2001). The physical disruption is usually carried out in the presence of the aforementioned chaotropic agents so that, post cell disruption, the released nucleic acids (total RNA) is not subject to RNase mediated degradation. Addition of diethyl pyrocarbonate (DEPC), a potent inhibitor of RNases, directly in the extraction buffer, has also been described for matrices which are known to contain high amounts of endogenous RNases (Bleve *et al.*, 2003). While there are many 'home brew' recipes for RNA extraction from environmental matrices (Ogram *et al.*, 1995; Bürgmann *et al.*, 2001; Hurt *et al.*, 2001), many commercial kits are now increasingly becoming available which work on a pre-defined environmental matrix; for example, the Mo Bio water and soil RNA extraction systems from Mo Bio Laboratories (Carlsbad, CA, USA) or the ZR soil/faecal RNA extraction kit from Zymo Research Corp (Irvine, CA, USA), amongst few others. Owing to proprietary and intellectual concerns, the exact description of the composition of the solutions used in most commercial kits is not always available.

Quality of RNA

The RNA obtained following extraction is highly unstable and susceptible to degradation by endogenous RNase's (Bustin, 2002, 2005), especially mRNAs which are several kilobases long. Apart from this, degradation could also occur due to handling errors and improper storage of the sample under suboptimal temperature conditions (Schoor *et al.*, 2003). Owing to the inherent chemical instability and its ability to get degraded due to base- or enzyme-catalysed hydrolysis, assessment of integrity of the extracted RNA is the first step in ensuring that the RNA submitted to RT-qPCR analysis would result in a reproducible data. For that, the general criteria for RNA integrity, irrespective from which environmental matrix it has been extracted from, could be summarized as:

- $A_{260\,nm}/A_{280\,nm}$ spectrophotometric absorbance value > 1.8;
- free of or extremely low genomic DNA contamination;
- free of enzymatic inhibitors of the reverse transcriptase enzyme and other enzymatic components of the RT-qPCR reaction;
- free of divalent ions and other co-factors which affect PCR reaction;
- low amounts/free of endogenous nucleases.

Conventional methods for RNA assessment are often not sensitive enough and also not specific enough for single-stranded RNA (Fleige and Pfaffl, 2006). This makes it difficult to assess the quality of extracted RNA from a wide range of environmental matrices. The interference from other non-target components during spectrophotometric assessment which include genomic DNA, proteins and other chemical moieties, also confound the true quality of the RNA. The conventional methods to assess RNA integrity can be done by classical denaturing agarose gel assessment, spectrophotometric assessment at multiple UV/VIS range wavelengths using a conventional spectrophotometer or a NanoDrop™ instrument, to a more modern lab-on-chip microfluidic platforms like the Bioanalyzer™ 2100 (Agilent Technologies, USA) and Experion™ (Bio-Rad Laboratories, USA).

Owing to the invariable low yields in extracting RNA from environmental samples, quantification and assessment of nucleic acids are more pre-disposed towards protocols which consume fewer amounts of the sample. This makes logical sense as there are always inherent procedural difficulties in extracting high quality RNA from environmental samples, forcing the end user to judiciously make use of the nucleic acid sample. Of the methods available for RNA assessment, the spectrophotometric methods using UV/VIS are the most popular and are performed at multiple wavelengths to assess truly the quality of the extracted RNA. Being non-discriminative, spectrophotometric quantification is done at 240 nm (background absorption and possible contamination), 260 nm (specific for nucleic acids), 280 nm (specific for proteins) and 320 nm (background absorption and possible contamination). On the basis of absorption at 260 nm, the quantity, and by the ratio at other wavelength values, namely $A_{260\,nm}/A_{280\,nm}$, $A_{260\,nm}/A_{240\,nm}$ and $A_{260\,nm}/A_{320\,nm}$, the purity can be assessed (Fleige and Pfaffl, 2006). $A_{260\,nm}/A_{280\,nm}$ ratio > 1.8 is considered an acceptable indicator of good RNA yield, suitable for downstream RT-qPCR analysis. However, it cannot determine the presence of contaminating inhibitors, which has to be empirically determined. This method however, has one major drawback- its inability to discern between RNA and carryover genomic DNA which, also contributes to the absorption value at 260 nm. This contamination could lead to serious overestimation (less than 50% accuracy) of the final RNA amounts in samples containing carryover DNA, bringing into question the reliability of this widely used estimation parameter (Bustin and Nolan, 2004). A more modern approach to correct this anomaly has been the use of advanced instrumentation in conjunction with RNA specific dyes. The NanoDrop ND-3300 instrument from NanoDrop Technologies, is an example of this approach and is increasingly becoming popular amongst researchers. RNA specific dyes like the RiboGreen™ dye (Molecular Probes, USA) can be used to detect RNA in conjunction with the NanoDrop quantification instrument. The main advantage of this quantification approach being the low requirement of sample (1–2 µl) and the ability of the instrument to read spectra over a wide-range (240–700 nm) on a single run.

The aforementioned spectrophotometric assessment methods, both conventional and modern, have one major drawback- the difficulty in assessing the integrity of the RNA preparation in the sample. Assessment by resolving the RNA sample on a denaturation agarose gel, to visually inspect the integrity of the 18S/28S or 16S/25S ribosomal RNA, is one possible approach. This however is extremely impractical if the experimental set-up is high throughput or if the sample amount is limiting. Both these are woefully unsuitable for assessing RNA extracted from environmental samples which generally are high-throughput and precious.

The modern lab-on-chip technologies holds much promise and is able to alleviate all the aforementioned drawbacks inherent to conventional quantification methodologies. The Agilent 2100 Bioanalyzer and the Experion system are part of this new generation of technologies which allows the end user to assess the RNA sample's quality, quantity and most importantly, its integrity (18S/28S or 16S/25S rRNA). All the quantifications are performed on a micro-fabricated chip which requires extremely small amounts of starting sample. The development of a RNA assessment quality parameter called the RNA integrity number (RIN) for the Agilent 2100 Bioanalyzer (Agilent Technologies, 2004; Schroeder *et al.*, 2006), is a step in the right direction to develop parameter(s) which, could be reliably used to compare the quality of starting samples. Since it is easy to directly compare samples which are submitted to RT-qPCR analysis, such universal parameters serve another important function- make it simpler for laboratories to cross collaborate and compare results. These quantification platforms although extremely accurate and promising, are however expensive. This makes it difficult to implement them in experimental situations which require high throughput quantification of the starting sample or for laboratories operating on a modest research budgets.

Inhibitors from environmental matrices: diversity and mitigation

Notwithstanding the difficulties in extracting high quality intact RNA from environmental matrices, one invariant factor an environmental microbiologist has to always contend with is the presence of substrate derived chemical inhibitors (reviewed by de Franchis *et al.*, 1988; Wilson 1997; Filion *et al.*, 2003; Sutlovic *et al.*, 2005). These chemical compounds are natural components of the environmental matrices and are invariably co-extracted with RNA preparation. The passive presence of these chemical moieties in the final RNA preparation has one unintended consequence-inhibition of the downstream RT-qPCR enzymatic reaction, including the reverse transcription reaction which is required for generating the cDNA template (Johnson *et al.*, 2005). Practically speaking, inhibition *per se* during the whole RT-qPCR work flow could also be the end consequence of other factors which are of 'non-sample' origin, for example poorly controlled reaction conditions, contaminants in reagents, user derived introduction of contamination or some complex interactions that, at best are difficult to describe or distinguish (Wilson, 1997). In the present context, we have sought to focus on the substrate (i.e. environmental matrix) derived inhibitors and their effect on downstream enzymatic reaction.

Till date, a wide range of these endogenous chemical inhibitors have been identified from different environmental matrices (reviewed by Wilson, 1997). The list of known inhibitors presented in this review, however, does not include any process/protocol introduced chemical moieties like phenol, EDTA, CTAB etc., which are also inhibitory to the enzymatic components of RT-qPCR (Demeke and Jenkins, 2010). Empirical extraction of nucleic acids from various environmental substrates carried out till date has made it fairly possible to predict, assuming the elimination is not total, the nature of inhibitor(s) which, may be present in the final RNA preparation. For example, one can expect the presence of 'brown contaminating' and humic/fulvic acids-based substances in RNA preparations originating from organic soil and sediments (Watson and Blackwell, 2000). Similarly, polyphenols and tannins based inhibitors are endemic to wines and fermented beverages (Zoecklein *et al.*, 1999; Phister and Mills, 2003; Delaherche *et al.*, 2004; Martorell *et al.*, 2005) and therefore, expected to get co-extracted during the RNA extraction protocol. As a result, an intuitive decision making process goes into selecting an optimal RNA extraction protocol, for a particular target environmental matrix. If an optimized protocol is unavailable, a specific RNA extraction protocol has to be 'tailored' for the target matrix, so that an efficient mitigation of the expected inhibitory factor(s) known to be prevalent within the target matrices could be achieved.

Mitigating the effects of endogenous inhibitors on RT-qPCR, co-extracted during RNA extraction, is one of the major challenges faced by researchers. Since it is difficult to describe the quantitative levels of a particular inhibitor(s) present within a target matrix sample, especially those sourced naturally, a two pronged inhibitor mitigation strategy is generally employed by most researchers. As a first part of this strategy, and the most logical, is the development of RNA extraction protocol(s) which, not only are capable of extracting high-quality, intact RNA, but are also able to efficiently sequester any matrix originating inhibitors (Bürgmann *et al.*, 2003; Monnet *et al.*, 2008; Ulve *et al.*, 2008). Considering the wide range and complexities amongst environmental matrices *per se*, tailoring any extraction protocol to yield high quality RNA, which is not only intact, but also inhibitor-free, has proven to be an extremely challenging undertaking.

Based on literature evidence, the most common strategy has been to use specific chemicals/additives in the RNA extraction protocol itself which, bind/adsorb to the inhibitor(s). In this way, a large load of inhibitors are removed during the extraction process itself, making the final RNA preparations virtually free of or having a very minimum amounts of chemical inhibitors. This 'pre-emptive' strategy is widely employed and some of the common examples of the chemicals which aid in reducing/eliminating the endogenous load of inhibitors are: acid phenol–chloroform–isoamyl alcohol (25:24:1) to remove inhibitors from sediments and soils (Tsai and Olson, 1991), $AlNH_4(SO_4)_2$ to reduce co-purification of PCR inhibitors (Braid *et al.*, 2003), a combination of 5% polyethylene glycol 8000 (PEG), polyvinylpolypyrrolidone (PVPP) and Sepharose 4B-polyvinylpolypyrrolidone spin columns (Arbeli and Fuentes, 2008), polymeric adsorbent Superlite DAX-8 (Schriewer *et al.*, 2011), flotation of substrate (Jacobson *et al.*, 2009), commercial glass milk, Ionex and membrane-based spin columns (Tsai and Olson, 1991; Wiedbrauk *et al.*, 1995; Braid *et al.*, 2003). These approaches have been found to be optimal for those environmental matrices where the level of target transcript(s) is relatively high. This being important as the use of these inhibitor adsorption chemicals requires one to use multiple phase extraction/elution steps. Such multistep recovery processes, result in a considerable loss of nucleic acids, making it difficult to detect moderate to rare transcripts.

The aforementioned mitigation strategies, although successful in most of the cases, work especially well when the target matrices have a simpler composition, for example water, or if the extraction matrices originate from highly controlled *in vitro* experimental set-ups. When highly complex environmental matrices like natural organic soils; organic sediments/sludge are used, which are known to carry a very heavy load of endogenous inhibitors, this mitigation approach is generally found to be inadequate. As a result, additional steps are deemed necessary to completely alleviate the effects of the inhibitor(s) which, do get co-extracted in the final RNA preparation. A very common strategy to achieve this goal is through the use of additives which bind/chelate the inhibitors in the RT-qPCR reaction itself. However for this strategy to work, the additive should be compatible with the enzymatic components of the RT-qPCR reaction itself. Identifying such novel additives, which are not only highly efficient in quenching the actions of an inhibitor, but compatible with the other enzymatic components of the RT-qPCR reaction, has been a major challenge. Some of the amplification facilitators/additives identified till date, which are able to satisfy these stringent requirements include the use of 'inhibitor tolerating' DNA polymerases (Al-Soud and Rådström, 1998), like bovine serum albumin (BSA) and single-stranded DNA binding T4 gene 32 protein (Kreader, 1996), bovine lacto transfer technique optimizer (BLOTTO) and spermidine, amongst others (Wang *et al.*, 2007). Irrespective of which one of these aforementioned strategies is employed; one of the most basic approach has always been to serially dilute the RNA sample to overcome the inhibitory effects of the endogenous inhibitor (Tsai and Olson, 1992; Queipo-Ortuno *et al.*, 2008). This simple approach, in conjunction to the aforementioned approaches, works well in most of the cases, however, this strategy is feasible only when the target transcript is present in high copy numbers. Excess dilution can lead to false negatives when targets are diluted below their detection limits or when the inhibitor(s) cannot be diluted to levels below their interference threshold (Queipo-Ortuno *et al.*, 2008).

Of the plethora of inhibitory moieties identified during RNA extraction from environmental matrices, very little is known about the actual mechanism through which this inhibition takes place. It is generally assumed that an inhibitor generally acts at multiple steps of the whole RT-qPCR analysis process. These steps could broadly be divided into three categories: (1) the nucleic acid extraction step, where it prevents complete cell lysis leading to an unrepresentative sampling; (2) the capture step, where RNA is recovered leading to an inefficient recovery or degradation; and, finally, (3) at the individual enzymatic step(s) of the RT-qPCR protocol which results in signal dampening. Since it is sometimes difficult to define an inhibition phenomenon at both the lysis and capture phases, the most logical target to characterize an inhibitor(s), has been to assay for any changes (usually negative), in the fidelity of the RT-qPCR enzymes, including the reverse transcriptase (RT) enzyme used for generating the cDNA.

Based on the multistep enzymatic cascade of RT-qPCR, the interaction of the inhibitor could itself be elucidated into three potential steps: (1) binding of the inhibitor to the enzymes of RT-qPCR namely: reverse transcriptase and DNA polymerase; (2) interaction of the inhibitor with the DNA; and finally (3) the interaction with the polymerase during primer extension cycle. This categorization is by no means absolute, since an inhibitor may act in more than one way and relationships between chemical, enzymatic, and physical factors often cannot be distinguished (Wilson, 1997). It is also probable that many inhibitors act through various physical and chemical means by interfering with the interaction between DNA and polymerase, for example 1 ng of humic acid is known to inhibit Taq DNA polymerase and reverse transcriptase (Opel *et al.*, 2010). Other factors which directly affect the components of the RT-qPCR are nucleases which hydrolyse DNA, divalent ions which bind to the co-factors required for the functioning of the polymerase, polyamines which prevent the DNA polymerase to access the template, amongst others (reviewed by Wilson, 1997; Hugget *et al.*, 2008).

Examples of common environmental matrices

A quick meta-analysis of those studies where qPCR and RT-qPCR techniques have been implemented to quantify microorganisms from environmental matrices, reveal a striking disparity- a much larger number of studies used DNA as a target instead of mRNA. This large scale bias in the use of DNA rather than RNA based targets could be attributed to the innate difficulties in extracting inhibitor-free RNA, as described in great detail in the earlier sections. Specifically, the compounded inefficiencies usually associated with RNA extraction from environmental matrices, the problem of accessing the cells enmeshed within the matrix component (for lysis to release of nucleic acids), high prevalence of endogenous RNase's, the extreme labiality of RNA species and co-extraction of inhibitors, could be some of the technical difficulties which in our opinion tip a researcher to favour DNA based targets. Despite these 'substrate defined' technical bottlenecks, it's highly satisfying to note that many attempts have indeed been made till date, in analysing the *in situ* microbial gene expression from various environmental matrices.

Amongst the host of RNA based gene transcripts as a target for RT-qPCR analysis, the choice of ribosomal RNA (rRNA) to infer the physiological status of a microorganism has certainly been very controversial (Smith *et al.*, 2006). Though, it is difficult for us to justify or debunk the validity of any study carried out till date where, modulation of a particular rRNA transcript(s) has successfully been used to describe the physiological functioning of the target microorganism(s), arguments certainly compelling against this approach, cannot be ignored. One of primary argument being the well known fact that the copy number of rRNA operons varies widely within a genome, therefore measuring the rRNA transcript numbers is not an accurate 'reflection of the functional processes' carried out by the target organism (Smith *et al.*, 2006). Also, the transcription rates of the rRNA is regulated by resource availability (Klappenbach *et al.*, 2000), for example starved cells tend to keep their ribosomal machinery active for a longer time (Wagner *et al.*, 2003), thereby potentially skewing the actual gene expression results. To circumvent such technical problems, functional genes are now being seen as more logical targets in other areas of microbiology, including environmental microbiology (Bælum *et al.*, 2008; Nicolaisen *et al.*, 2008). Keeping these caveats in mind and avoiding any unnecessary complications, we have in the future sections, only considered those studies where mRNA transcripts have been used as targets to analyse endogenous microbial gene expression from environmental matrices using RT-qPCR. Some of these common environmental matrices are described below:

Soil

Soil is home to highly diverse, complex microbial communities, which provide immense ecological services to mankind and its functioning, both direct and indirect, deeply impact many biological and bio-geochemical processes (Gutknecht *et al.*, 2006). Because of this importance, amongst all the known environmental matrices where the RT-qPCR technology has been implemented, it either pertains to soil or matrices derived from it. This is despite severe technical handicaps usually associated with working with this matrix, for example heavy compositional complexity and presence of chemical inhibitors, amongst others. Since only 1% of the endogenous soil microbial population is cultivable (Davis *et al.*, 2005), the acute gap in our information pertaining to the exact genetic makeup of the target microorganism(s), makes it even more challenging to design specific RT-qPCR primers/probes when soil is the substrate in question.

With specific reference to soils, one overarching question molecular microbiologists have constantly been interested in is the possibility of developing a quantitative understanding of both the abilities and diversity of complex natural soil microbial communities, preferably under *in situ* conditions (Bælum *et al.*, 2006; Bælum and Jacobsen, 2009). While some headway has been made in addressing this question, the introduction of RT-qPCR, a molecular tool with an undeniable ability to analyse target microorganism(s) with an extreme high specificity, sensitivity and rapidity, has given a tremendous boost to one's ability to address this seemingly simple, yet technically intractable question. The latter being primarily attributed to the high compositional complexity and intrinsic diversity amongst natural soil matrices *per se*. Notwithstanding this technical challenge, such understanding is extremely important if one needs to define the precise 'structure–function' relationship amongst soil based microbial communities. Of all the studies carried out till date, two distinct experimental approaches could be seen: (a) *in situ* gene expression experiments, where the expression of the target gene(s) is analysed directly from the natural soils and (b) seeding experiments, where the natural soil samples are seeded with a specific test microbe and its expression studied under defined test conditions.

An *in situ* analysis of microbial gene expression is seen as the 'holy grail' by many interested in understanding the endogenous microbial structure–function relationship (Bælum *et al.*, 2006). Owing to low efficiency of RNA extraction, high prevalence of endogenous inhibitors and rapid degradation of mRNA, especially when working with natural field soils, realization of this goal has not been possible as one would have expected (Bælum *et al.*, 2006). As a result, most gene expression studies have tended to use soil matrices with low concentrations of co-extractable enzymatic inhibitors, organic radicals or nucleases (Holmes 2004; Steunou *et al.*, 2006; Smith *et al.*, 2007). Despite this, researchers still have attempted, and in some cases succeeded, in analysing soils matrices, e.g. peat soil for microbial gene expression which, earlier were thought to be intractable due to their harsh endogenous constituents (Freitag and Prosser, 2009).

Understanding the *in situ* temporal expression of a gene has a special significance in certain area of soil microbiology, for example bio-remediation where such understanding is important to implicate the role of a gene in the actual mineralization/degradation process. An illustrative example could be the work of Holmes *et al.* (2004), who showed that it's possible to monitor the *in situ* changes in the expression pattern of the *nifD* gene in Geobacteraceae present in aquifer sediment in response to fluctuating sources of nitrogen. When these *in situ* results were compared with the highly controlled laboratory experimental set-ups, the *in situ* expression rate of the *nifD* gene was close to 100-fold higher. The authors suggested that there are 'unknown factors' operating in natural systems which, cannot be mimicked under controlled conditions. In other words, if an effective bio-remediation strategy has to be formulated using Geobacteraceae; one needs to understand the expression pattern of the *nifD* gene preferably under *in situ* conditions. This study underscores the utility of *in situ* systems to understand the exact metabolic state of the target microorganism. Similar studies where transcription dynamics of microbial genes have been assessed *in situ* using RT-PCR targeted *atzA* and *atzD* genes expression involved in the bio-degradation of the pesticide atrazine in the soil drilosphere, which essentially is a soil influenced by earthworm secretions (Monard *et al.*, 2010), *alkB* (alkane monooxygenase) involved in the microbial degradation of hydrocarbons in Antarctic soils (Powell *et al.*, 2006) and the *mcrA* (methyl coenzyme M reductase) and *pmoA* (methane monooxygenase) genes involved in the methanogenesis process in peat soils (Freitag and Prosser, 2009; Freitag *et al.*, 2010) amongst others (Bælum *et al.*, 2008; Chin *et al.*, 2008; O'Neil *et al.*, 2008; Mårtensson *et al.*, 2009; Mouser *et al.*, 2009; DiDonato *et al.*, 2010; Liu *et al.*, 2010a,b; N'Guessan *et al.*, 2010; Paulin *et al.*, 2010; Yuan *et al.*, 2011). The methanogenesis study in peat soils (Freitag and

Prosser, 2009; Freitag *et al.*, 2010) is especially notable, considering the extreme difficulties in working with peat-based substrates. Components of peat soil are detrimental for RT-qPCR analysis since it is difficult to extract high-quality nucleic acids, owing to its high acidity (pH 3.8), high water content (80–90%), high concentrations of co-extractable organic matter, high capacity of complexing nucleic acids (Crecchio and Stotzky, 1998) and high concentrations of stable free radicals (Shinozuka *et al.*, 2001) and phenolics (Vaughan and Ord, 1982).

Limited numbers of spiking experiments have also been reported whereby; natural soil is first seeded with a test microorganism (containing the gene(s) of interest) and then, transcriptional modulation of the(se) gene(s) monitored using RT-qPCR. Subsequent work where such type of approach has been taken include quantification of *invA* mRNA in *Salmonella* spp. seeded in soil and chicken manure (Jacobsen and Holben, 2007; García *et al.*, 2010); herbicide degradation genes in *Cupriavidus necator* (Nicolaisen *et al.*, 2008), subsurface sediment (Holmes *et al.*, 2004; Elifantz *et al.*, 2010), *bphAa* gene involved in the degradation of polychlorinated biphenyl (PCB) in *Rhodococcus* sp. strain RHA1 (Wang *et al.*, 2008), *tfdC* and C230 genes involved in the cleavage of the of the pollutant 2,4-D (Lillis *et al.*, 2010), and more recently, the expression of two antibiotic encoding biosynthetic genes in the bacterium *Pseudomonas* sp. LBUM300 of biocontrol interest (DeCoste *et al.*, 2011; Novinscak and Filion, 2011).

Food and other edible matrices

Microbial gene expression using RT-PCR in food and water based matrices (e.g. beef, chicken, carbonated beverages, fresh produce, wine, drinking water) is mainly performed for biosafety and surveillance purposes (Fratamico, 2001; Straub and Chandler, 2003; Rowan, 2011); as part of mandates usually delegated to governmental regulatory authorities (USFDA, 2001; WHO, 2002). In the interest of public safety at large and the societal importance of food and water, microbiological testing of these matrices is not only expected to be very rapid, but also extremely accurate (Jaykus, 2003). With the ability to acquire data in real-time, coupled with no post-sample processing, the RT-qPCR technology is an ideal technique to further this goal. Despite some initial cost concerns *viz à viz* the traditional protocols which are generally slower and cumbersome to perform, the RT-qPCR technology is now a widely accepted analytical tool in determining food microbial biosafety (Jaykus, 2003; Hanna *et al.*, 2005). The implementation of the RT-qPCR technology, specifically addresses the accuracy factor also interpreted as the 'zero tolerance' policy in the food microbial biosafety industry. Departure from such excruciatingly high verification standards could result in an erosion of consumer confidence, leading to potential severe economic consequences to both- producer and the end consumer.

Viability status of the target microorganism is extremely critical to accurately determine the biosafety status of a food sample (Keer and Birch, 2003). The conventional cultivation protocol even though highly cumbersome, specifically addresses this aspect, and therefore continues to be a mainstay technique in food microbiology (Harrigan, 1998). The relatively non-specific and time consuming aspect of this approach affects the decision making process in determining the biosafety status of food samples. As a result, alternatives are constantly being sought, which do not suffer from some of the logistical drawbacks. The advent of PCR changed this and offered a viable alternative to these cultivation based protocols (Krascsenicsova *et al.*, 2008). Initially, the PCR based protocols made extensive use of DNA based targets, including ribosomal (rDNA) in microbial food biosafety sciences (Vaitilingom *et al.*, 1998; Fortin *et al.*, 2001; Rijpens and Herman, 2002; Yaron and Matthews, 2002; Rodriguez-Lazaro *et al.*, 2003; Malorny *et al.*, 2007). Though able to detect the presence/absence of the target microorganism, this DNA based approach falls short in determining the viability status of the target microorganism. As viability *per se* is critical in accurately determining the biosafety status of food/water, decisions which have solely relied on DNA-based data, have attracted a great deal of controversy (Drahovska *et al.*, 2001) and continues to do so (Fey *et al.*, 2004). One major criticism against this approach is the well known fact that

DNA is able to persist in dead cells, and therefore any positive PCR signal could potentially be contaminated, or at worst originate wholly from these non-viable, dead cells (Gonzalez-Escalona *et al.*, 2009). To avoid such confounding situations, non-DNA targets have been proposed and for that matter, less stable mRNA offers a much more attractive alternative (Sheridan *et al.*, 1998).

As mentioned earlier, expression analysis of functional gene transcripts allows one not only to detect the presence/absence of the target microorganism within a food sample, but also to infer its viability (Fey *et al.*, 2004). In food-based matrices, for mRNA *per se* to be a reliable target for RT-qPCR and an accurate indicator of cell metabolic status, it should meet certain criteria. These criteria could be summarized as:

1. For specificity purposes, the target mRNA transcription should be species or strain specific.
2. The target mRNA transcription should have a sufficiently brief half-life, so as to allow its detection.
3. The target mRNA transcription should ideally be expressed at sufficiently high copy numbers.
4. Its expression pattern should have some correlation with well-defined physiological parameters (typically spoilage).

Identifying candidate transcripts which are able to satisfy all the aforementioned criteria's, has proven to be an extremely difficult undertaking for food microbiologists. Thus, it is not surprising to note that there aren't many studies in the literature where RT-qPCR has been used to assess microbial gene expression in food/water based matrices. This difficulty in identifying a candidate mRNA could also be one reason where researchers have been forced to use the rRNA targets for determining the biosafety status of food (Hierro *et al.*, 2006; Wan *et al.*, 2006), despite the well known fact that rRNA expression is not too tightly linked to the physiological status of an organism (Ruimy *et al.*, 1994; Gourse *et al.*, 1996) and hence, not an ideal reliable indicator of cell viability status (Fey *et al.*, 2004, Gonzalez-Escalona *et al.*, 2009). Most, if not all the studies which have sought to assess the biosafety status of a particular food sample using mRNA based targets, have invariably made use of the 'endpoint reporting method' i.e. conventional agarose gel electrophoresis (Vaitilingom *et al.*, 1998; Fortin *et al.*, 2001; Yaron and Matthews, 2002). Though highly invaluable from a scientific stand point, these reverse transcription-PCR (RT-PCR) studies would have spurred, as one might expect, a rapid adaptation of the RT-qPCR technology, to further enhance the dynamics and accuracy of detecting the target microorganism(s), in the area of food microbial biosafety. However, looking at the number of studies in the literature, this 'natural progression' leading to a wide-scale adaptation of the RT-qPCR technology, has not yet happened. The main reasons cited behind this application gap, is the high set up and implementation costs of the RT-qPCR platform (Jaykus, 2003; Hanna *et al.*, 2005) and the exceptionally low tolerance for false-positive results in the area of food biosafety which is not totally difficult to rule out, considering the highly sensitive nature of this fluorescence-based detection technique.

Bleve *et al.* (2003), for the first time implemented the RT-qPCR technique along with the conventional end-point RT-PCR technique to quantify yeast and mould contaminants in yogurts and other pasteurized fruit-derived products (fruit juices and preserves). Yeast and mould based microorganisms have been heavily implicated in food spoilage and contamination. Using the fungal specific actin (*act*) transcripts as target, they were able to show that it is possible to detect viable and viable but not culturable (VBNC) cells from food samples. No signal was however detected from dead cells. The dynamics of this detection was very interesting and depended on the nature of the detection system employed. Specifically, the RT-PCR was able to detect 10 CFU/ml and 10^2–10^3 CFU/ml of the fungi in pure culture and artificially contaminated food products, respectively. However, when the RT-qPCR system was used, the detection limit was raised 10- to 1000-fold higher, for the same target and extraction matrices. The authors noted that the extremely large dynamic range of detection, typical for RT-qPCR, is the key towards offsetting some of the limitations of the RT reaction, which is inherently vulnerable to inhibition by some of the matrix components (e.g. nucleases, fats, low pH, and chelating Mg^{2+} ions), used in this particular study. Other examples of RT-qPCR-based assessment of endogenous microbial gene expression in processed foods include: quantification of the *sea* and *sed* gene transcripts encoding enterotoxins A and D to quantify *Staphylococcus aureus* in cheese and milk (Duquenne, 2010), meat (Dolan *et al.*, 2010), pork (Techathuvanan *et al.*, 2010) and finally, *Salmonella* spp. in chicken and eggs (Wang and Mustapha, 2010).

In the realm of unprocessed foods/water-based matrices, there are few examples to be found where RT-qPCR has been used to determine the endogenous microbial gene expression. Excellent examples of unprocessed matrices are fresh vegetable products (e.g. raw salads) and drinking water. While fresh products are consumed raw with minimal pre-treatment, water is subject to microbial contamination either at source or during distribution to end consumers. In other words, it is an important vehicle for transmission of a large number of microbial borne diseases and therefore a subject of constant surveillance by regulatory authorities (Straub and Chandler, 2003; Howard *et al.*, 2006; Fong *et al.*, 2007). Fey *et al.* (2004) showed for the first time that it was possible to detect *Salmonella enterica* serovar Typhimurium using RT-qPCR in environmental water samples, using two RNA targets: the mRNA encoding for the *invA* (invasion protein A) gene and 16S rRNA. Notwithstanding the successful demonstration by these authors of RT-qPCR to detect microbial gene expression in water, there have been no further reported studies on similar lines, except for the detection of water based RNA viruses (Baert *et al.*, 2008; Parshionikar *et al.*, 2010). One major reason for this could be the technical difficulties in concentrating the target microorganisms from very large volumes of water (typically 100 ml to 1 litre), so as to get enough biomass for a representative and reproducible extraction of RNA (Rutjes *et al.*, 2005). Other examples of RT-qPCR in water based matrices are for the presence of yeast in wine (Hierro *et al.*, 2006) and juice based products (Connor *et al.*, 2005; Wan *et al.*, 2006). Both these studies made use of rRNA as targets to detect the viability status of the targeted microorganisms. Similarly, RT-qPCR has been used for detection of various serovars of *S. enterica* in raw vegetables and fresh products, specifically spinach (Gonzalez-Escalona *et al.*, 2009) and peppers (Miller *et al.*, 2010), targeting the *invA* gene transcripts.

Environmental biosolids/sludge

Analysis of microbial gene expression in biosolids/faecal samples is done primarily as public biosafety measures to detect the active presence of microbes which are known to affect human health. Most of the studies done till date on these environmental matrices have been qualitative, i.e. detect the presence/absence of the target microbial pathogen (Novinscak *et al.*, 2008; Karpowicz *et al.*, 2009). Recovery of RNA from biosolids/sludge is difficult due to endemic presence of array of inhibitors like humic acids, fulvic acids, fats, and proteins (Rosen *et al.*, 1992; Tsai and Olson, 1992; Whitehouse and Hottel, 2007). Environmental biosolids, especially urban sludge, is known to contain a heavy load of these substances, in addition to numerous organic/inorganic compounds, such as polyphenols and heavy metals (USEPA, 2000). These compounds have been found to get concentrated with microbial activity, resulting in its co-extraction during RNA extraction (Rock *et al.*, 2010). Owing to these reasons, analysis of microbial gene expression in biosolids using RT-qPCR is extremely difficult and predictably very few studies on this substrate can be found in literature (Dunaev *et al.*, 2008; Matsuda *et al.*, 2009; Kubota *et al.*, 2010; Rock *et al.*, 2010).

Conclusions

It is beyond doubt that there has been resurgence in the application of nucleic acid targeting tools in every aspect of biology, owing to reduced technology costs, constant development of effective nucleic acids recovery protocols, which also include commercial options, and last but not the least, the ever increasing access to post-analysis processing systems, i.e. software, online databases. These aforementioned advancements have however failed to provide the necessary traction for researchers working in the area of environmental microbiology, owing to certain difficulties endemic to this area of research. Of several, two major issues which have made it difficult to rapidly introduce advanced molecular tools in this area: (a) limited genetic information of environmental microbes due to extremely low cultivability and (b) practical difficulties in extracting high quality nucleic acids from environmental matrices, suitable for sensitive application like RT-qPCR. Despite these 'cultivation-defined' and 'substrate-dependent' technical bottlenecks, DNA/RNA targeting molecular tools, mainly end-point PCR-based, have extensively been used to analyse the expression status of targeted microbial transcripts. Though qualitative, these preliminary studies have given us invaluable insights into the functioning of the target microbe under defined experimental conditions.

To overcome some of the well known technical limitations of end-point PCR, detection platforms like RT-qPCR have proven to be an ideal solution. Using this highly versatile technology, it is now possible to simultaneously detect and quantify in real-time, the expression pattern of a microbial transcripts isolated from environmental substrates. Practical difficulties in working with environmental matrices have however impeded our ability to robustly apply this versatile technology. Technical issues such as extraction of inhibitor-free, structurally intact, total RNA from environmental substrates which is suited for a sensitive technique like RT-qPCR, continues to be a major impediment. It is however satisfying to note that despite these severe 'substrate defined' limitations, several attempts have indeed been made to implement the RT-qPCR technology in environmental microbiology. Efforts like these have given us invaluable insights into the expression status of a particular transcript and hence, the biological functioning of microbes, specifically under natural *in situ* conditions. In summary, a better understanding of the technical difficulties faced by microbiologist in working with environmental matrices, especially those sourced naturally, should go a long way in developing efficient mitigation strategies.

References

Agilent Technologies. (2004). DNA, RNA, protein and cell analysis. Agilent application note, publication number 5989-1197EN.

Al-Soud, W.A., and Rådström, P. (1998). Capacity of nine thermostable DNA polymerases to mediate DNA amplification in the presence of PCR-inhibiting samples. Appl. Environ. Microbiol. *64*, 3748–3753.

Arbeli, Z., and Fuentes, C.L. (2008). Improved purification and PCR amplification of DNA from environmental samples. Int. Biodeterior. Biodegrad. *62*, 14–14.

Bælum, J., and Jacobsen, C.S. (2009). TaqMan probe-based real-time PCR Assay for detection and discrimination of class I, II, and III *tfdA* genes in soils treated with phenoxy acid herbicides. Appl. Environ. Microbiol. *75*, 2969–2972.

Bælum, J., Henriksen, T., Hansen, H.C.B., and Jacobsen, C.S. (2006). Degradation of 4-chloro-2-methylphenoxyacetic acid in top- and subsoil is quantitatively linked to the class III *tfdA* gene. Appl. Environ. Microbiol. 72, 1476–1486.

Bælum, J., Nicolaisen, M.H., Holben, W.E., Strobel, B.W., Sørensen, J., and Jacobsen, C.S. (2008). Direct analysis of *tfdA* gene expression by indigenous bacteria in phenoxy acid amended agricultural soil. ISME J. 2, 677–687.

Baert, L., Wobus, C.E., Van Coillie, E., Thackray, L.B., Debevere, J., and Uyttendaele, M. (2008). Detection of murine noro virus 1 by using plaque assay, transfection assay, and real-time reverse transcription-PCR before and after heat exposure. Appl. Environ. Microbiol. *74*, 543–546.

Bakken, L.R., and Frostegård, Å. (2006). Nucleic acid extraction from soil. In Soil Biology, Nannipieri, P., and Smalla, K., eds (Springer, Berlin, Germany), pp. 49–73.

Bakken, L.R., and Lindahl, V.S. (1995). Recovery of bacterial cells from soil. In Nucleic Acids in the Environment, Methods and Application: Laboratory Manual, Trevors, J.T., and Van Elsas, J.D., eds (Springer, Berlin, Germany), pp. 8–27.

Bleve G., Rizzotti, L., Dellaglio, F., and Torriani, S. (2003). Development of reverse transcription (RT)-PCR and real-time RT-PCR assays for rapid detection and quantification of viable yeasts and molds contaminating yogurts and pasteurized food products. Appl. Environ. Microbiol. *69*, 4116–4122.

Braid M.D., L.M. Daniels and C.L. Kitts, (2003) Removal of PCR inhibitors from soil DNA by chemical flocculation, J. Microbiol. Methods 52, 389–393.

Bürgmann, H., Widmer, F., Sigler, W.V., and Zeyer. J. (2003). mRNA extraction and reverse transcription-PCR protocol for detection of *nifH* gene expression by *Azotobacter vinelandii* in soil. Appl. Environ. Microbiol. *69*, 1928–1935.

Bustin, S.A. (2002). Quantification of mRNA using real-time reverse transcription PCR (RT-PCR): trends and problems. J. Mol. Endocrinol. *29*, 23–39.

Bustin, S.A., and Nolan, T. (2004). Template handling, preparation, and quantification, In The Real-Time PCR Encyclopedia A–Z of Quantitative PCR, Bustin, S.A., ed. (International University Line, La Jolla, CA), pp. 87–120.

Bustin, S.A., Benes, V., Nolan, T., and Pfaffl, M.W. (2005). Quantitative real-time RT-PCR-a perspective. J. Mol. Endocrinol. *34*, 597–601.

Chin, K.J., Sharma, M.L., Russell, L.A., O'Neill, K.R., and Lovley, D.R. (2008). Quantifying expression of a dissimilatory (bi) sulfite reductase gene in petroleum-contaminated marine harbor sediments. Microb. Ecol. *55*, 489–499.

Connor, C.J., Luo, H., Gardener, B.B., and Wang, H.H. (2005). Development of a real-time PCR-based system targeting the 16S rRNA gene sequence for rapid detection of *Alicyclobacillus* spp. in juice products. Int. J. Food Microbiol. *99*, 229–235.

Crecchio, C., and Stotzky, G. (1998). Binding of DNA on humic acids: effect on transformation of *Bacillus subtilis* and resistance to DNase. Soil Biol. Biochem. *30*, 1061–1067.

Davis, K.E., Joseph, S.J., and Janssen, P.H. (2005). Effects of growth medium, inoculum size, and incubation time on culturability and isolation of soil bacteria. Appl. Environ. Microbiol. *71*, 826–834.

DeCoste, N.J., Gadkar, V.J., Filion, M. (2011). Relative and absolute quantitative real-time PCR-based quantifications of *hcnC* and *phlD* gene transcripts in natural soil spiked with *Pseudomonas* sp. strain LBUM300. Appl. Environ. Microbiol. *77*, 41–47.

Deer, D.M., Lampel, K., and González-Escalona, N. (2010). A versatile internal control for use as DNA in real-time PCR and as RNA in real-time reverse transcription PCR assays. Lett. Appl. Microbiol. *50*, 366–372.

Delaherche, A., Claisse, O., and Lonvaud-Funel, A. (2004). Detection and quantification of *Brettanomyces bruxellensis* and 'ropy' *Pediococcus damnosus* strains in wine by real-time polymerase chain reaction. J. Appl. Microbiol. *97*, 910–915.

Demeke, T., and Jenkins, G.R. (2010). Influence of DNA extraction methods, PCR inhibitors and quantification methods on real-time PCR assay of biotechnology-derived traits. Anal. Bioanal. Chem. *396*, 1977–1990.

DiDonato, R.J. Jr., Young, N.D., Butler, J.E., Chin, K-J., Hixson, K.K., Mouser, P., Lipton, M.S., DeBoy, R., and Methé, B.A. (2010). Genome sequence of the delta proteobacterial strain NaphS2 and analysis of differential gene expression during anaerobic growth on naphthalene. PLoS ONE Genome *19*, e14072.

Dolan, A., Burgess, C.M., Fanning, S., and Duffy, G. (2010). Application of quantitative reverse-transcription PCR (qRT-PCR) for the determination of the total viable count (TVC) on meat samples. J. Appl. Microbiol. *109*, 91–98.

van Doorn, R., Klerks, M.M., van Gent-Pelzer, M.P., Speksnijder, A.G., Kowalchuk, G.A., and Schoen, C.D. (2009). Accurate quantification of microorganisms in PCR-inhibiting environmental DNA extracts by a novel internal amplification control approach using Biotrove OpenArrays. Appl. Environ. Microbiol. *75*, 7253–7260.

Drahovska, H., Turna, J., Piknova, E., Kuchta, T., Szitasova, I., Skarkova, A., and Sasik, M. (2001). Detection of *Salmonella* by polymerase chain reaction targeted to *fimC* gene. Biologia 56, 611–616.

Dunaev, T., Sevda, A., and Metin, D. (2008). Gene expression monitoring for detection and quantification of viable but non-culturable pathogens in biosolids. Proceedings of the Water Environment Federation, Residuals and Biosolids *11*, 1053–1063.

Duquenne, M., Fleurot, I., Aigle, M., Darrigo, C., Borezee-Durant, E., Derzelle, S., Bouix, M., Deperrois-Lafarge, V., and Delacroix-Buchet, A. (2010). Tool for quantification of Staphylococcal enterotoxins gene expression in cheese. Appl. Environ. Microbiol. *76*, 1367–1374.

Elifantz, H., N'Guessan, L.A., Mouser, P.J., Williams, K.H., Wilkins, M.J., Risso, C., Holmes, D.E., Long, P.E., and Lovley, D.R. (2010). Expression of acetate permease-like (apl) genes in subsurface communities of *Geobacter* species under fluctuating acetate concentrations. FEMS Micro. Ecol. 73, 441–449.

Fey, A., Eichler, S., Flavier, S., Christen, R., Hofle, M.G., and Guzman, C.A. (2004). Establishment of a real-time PCR-based approach for accurate quantification of bacterial RNA targets in water, using *Salmonella* as a model organism. Appl. Environ. Microbiol. *70*, 3618–3623.

Filion, M., St-Arnaud, M., and Jabaji-Hare, S.H. (2003). Direct quantification of fungal DNA from soil substrate using real-time PCR. J. Microbiol. Methods 53, 67–76.

Fleige, S., and Pfaffl, M.W. (2006). RNA integrity and the effect on the real-time qRT-PCR performance. Mol. Asp. Med. 27, 126–139.

Fleming, J.T., and Sayler, G.S. (1995). Assessment of gene expression in the environment: quantitative mRNA analysis in contaminated soils. In Applications of Molecular Biology in Environmental Chemistry, Minear, R.A., Ford, A.M., Needham L.L., and Karch, N.J., eds (Lewis Publishers, New York), pp. 45–58.

Fong, T.T., Mansfield, L.S., Wilson, D.L., Schwab, D.J., Molloy, S.L., and Rose, J.B. (2007). Massive microbiological groundwater contamination associated with a waterborne outbreak in Lake Erie, South Bass Island, Ohio. Environ. Health Perspect. *115*, 856–864.

Fortin, N.Y., Mulchandani, A., and Chen, W. (2001). Use of real-time polymerase chain reaction and Molecular beacons for the detection of *Escherichia coli* O157:H7. Anal. Biochem. *289*, 281–288.

de Franchis, R., Cross, N.C., Foulkes, N.S., and Cox, T.M. (1988). A potent inhibitor of Taq polymerase co-purifies with human genomic DNA. Nucleic Acids Res. *16*, 10355.

Fratamico, P.M. (2001). Applications of the polymerase chain reaction for detection, identification, and typing of food borne microorganisms. In Microbial Food Contamination, Wilson, C.L., and Droby, S., eds (CRC Press, New York), pp. 95–99.

Freitag, T.E., and Prosser, J.I. (2009). Correlation of methane production and functional gene transcriptional activity in a peat soil. Appl. Environ. Microbiol. 75, 6679–6687.

Freitag, T.E., Toet, S., Ineson, P., and Prosser, J.I. (2010). Links between methane flux and transcriptional activities of methanogens and methane oxidizers in a blanket peat bog. FEMS Microbiol. Ecol. 73, 157–165.

Frostegård, A., Courtois, S., Ramisse, V., Clerc, S., Bernillon, D., Le Gall, F., Jeannin, P., Nesme, X., and

Simonet, P. (1999). Quantification of bias related to the extraction of DNA directly from soils. Appl. Environ. Microbiol. *65*, 5409–5420.

García, R., Bælum, J., Fredslund, L., Santorum, P., and Jacobsen, C.S. (2010). Influence of temperature and predation on survival of *Salmonella enterica* Serovar Typhimurium and expression of *invA* in soil and manure-amended soil. Appl. Environ. Microbiol. *76*, 5025–5031.

González-Escalona, N., Hammack, T.S., Russel, M., Jacobson, A.P., De Jesús, A.J., Brown, E.W., and Lampel, K.A. (2009). Detection of live *Salmonella* sp. cells in produce by a TaqMan-Based quantitative reverse transcriptase real-time PCR targeting *invA* mRNA. Appl. Environ. Microbiol. *75*, 3714–3720.

Gourse, R.L., Gaal, T., Bartlett, M.S., Appleman, J.A., and Ross, W. (1996). rRNA transcription and growth rate-dependent regulation of ribosome synthesis in *Escherichia coli*. Annu. Rev. Microbiol. *50*, 645–677.

Gutknecht, J.L.M., Goodman, R.M., and Balser, T.C. (2006). Linking soil process and microbial ecology in freshwater wetland ecosystems. Plant and Soil. *289*, 17–34.

Hahn, D., Kester, R., Starrenburg, M.J.C., and Akkermans, A.D.L. (1990). Extraction of ribosomal RNA from soil for detection of *Frankia* with oligonucleotide probes. Arch. Microbiol. *154*, 329–335.

Hanna, S.E., Connor, C.J., and Wang, H.H. (2005). Real-time polymerase chain reaction for the food microbiologist: technologies, applications, and limitations. J. Food Sci. *70*, R49-R53.

Harrigan, W.F. (1998). Laboratory Methods in Food Microbiology, 3rd edition (Academic press, New York).

Hierro, N., Esteve-Zarzoso, B., Gonzalez, A., Mas, A., and Guillamón, J.M. (2006). Real-time quantitative PCR (QPCR) and reverse transcription-QPCR for detection and enumeration of total yeasts in wine. Appl. Environ. Microbiol. 72, 7148–7155.

Holmes, D.E., Nevin, K.P., and Lovley, D.R. (2004). *In situ* expression of *nifD* in Geobacteraceae in subsurface sediments. Appl. Environ. Microbiol. *70*, 7251–7259.

Hoorfar, J., Cook, N., Malorny, B., Wagner, M., De Medici, D., Abdulmawjood, A., and Fach, P. (2003). Making internal amplification control mandatory for diagnostic PCR. J. Clin. Microbiol. *41*, 5835.

Hoorfar, J., Cook, N., Malorny, B., Wagner, M., De Medici, D., Abdulmawjood, A., and Fach, P. (2004). Diagnostic PCR: making internal amplification control mandatory. Lett. Appl. Microbiol. *38*, 79–80.

Howard, G., Bartram, J., Pedley, S., Schmoll, O., Chorus, I., and Berger, P. (2006). Groundwater and public health. In Protecting groundwater for health: managing the quality of drinking-water sources, Schmoll, O., Howard, G., Chilton, J., and Chorus, I., eds (IWA Publishing, London), pp. 3–19.

Hugget, J.F., Novak, T., Garson, J.A., Green, C., Morris-Jones, S.D., Miller, R.F., and Zumla, A. (2008). Differential susceptibility of PCR reactions to inhibitors: an important and unrecognized phenomenon. BMC Res. Notes. *1*, 70.

Hurt, R.A., Qiu, X., Wu, L., Roh, Y., Palumbo, A.V., Tiedje, J.M., and Zhou, J. (2001). Simultaneous recovery of RNA and DNA from soils and sediments. Appl. Environ. Microbiol. *67*, 4495–4503.

Ijzerman, M.M., Dahling, D.R., and Fout, G.S. (1997). A method to remove environmental inhibitors prior to the detection of waterborne enteric viruses by reverse transcription-polymerase chain reaction. J. Virol. Meth. *63*, 145–153.

Imbeaud, S., Graudens, E., Boulanger, V., Barlet, X., Zaborski, P., Eveno, E., Mueller, O., Schroeder, A., and Auffray, C. (2005). Toward standardization of RNA quality assessment using user-independent classifiers of microcapillary electrophoresis traces. Nucleic Acids Res. *33*, e56.

Jacobsen, C.S., and Holben, W.E. (2007). Quantification of mRNA in *Salmonella* sp. seeded soil and chicken manure using magnetic capture hybridization RT-PCR. J. Microbiol. Methods *69*, 315–321.

Jacobson, M., Norling, B., Gunnarson, A., and Aspan, A. (2009). Flotation-a new method to circumvent PCR inhibitors in the diagnosis of *Lawsonia intracellularis*. Int. J. Microbiol. *2009*, 410945.

Jaykus, L. (2003). Challenges to developing real-time methods to detect pathogens in foods. ASM News *69*, 341–347.

Johnson, D.R., Lee, P.K.H., Holmes, V.F., and Alvarez-Cohen, L. (2005). An internal reference technique for accurately quantifying specific mRNAs by real-time PCR with application to the *tceA* reductive dehalogenase gene. Appl. Environ. Microbiol. *71*, 3866–3871.

Karpowicz, E., Novinscak, A., Bärlocher, F., and Filion, M. (2010). qPCR quantification and genetic characterization of *Clostridium perfringens* populations in biosolids composted for 2 years. J. Appl. Microbiol. *108*, 1365–2672.

Keer, J.T., and Birch, L. (2003). Molecular methods for the assessment of bacterial viability. J. Microbiol. Methods *53*, 175–183.

Klappenbach, J.A., Dunbar, J.M., and Schmidt, T.M. (2000). rRNA operon copy number reflects ecological strategies of bacteria. Appl. Environ. Microbiol. *66*, 1328–1333.

Krascsenicsova, K., Piknova, L., Kaclikova, E., and Kuchta, T. (2008). Detection of *Salmonella enterica* in food using two-step enrichment and real-time polymerase chain reaction. Lett. Appl. Microbiol. *46*, 483–487.

Kreader, C.A. (1996). Relief of amplification in PCR with bovine serum albumin or T4 gene 32 protein. Appl. Environ. Microbiol. *62*, 1102–1106.

Kubota, H., Tsuji, H., Matsuda, K., Kurakawa, T., Asahara, T., and Nomoto, K. (2010). Detection of human intestinal catalase-negative, gram-positive cocci by rRNA-targeted reverse transcription-PCR. Appl. Environ. Microbiol. *76*, 5440–5451.

Lillis, L., Clipson, N., and Doyle, E. (2010). Quantification of catechol dioxygenase gene expression in soil during degradation of 2, 4-dichlorophenol. FEMS. Microbiol. Ecol. 73, 363–369.

Liu, B., Mørkved, P.T., Frostegård, Å., and Bakken, L.R. (2010a). Denitrification gene pools, transcription and kinetics of NO, N_2O and N_2 production as affected by soil pH. FEMS Microbiol. Ecol. 72, 407–417.

Liu, Y.J., Zaprasis, A., Liu, S.J., Drake, H.L., and Horn, M.H. (2010b). The earthworm *Aporrectodea caliginosa* stimulates abundance and activity of phenoxyalkanoic acid herbicide degraders. The ISME J. 5, 473–485.

Malorny, B., Bunge, C., and Helmuth, R. (2007). A real-time PCR for the detection of *Salmonella enteritidis* in poultry meat and consumption eggs. J. Microbiol. Methods *70*, 245–251.

Mårtensson, L., Díez, B., Wartiainen, I., Zheng, W., El-Shehawy, R., and Rasmussen, U. (2009). Diazotrophic diversity, *nifH* gene expression and nitrogenase activity in a rice paddy field in Fujian, China. Plant and Soil 325, 207–218.

Martorell, P., Querol, A., and Fernández-Espinar, M.T. (2005). Rapid identification and enumeration of *Saccharomyces cerevisiae* species directly from wine by real-time PCR. Appl. Environ. Microbiol. *71*, 6823–6830.

Matsuda, K., Tsuji, H., Asahara, T., Matsumoto, K., Takada, T., and Nomoto, K. (2009). Establishment of an analytical system for the human fecal microbiota, based on reverse transcription-quantitative PCR targeting of multicopy rRNA molecules. Appl. Environ. Microbiol. 75, 1961–1969.

Mouser, P.J., N'Guessan, A.L., Elifantz, H., Holmes, D.E., Williams, K.H., Wilkins, M.J., Long, P.E., and Lovley, D.R. (2009). Influence of heterogeneous ammonium availability on bacterial community structure and the expression of nitrogen fixation and ammonium transporter genes during *in situ* bioremediation of uranium-contaminated groundwater. Environ. Sci. Tech. 43, 4386–4392.

Miller, N.D., Draughon, F.A., and D'Souza, D.H. (2010). Real-time reverse-transcriptase-polymerase chain reaction for *Salmonella enterica* detection from Jalapeño and Serrano peppers. Foodborne Pathogens and Disease. 7, 367–373.

Monard, C., Martin-Laurent, F., Devers-Lamrani, M., Lima, O., Vandenkoornhuyse, P., and Binet, F. (2010). atz gene expressions during atrazine degradation in the soil drilosphere. Mol. Ecol. *19*, 749–759.

Monnet, C., Ulve, V., Sarthou, A.S., and Irlinger. F. (2008). Extraction of RNA from cheese without prior separation of microbial cells. Appl. Environ. Microbiol. *74*, 5724–5730.

N'Guessan, A.L., Elifantz, H., Nevin, K.P., Mouser,.PJ., Methé, B., Woodard, T.L., Manley, K., Williams, K.H., Wilkins, M.J., Larsen, J.T., *et al.* (2010). Molecular analysis of phosphate limitation in Geobacteraceae during the bioremediation of a uranium-contaminated aquifer. ISME J. *4*, 253–266.

Nicolaisen, M.H., Bælum, J., Jacobsen, C.S., and Sørensen, J. (2008). Transcription dynamics of the functional *tfdA* gene during MCPA herbicide degradation by *Cupriavidus necator* AEO106 (pRO101) in agricultural soil. Environ. Microbiol. *10*, 571–579.

Nolte, F.S. (2004). Novel internal controls for real-time PCR assays. Clin. Chem. *50*, 801–802.

Novinscak, A., and Filion, M. (2011). Effect of soil clay content on RNA isolation and detection and quantification of bacterial gene transcripts in soil using qRT-PCR. Appl. Environ. Microbiol. 77, 6249–6252.

Novinscak, A., Surette, C., Allain, C., and Filion, M. (2008). Application of molecular technologies to monitor the microbial content of biosolids and composted biosolids. Water Sci. Tech. 57, 471–477.

Ogram, A., Sun, W., Brockman, F.J., and Fredrickson, J.K. (1995). Isolation and characterization of RNA from low-biomass deep-subsurface sediments. Appl. Environ. Microbiol. *61*, 763–768.

O'Neil, R.A., Holmes, D.E., Coppi, M.V., Adams, L.A., Larrahondo, M.J., Ward, J.E., Nevin, K.P., Woodard, T.L., Vrionis, H.A., N'Guessan, A.L., and Lovley, D.R. (2008). Gene transcript analysis of assimilatory iron limitation in Geobacteraceae during groundwater bioremediation. Environ. Microbiol. *10*, 1218–1230.

Opel, K.L., Chung, D., and McCord, B.R. (2010). A study of PCR inhibition mechanisms using real time PCR. J. Forensic Sci. 55, 25–33.

Parshionikar, S., Laseke, I., and Fout, G.S. (2010). Use of propidium monoazide in reverse transcriptase PCR to distinguish between infectious and noninfectious enteric viruses in water samples. Appl. Environ. Microbiol. *76*, 4318–4326.

Paulin, M.M., Nicolaisen, M.H., and Sørensen, J. (2010). Abundance and expression of enantioselective *rdpA* and *sdpA* dioxygenases genes during degradation of the racemic herbicide (R,S)-2-(2,4-Dichlorophenoxy) propionate in soil. Appl. Environ. Microbiol. *76*, 2873–2883.

Phister, T.G., and Mills, D.A. (2003). Real-time PCR assay for detection and enumeration of *Dekkera bruxellensis* in wine. Appl. Environ. Microbiol. *69*, 7430–7434.

Powell, S.M., Ferguson, S.H., Bowman, J.P., and Snape, I. (2006). Using real-time PCR to assess changes in the hydrocarbon-degrading microbial community in Antarctic soil during bioremediation. Microbial Ecol. *52*, 523–532.

Queipo-Ortuno, M.I., Colmenero, J.D.D., Macias, M., Bravo, M.J., and Morata, P. (2008). Preparation of bacterial DNA template by boiling and effect of immunoglobulin G as an inhibitor in real-time PCR for serum samples from patients with brucellosis. Clin. Vacc. Immunol. *15*, 293–296.

Raeymarkers, L. (1993). Quantitative PCR: theoretical consideration with practical implication. Anal. Biochem. *214*, 582–585.

Rijpens, N.P., and Herman, L.M. (2002). Molecular methods for identification and detection of bacterial food pathogens. J. AOAC Int. *85*, 984–995.

Rock, C., Alum, A., and Abbaszadegan, M. (2010). PCR inhibitor levels in concentrates of biosolid samples predicted by a new method based on excitation-emission matrix spectroscopy. Appl. Env. Microbiol. *76*, 8102–8109.

Rodriguez-Lazaro, D., Hernandez, M., Esteve, T., Hoorfar, J., and Pla, M. (2003). A rapid and direct real time

PCR-based method for identification of *Salmonella* spp. J. Microbiol. Methods 54, 381–390.

Rossen, L., Norskov, P., Holmstrom, K., and Rasmussen, O.F. (1992). Inhibition of PCR by components of food samples, microbial diagnostic assays and DNA-extraction solutions. Int. J. Food Microbiol. *17*, 37–45.

Rowan, N.J. (2011). Defining established and emerging microbial risks in the aquatic environment: current knowledge, implications, and outlooks. Int. J. Microbiol. *46*, 2832.

Ruimy, R., Breittmayer, V., Boivin, V., and Christen, R. (1994). Assessment of the state of activity of individual bacterial cells by hybridization with a ribosomal RNA targeted fluorescently labeled oligonucleotidic probe. FEMS Microbiol. Ecol. *15*, 207–213.

Rutjes, S.A., Italiaander, R., van den Berg, H.H.J.L., Lodder, W.J., and de Husman, R.A.M. (2005). Isolation and detection of enterovirus RNA from large-volume water samples by using the NucliSens miniMAG System and real-time nucleic acid sequence-based amplification. Appl. Environ. Microbiol. *71*, 3734–3740.

Sambrook, J., and Russel, D.W. (2001). Molecular Cloning: A Laboratory Manual (CSH Laboratory Press, Cold Spring Harbor, New York).

Schoor, O., Weinschenk, T., Hennenlotter, J., Corvin, S., Stenzl, A., Rammensee, H.-G., and Stevanovic, S. (2003). Moderate degradation does not preclude microarray analysis of small amounts of RNA. Biotechniques *35*, 1192–1201.

Schriewer A., Wehlmann, A., and Wuertz, S. (2011). Improving qPCR efficiency in environmental samples by selective removal of humic acids with DAX-8. J. Microbiol. Methods *85*, 16–21.

Schroeder, A., Mueller, O., Stocker, S., Salowsky, R., Leiber, M., Gassmann, M., Lightfoot, S., Menzel, W., Granzow, M., and Ragg, T. (2006) The RIN: an RNA integrity number for assigning integrity values to RNA measurements. BMC Mol. Biol. *7*, 3.

Sharkey, F.H., Banat, I.M., and Marchant, R. (2004). Detection and quantification of gene expression in environmental bacteriology. Appl. Environ. Microbiol. *70*, 3795–3806.

Sheridan, G.E.C., Masters, C.I., Shallcross, J.A., and Mackey, B.M. (1998). Detection of mRNA by reverse transcription-PCR as an indicator of viability in *Escherichia coli* cells. Appl. Environ. Microbiol. *64*, 1313–1318.

Shinozuka, T., Enomoto, Y., Hayashi, H., Andod, H., and Yamaguchi, T. (2001). Variation of free radical concentrations of peat humic acid by methylation. In Humic Substances, Structures, Models and Functions, Ghabbour, E.A., and Davies, G., eds (RSC Publishing, Cambridge, UK), pp. 83–93.

Smith, C.J., Nedwell, D.B., Dong, L.F., and Osborn, A.M. (2006). Evaluation of quantitative polymerase chain reaction-based approaches for determining gene copy and gene transcript numbers in environmental samples. Environ. Microbiol. *8*, 804–815.

Smith, C.J., Nedwell, D.B., Dong, L.F., and Osborn, A.M. (2007). Diversity and abundance of nitrate reductase genes (*narG* and *napA*), nitrite reductase genes (*nirS* and *nrfA*), and their transcripts in estuarine sediments. Appl. Environ. Microbiol. *73*, 3612–3622.

Steunou, A.-S., Bhaya, D., Bateson, M.M., Melendrez, M.C., Ward, D.M., Brecht, E., Peters, J.W., Kühl, M., and Grossman, A.R. (2006). *In situ* analysis of nitrogen fixation and metabolic switching in unicellular thermophilic cyanobacteria inhabiting hot spring microbial mats. Proc. Natl. Acad. Sci. U.S.A. *103*, 2398–2403.

Straub, T.M., and Chandler, D.P. (2003). Towards a unified system for detecting waterborne pathogens. J. Microbiol. Methods 53, 185–197.

Suslov, O., and Steindler, D.A. (2005). PCR inhibition by reverse transcriptase leads to an overestimation of amplification efficiency. Nucleic Acids Res. *33*, e181.

Sutlović, D., Definis, G.M, Andelinović, S., Gugić, D., and Primorac, D. (2005). Taq polymerase reverses inhibition of quantitative real time polymerase chain reaction by humic acid. Croat. Med. J. *46*, 556–562.

Takayama, K., and Kjelleberg, S.A. (2000). The role of RNA stability during bacterial stress responses and starvation. Environ. Microbiol. 2, 355–365.

Techathuvanan, C., Draughon, F.A., and D'Souza, D.H. (2010). Real-time reverse transcriptase PCR for the rapid and sensitive detection of *Salmonella typhimurium* from pork. J. Food Prot. 73, 507–514.

Tsai, Y.L., and Olson, B.H. (1992). Rapid method for separation of bacterial DNA from humic substances in sediments for polymerase chain reaction. Appl. Environ. Microbiol. *58*, 2292–2295.

Ulve, V.M., Monnet, C., Valence, F., Fauquant, J., Falentin, H., and Lortal, S. (2008). RNA extraction from cheese for analysis of *in situ* gene expression of *Lactococcus lactis*. J. Appl. Microbiol. *105*, 1327–1333.

USFDA U.S. Food and Drug Administration. (2001). Food Code. Washington, D.C.: Food and Drug Administration. Available from: <http://www.cfsan.fda.gov/~dms/fc01-toc.html

U.S. Environmental Protection Agency. (2000). Supplementary guidance for conducting health risk assessment of chemical mixtures. Report no. USEPA/630/R-00/002. National Center for Environmental Assessment, Office of Research and Development, U.S. (EPA, Washington, DC).

Vaitilingom, M., Gendre, F., and Brignon, P. (1998). Direct detection of viable bacteria, molds and yeasts by reverse transcriptase PCR in contaminated milk samples after heat treatment. Appl. Environ. Microbiol. *64*, 1157–1160.

Vaughan, D., and Ord, B.G. (1982). An *in vitro* effect of soil organic matter fractions and synthetic humic acids on the generation of superoxide radicals. Plant Soil *66*, 113–116.

Wagner, M., Horn, M., and Daims, H. (2003). Fluorescence *in situ* hybridization for the identification and characterization of prokaryotes. Curr. Opin. Microbiol. *6*, 302–309.

Wan, K., Yousef, A.E., Schwartz, S.J., and Wang, H.H. (2006). Rapid, specific, and sensitive detection of spoilage molds in orange juice using a real-time TaqMan PCR assay. J. Food Prot. *69*, 385–390.

Wang L., and Mustapha, A. (2010). EMA-real-time PCR as a reliable method for detection of viable *Salmonella* in chicken and eggs. J. Food Sci. *75*, M134–M139.

Wang, M.Y., Olson, B.H., and Chang, J.S. (2007). Improving PCR and qPCR detection of hydrogenase A (*hydA*) associated with Clostridia in pure cultures and environmental sludges using bovine serum albumin, Appl. Microbiol. Biotechnol. *77*, 645–656.

Wang, Y., Shimodaira, J., Miyasaka, T., Morimoto, S., Oomori, T., Ogawa, N., Fukuda, M., and Fujii, T. (2008). Detection of *bphAa* gene expression of *Rhodococcus* sp. strain RHA1 in soil using a new method of RNA preparation from soil. Biosci. Biotechnol. Biochem. *72*, 694–701.

Watson, R.J., and Blackwell, B. (2000). Purification and characterization of a common soil component which inhibits the polymerase chain reaction. Can. J. Microbiol. *46*, 633–642.

Whitehouse, C.A., and Hottel, H.E. (2007). Comparison of five commercial DNA extraction kits for the recovery of *Francisella tularensis* DNA from spiked soil samples. Mol. Cell. Probes *21*, 92–96.

Wiedbrauk, D.L., Werner, J.C., and Drevonc, A.M. (1995). Inhibition of PCR by aqueous and vitreous fluids. J. Clin. Microbiol. *33*, 2643–2646.

Wilson, I.G. (1997). Inhibition and facilitation of nucleic acid amplification. Appl. Environ. Microbiol. *63*, 3741–3751.

World Health Organization. (2002). The World Health Report. Reducing Risks, Promoting Healthy Life (World Health Organization, Geneva, Switzerland).

Yaron, S., and Matthews, K.R. (2002). A reverse transcriptase-polymerase chain reaction assay for detection of viable *Escherichia coli* O157:H7: investigation of specific target genes. J. Appl. Microbiol. *92*, 633–640.

Yuan, Y., Conrad, R., and Lu, Y. (2011). Responses of methanogenic archaeal community to oxygen exposure in rice field soil. Environ. Micro. Rep. *1*, 345–347.

Zhou, J., Bruns, M.A., and Tiedje, J.M. (1996). DNA recovery from soils of diverse composition. Appl. Environ. Microbiol. *62*, 316–322.

Zoecklein, B.W., Fugelsang, K.C., Gump, B.H., and Nery, F.S. (1999). Wine Analysis and Production (Aspen Publishers, New York).

Quantitative Reverse Transcriptase Polymerase Chain Reaction for Validating Microbial Microarray Data

10

Dan Tulpan, Michelle Davey and Mark Laflamme

Abstract

The ability of DNA microarray technology to identify and quantify microbial entities and genes of interest in various environments, such as soil, water, air, compost, and blood, propelled biological, environmental and clinical research into the post-genomic era. Nevertheless, as it is valid for any new technology, errors may occur at different stages along the experimental process. Three sources of errors associated with DNA microarray utilization have been identified by Taniguchi *et al.* (2001), namely (i) the microarray fabrication, (ii) the microarray experiment and (iii) the interpretation of results (data analysis). Validation strategies are typically required to alleviate and eventually repair the undesired errors that may arise in a microarray experiment. One of the validation techniques widely accepted and used worldwide is the quantitative reverse transcriptase polymerase chain reaction (RT-qPCR).

This chapter will provide succinct introductions to microarray technologies applied to microbial research and fundamental notions regarding RT-qPCR and its use to validate microarray results. A discussion including advantages and disadvantages of microbial microarray validation using RT-qPCR will be presented and current and future trends and research directions will be summarized towards the end of the chapter.

Introduction

Fundamentals of microbial microarray technology

Over the past decade, microarrays have revolutionized researchers' ability to study simultaneously large numbers of biologically relevant molecules. While being used with the aim of identifying, quantifying or characterizing specific bio entities in complex mixtures, the microarrays functionality principle is mostly unchanged across applications. Typically a large collection of probing molecules are affixed on solid surfaces at specific locations in a matrix pattern and exposed to a sample consisting of a mixture of target components. Multiplexed hybridization processes occur at various sites across the solid surface, where probes capture corresponding targets and specific detection methods (e.g. florescence) are applied to those locations. While most arrays in use today are based on DNA oligonucleotides (Ehrenreich, 2006; Petersen and Kawasaki, 2007), arrays containing antibodies (Wingren and Borrebaeck, 2006), enzymes (Reymond and Babiak, 2007), proteins (Kricka *et al.*, 2006) or live cells (Elad *et al.*, 2008) are slowly finding their ways into exploratory research.

Largely owing to advances in biochemistry, robotics and bioinformatics, microarray fabrication has diversified, therefore leading to complex technologies that could be used to further group microarray platforms based on characteristics such as solid-surface support, probe sequence types, and target detection. A recent review paper by Miller and Tang (2009) provides a detailed overview of the basic concepts of microarray technologies and their potential applications in clinical microbiology. They identify and describe underlining technologies and major producers of 5 categories of microarrays, namely (i) printed (custom), (ii) *in situ*-synthesized (Affymetrix Inc., Roche NimbleGen, Agilent Technologies, TessArae), (iii) high-density bead-based (Illumina), (iv) electronic (NanoGen) and (v) suspension bead-based (Luminex Molecular Diagnostics). A selection of existing commercial microarrays for microbial studies is summarized in Table 10.1.

Areas of life sciences where microarray based microbial studies are pursued include gene expression, environmental microbiology and microbial ecology. An excellent review of general and specific applications of DNA microarrays for the study of microbial systems is presented in Ye *et al.* (2001) and a more detailed collection of applications is summarized in Table 10.2.

Microbial ecological microarrays typically use three types of probes: oligonucleotides (20–70 nt), gene fragments (200–1000 nt) and partial or entire microbial genome DNAs (3–7 Mnt). Starting with the study of Guschin *et al.* in 1997, oligonucleotide arrays have been successfully used to detect, for example, various ammonia-oxidizing bacteria using as substrate

Table 10.1 Commercial microarrays used for microbial studies and pathogen identification

Producer	Microarray purpose	Microarray name/application
Affymetrix	Tiling	*S. cerevisiae* Tiling 1.0R Array *S. pombe* Tiling 1.0FR Array
	3′ Gene Expression Analysis Arrays	Yeast Genome 2.0 Array Yeast Genome S98 Array
	Prokaryotic Gene Expression Analysis Arrays	*E. coli* Antisense Genome Array *E. coli* Genome 2.0 Array *P. aeruginosa* Genome Array *S. aureus* Genome Array
	DNA Analysis Arrays	SARS Resequencing Array
Agilent	Model Organism Chip-on-chip	*S. pombe* Genome Microarray, 1x244K Yeast Genome Microarray, 1x244K Yeast Genome Microarray, 4x44K
	Gene expression	*E. coli* Gene Expression Microarray, 8x15K Yeast (V2) Gene Expression Microarray, 8x15K
Akonni Biosystems	Diagnostic arrays	TruArray HE (HSV-1, HSV-2, VZV) TruArray MRSA (*S.aureus*) TruArray BTA (*B.anthracis*, *Y.pestis*, Orthopoxvirus and Venezuelan equine encephalitis) TruArray BER (35 bacteria species)
AutoGenomics Inc.	Molecular diagnostics	BioFilmChip microarrays for 50+ pathogens
Qiagen	Pathogen detection	18 HPV types, BK virus, *Chlamydia trachomatis*, *Neisseria gonorrhoeae*, multiple bacteria and adenoviruses
Oxford Gene Technology	Gene expression	48 species specific microbial arrays
Roche NimbleGen	Whole-genome tiling	*E. coli* ChIP 385K *S. cerevisiae* ChIP 385K
	Gene-expression	26 Archaea and 100+ Bacteria chips
The Forsyth Institute	Diagnostic array	HOMIM – Human oral microbe identification array (272 bacterial species)
Veredus Laboratories	Diagnostic arrays	VereFlu (Human Influenza A and B) VereThreat (Anthrax, Plague, Small Pox and Tularemia)

a polyacrylamide gel with nine probes per interrogation spot. Five years later, a high density Affymetrix GeneChip, containing over 30,000 oligonucleotide probes was used by Wilson *et al.* (2002) to identify and phylogenetically characterize populations of airborne bacterial species. A novel approach based on random sequence probes was introduced by Belosludtsev *et al.*

Table 10.2 Applications of DNA microarrays in microbial systems

Research area	Organism	References
Genome-wide transcriptional activity	*E. coli*	Selinger *et al.* (2000), Wei *et al.* (2001)
	C. crescentus	Laub *et al.* (2000), Stephens (2001)
	B. subtilis	Ye *et al.* (2000); Kort *et al.* (2008)
	S. cerevisiae	Iyer *et al.* (2001), Fay *et al.* (2004), Mira *et al.* (2010)
Regulon elucidation	*E. coli*	Zimmer *et al.* (2000)
	B. subtilis	Fawcet *et al.* (2000), Hecker and Engelmann (2000), Ye et al. (2000)
	S. pneumoniae	De Saizieu *et al.* (2000), Rimini *et al.* (2000), Peterson *et al.* (2000)
Studies of unknown DNA regions and species identification	*M. tuberculosis*	Troesch *et al.* (1999)
	Multiple species	Bavykin *et al.* (2001)
	Multiple species of respiratory viruses	Brunstein *et al.* (2006), Li *et al.* (2007), Takashi *et al.* (2008), Raymond *et al.* (2009)
Nucleic acids – protein interactions	*S. cerevisiae*	Ren *et al.* (2000), Townsend *et al.* (2003), Gerber *et al.* (2004)
Comparative genomics and genotyping	*M. tuberculosis*, *M. bovis* and BCG strains	Behr *et al.* (1999)
	H. pylori and *M. tuberculosis*	Salama *et al.* (2000), Kato-Maeda *et al.* (2001)
	S. pneumoniae	Hakenbeck *et al.* (2001)
	E. coli	Riehle *et al.* (2001)
Identification of virulence factors of microbial pathogens	*H. pylori*	Salama *et al.* (2000); Israel *et al.* (2001)
	E. coli	Chizhikov *et al.* (2001), Lloyd *et al.* (2007)
	Mycobacterium species	Ribeiro-Gimares and Passolani (2007)
Host response to pathogens or resident microflora	*B. pertussis*	Belcher *et al.* (2000)
	P. aeruginosa	Ichikawa *et al.* (2000)
	L. monocytogenes	Cohen *et al.* (2000)
	C. pneumoniae	Coombes and Mahony (2001)
	P. sojae	Moy *et al.* (2004)
	P. falciparium	Winzeler, (2006)
Effect of drugs, inhibitors and toxicity on gene expression	*M. tuberculosis*	Wilson *et al.* (1999), Fu and Tai (2009)
	E. coli	Shaw *et al.* (2003)
	S. pneumoniae	Ng *et al.* (2003)
	S. aureus	Utaida *et al.* (2003), Muthaiyan *et al.* (2008)
Microbial evolution and epidemiological studies	*M. tuberculosis*	Kato-Maeda *et al.* (2001)
Pathway engineering and bioprocess optimization	*E. coli*	Oh and Liao (2000a,b); Gill *et al.* (2001); Tao *et al.* (2001)
Functional studies	*S. cerevisiae*	Kastenmayer *et al.* (2006)

Table 10.2 Continued

Research area	Organism	References
Pathogen detection and biodefence	*Bacillus anthracis*	Sergeev *et al.* (2004), Doran *et al.* (2007)
	Clostridium botulinium	Wilson *et al.* (2002b), Raphael *et al.* (2010)
	Yersinia pestis	Zhou *et al.* (2004), Wojciechowski *et al.* (2010)
	Francisella tularensis	Lin *et al.* (2006), Postic *et al.* (2010)
	Ebola	Lin *et al.* (2006), Leski *et al.* (2009)
	Burkholderia pseudomallei	DeSantis *et al.* (2005), Schmoock *et al.* (2009)
	Staphylococcus enterotoxin B	Couzinet *et al.* (2005)
	Avian influenza	Anjum *et al.* (2006), Greninger *et al.* (2010)
	SARS	Cutler *et al.* (2001), Long *et al.* (2004)

(2004) for organism identification. They designed a 'universal microarray' (also known as a genome-independent microarray, anonymous microarray or generic microarray) consisting of 14,283 unique probes (12–13 nt each) whose sequences were randomly chosen so as to minimize their similarity while being organism independent. While their approach proved useful in fingerprinting microbial genomes, a large variation in signal intensities between replicates was observed, most probably due to the rather simple heuristic methods used in the design of the probes. Further development of universal microarray technologies applicable to the identification of unknown species is reported in recent publications (Cannon *et al.*, 2006; Chandler *et al.*, 2006; Royce *et al.*, 2007; Pasquer *et al.*, 2010). Another promising technology envisaged to improve microarray hybridization specificity consists of using gold nanoparticles coupled with specific oligonucleotide sequences for target detection (Taton *et al.*, 2000). This approach was successfully used for speciation of pathogenic strains in nosocomial infections and for the rapid detection of methicillin resistance in *S. aureus* from cultured cells (Storhoff *et al.*, 2004) and SNP detection (Bao *et al.*, 2005). However, this approach cannot be applied for high-density microarrays, since each gold nanoparticle is associated with one specific probe sequence (Avarre *et al.*, 2007).

Successful applications of microbial oligonucleotide microarrays have been reportedly used for the characterization of oral microbial diversity (Paster *et al.*, 2006; Preza *et al.*, 2008, 2009a,b; Colombo *et al.*, 2009), detection of functional genes in microbial communities (Bodrossy *et al.*, 2004; Majtan *et al.*, 2004; Rhee *et al.*, 2004; Chung *et al.*, 2005), host–microbe interactions (Cummings and Relman, 2000; Kato-Maeda *et al.*, 2001; Gomez *et al.*, 2010) and bacterial pathogenicity (Schoolnik, 2002). Nevertheless, applications of this technology to the characterization of microbial communities are still hampered mainly by the unknown and extreme variety of sequence composition in these samples (Call *et al.*, 2003).

The advantages of using oligonucleotide probes reside in the availability of data specific to a large number of uncultivated organisms, the option of using existing probe design software and the low fabrication cost for their synthesis (Bae *et al.*, 2006). The main disadvantages are the poor resolution at the species level for some bacterial groups that have between 2 and 13 rRNA operons in their genomes and the heterogeneity of 16S rRNA genes for intragenomic sequences, which affects the ability of designing high resolution probes.

By contrast with oligonucleotide-based microarrays, gene fragment and whole genome based arrays have led to advances in sensitivity, specificity and quantification of microbial species. While the length of probes increases from less than 100 nt for oligonucleotide arrays to hundreds and millions of nucleotides for gene fragment and whole genome arrays, their sensitivity also increases up to sub nano gram limits (0.25 ng) for the latter arrays (Bae *et al.*, 2006). In 2004, Zhang *et al.* adopted a 96-well high throughput working protocol to efficiently isolate high-quality genomic bacterial DNA and created a 'Library on a Slide' microarray technology using an *E. coli* reference collection allowing them to determine the absence or presence of a gene or part of a gene within the given species. More recent studies reported the successful use of whole genome microarrays in comparative genomic analysis of bacterioplankton gene transcripts (Poretsky *et al.*, 2005), *Streptococcus mutans* (Zhang *et al.*, 2009), and group B *Streptococcus* (Seo *et al.*, 2010). A major challenge with these increasingly promising microarray technologies resides in the limited understanding and the absence of experimentally validated computational models for nucleic acid hybridization chemistry of such extremely large probes. The current state-of-the-art hybridization models heavily rely on tens and hundreds of experimental parameters derived from studies of short DNA and RNA sequences (less than 100 nt) hybridized in solutions (Serra *et al.*, 1995; SantaLucia and Hicks, 2004), thus their applicability is limited to probe selection for nucleotide microarrays. Another challenge for genome microarrays resides in their inability to probe uncultivated bacteria. A suggested potential solution for this drawback consists of amplifying single flow-sorted bacterial cells via the multiple displacement amplification (MDA) procedure using phi 29 DNA polymerase, which reportedly produced a 5 billion-fold amplification of genomic DNA in *E. coli* (Raghunathan *et al.*, 2005) and cyanobacterium *Synechocystis* sp. strain PCC6803 (Yokouchi *et al.*, 2006).

One research direction where microbial microarrays play a critical role is pathogen detection and bio-defence. Based on the classification and prioritization of biological threat agents provided by the US Centre for Disease Control, three major categories including 36 bio-hazardous agents, ranging from *Bacillus anthracis* (Anthrax) and *Clostridium botulinium* (Botulism) to Avian influenza (H5N1) and SARS have been identified to pose a significant threat for public health. Uttamchandani *et al.* (2009) provide a succinct review of microarray applications for pathogen detection and bio-defence and highlight the importance and the impact of microarray technologies on shortening the time (from days to hours or minutes) required to detect and identify bio-threats with high accuracy.

In research areas that investigate higher level eukaryotic organisms, standard microarray platforms have been established based on well defined probe design principles. However, the microbial microarrays field relies heavily on custom designs given the new challenges specific to this research area, namely: smaller sample sizes, diverse environments (soil, water, air), and humongous numbers of unknown species. Thus a good understanding of microarray probe design technologies and their scalability and adaptability for microbial studies is highly desirable. From a probe design point of view, nucleotide microarray technologies are better supported by existing hybridization models and design methods. The specificity of hybridizing oligonucleotides with targets is a complex process that involves parallel competitive interactions, which in turn can be simulated using advanced mathematical and thermodynamic models based on very large systems of coupled quadratic equations. Tulpan *et al.* (2005) present a model for competitive hybridizations between every probe and two targets applied to sets of probes designed with the aid of a non-trivial stochastic local search algorithm that maximizes the energetic gap between perfect match duplexes and undesired cross-hybridizations. The model makes the simplifying assumption that the equilibria between any such triplets of strands are independent of each other. One of the most debated criteria for probe design is finding an optimal probe length for which both, sensitivity and specificity are maximized. While some studies suggest that the optimal nucleotide length that achieves such a compromise between specificity and sensitivity is 60 nt (Hughes *et al.*, 2001; Agilent Inc.), other studies suggest that 50-mers (Schroder *et al.*, 2001), 25-mers (Affymetrix Inc.) and 19-mers (Zhou and Thompson, 2002) provide the highest discrimination between targets with higher sequence similarities. Since finding a general rule in regard to the optimal probe length proves to be so difficult and causes major discrepancies between oligonucleotide melting temperatures for high density arrays (Southern *et al.*, 1999), alternative approaches have been suggested that downplay the importance of the probe length criterion and focus on selecting probes of various lengths with isothermal behaviours. One such approach was proposed by Blazewicz and Kasprzak (2006) for the reconstruction of DNA chains using sequencing by hybridization (SBH), where equal melting temperature rather than equal length probes are selected. Four years later, Høvik and Chen (2010) describe a comprehensive computational pipeline that employs isothermal probe selection for designing tiling probe sets to study microbial transcriptome profiles. A comprehensive summary of existing probe design techniques is provided in Table 10.3.

To date, the usage of microarrays as investigative and discovery based tools for microbial communities in complex samples is still in its infancy, mostly due to difficulties that arise from hybridizing genomic DNA, the absence of validated computational modelling techniques for long DNA sequence hybridization and the complexity of unknown microbial species and subspecies. While solutions for similar problems have been addressed in other fields, such as biomedical research (tiling arrays, complex analysis techniques), more efforts are still required to be pursued for ecological studies.

Fundamentals of RT-qPCR technology

Microarray technology is an extremely powerful tool for screening gene expression profiles of biological samples. However, microarray data can easily contain errors originating from probe interchange, array production, labelling reactions, hybridization, and data acquisition. After data mining, it is usually advisable to validate data of those genes considered to be the most important with reliable independent techniques that allow for precise quantification of transcriptional abundance.

In 1977, the Northern blot was introduced for RNA analysis (Alwine *et al.*, 1977) and named after the DNA Southern blot invented by Sir Edwin Southern. The technique uses denaturing gel electrophoresis and blotting with hybridization probe-dependent detection of target RNAs. RNA levels can be quantified and directly compared between multiple samples on a single membrane (Kevil *et al.*, 1997; Taniguchi *et al.*,

Table 10.3 A selection of popular probe design software packages and their availability as of December 2010

Program	Company/Author	Reference/URL	Availability
AlleleID	Premier Biosoft	http://www.premierbiosoft.com/bacterial-identification/index.html	WD
ARB	Ludwig *et al.* (2004)	http://www.arb-home.de/	WD
Array Designer 4.25	Premier Biosoft	http://www.softpedia.com/get/Science-CAD/Array-Designer.shtml	WD
ArrayOligoSelector	Bozdech *et al.* (2003)	http://arrayoligosel.sourceforge.net/	WD
ChipD	Dufour *et al.* (2010)	http://chipd.uwbacter.org/	WS
CommOligo v2.0	Li *et al.* (2005)	http://ieg.ou.edu/software.htm	WD
eArray	Agilent Inc.	http://www.agilent.com/genomics/earray	WS
Featurama/ ProbePicker	ProbePicker	http://probepicker.sourceforge.net/	NA
GenomePRIDE v1.0	Haas *et al.* (2003)	http://pride.molgen.mpg.de/genomepride.html	FA
GoArrays	Rimour *et al.* (2005)	http://www.isima.fr/bioinfo/goarrays/	NA
HPD	Chung *et al.* (2005)	http://brcapp.kribb.re.kr/HPD/	NA
MOPeD	Patel *et al.* (2010)	http://moped.genetics.emory.edu/	WD
MPrime	Rouchka *et al.* (2005)	http://kbrin.a-bldg.louisville.edu/Tools/Mprime/	WS

Table 10.3 Continued

Program	Company/Author	Reference/URL	Availability
Mprobe v2.0	Li and Ying (2006)	http://www.biosun.org.cn/mprobe/	WD
MYcroarray Probe Design	MYcroarray.com	http://www.mycroarray.com/technology/design.html	FA
OliCheck	Charbonnier *et al.* (2005)	http://www.genomic.ch/download.php	WD
Olid	Talla *et al.* (2003)	NA	FA
OligoArray v2	Rouillard *et al.* (2003)	http://berry.engin.umich.edu/oligoarray2/	WD
OligoDb	Mrowka *et al.* (2002)	http://bioinf.charite.de/oligodb/	NA
OligoDesign	Tolstrup *et al.* (2003)	http://oligo.lnatools.com/expression/	NA
OligoFaktory	Schretter and Milinkovitch (2006)	http://oligofaktory.lanevol.org/index.jsp	NA
OligoPicker	Wang and Seed (2003)	http://pga.mgh.harvard.edu/oligopicker/	WD
OligoSpawn	Zheng *et al.* (2006)	http://oligospawn.ucr.edu/	NA
OligoWiz	Nielsen *et al.* (2003)	http://www.cbs.dtu.dk/services/OligoWiz/	WD, WS
Oliz	Chen and Sharp (2002)	http://www.utmem.edu/pharmacology/otherlinks/oliz.html	NA
Osprey	Gordon and Sensen (2004)	http://osprey.ucalgary.ca/cgi-bin/oligo_array	NA
PhylArray	Militon *et al.* (2007)	http://fc.isima.fr/~rimour/phylarray/	WS
PICKY	Chou *et al.* (2004)	http://www.complex.iastate.edu/download/Picky/index.html	WD
PrimeArray	Raddatz *et al.* (2001)	http://pages.unibas.ch/wtt/Products/PrimeArray/primearray.html	NA
PRIMEGENS v2	Srivastava *et al.* (2008)	http://compbio.ornl.gov/structure/primegens/	NA
ProbeMaker	Stenberg *et al.* (2005)	http://probemaker.sourceforge.net/	WD
Probemer	Emrich *et al.* (2003)	http://probemer.cs.loyola.edu/	NA
Prober	Navin *et al.* (2006)	http://prober.cshl.edu/	WD
Probe-sel	Kaderali and Schliep (2002)	NA	FA
ProbeSelect	Li and Stormo (2001)	http://ural.wustl.edu/resources.html	NA
Probe Tool	University of Cologne	http://jakob.genetik.uni-koeln.de/bioinformatik/software/probedesign/	WD
ProbeWiz	Nielsen and Knudsen (2002)	http://www.cbs.dtu.dk/services/DNAarray/probewiz.php	WS
ProDesign	Feng and Tillier (2007)	http://www.uhnresearch.ca/labs/tillier/ProDesign/ProDesign.html	WS
ProMide	Rahmann (2003)	http://oligos.molgen.mpg.de/	FA
ROSO	Reymond *et al.* (2004)	http://pbil.univ-lyon1.fr/roso/	WS, NA
SEPON	Hornshoj *et al.* (2004)	http://www.agrsci.dk/hag/sepon/	NA
Teolenn	Jourdren *et al.* (2010)	http://transcriptome.ens.fr/teolenn/	WD
SLSDNADesigner	Tulpan *et al.* (2005)	http://www.cs.ubc.ca/labs/beta/Software/DnaCodeDesign/	WD
UPS v2.0	Chen *et al.* (2010)	http://array.iis.sinica.edu.tw/ups/	WS
YODA	Nordberg (2005)	http://pathport.vbi.vt.edu/YODA/	NA

WD, web download; NA, not available (web site); FA, from authors; WS, web server.

2001). However, non-specific hybridization can confound data interpretation and the use of radioactivity is frequently necessary (Maderazo, 2003). Furthermore, this method lacks the accuracy of qPCR. Serial analysis of gene expression (SAGE) is a sequence-based high-throughput technique that can more exactly quantify gene expression than microarray experiments (Zhang *et al.*, 1997), however the technique is cumbersome, time-consuming, requires multiple manipulations of the samples, and requires large amounts of input RNA. Although several variants have been developed since the technique was first introduced, such as LongSAGE (Saha *et al.*, 2002), RL-SAGE (Gowda *et al.*, 2004), and SuperSAGE (Matsumura *et al.*, 2005), the basis of these techniques, and hence their limitations, remains the same. The RNase protection assay is also used to detect, quantify, and characterize RNA species due to its sensitivity and specificity (Zinn *et al.*, 1983; Melton *et al.*, 1984). Hybridization of antisense RNA corresponding to a known complementary target sequence prevents target digestion by single strand-specific RNase activity. This process results in the degradation of all remaining single-stranded RNAs (i.e. those not hybridized to the probe sequence), enabling the accurate quantification of specific target sequences. However, these techniques are long to perform and involve protocols that are not amenable to the examination of different transcripts in a high-throughput manner. Furthermore, the required amount of RNA can be quite large (VanGuilder *et al.*, 2008).

Since its inception more than 25 years ago (Saiki *et al.*, 1985), the polymerase chain reaction (PCR) has undergone multiple improvements and adaptations for a plethora of molecular biology applications. Although a number of quantitative PCR methods relying on end point analysis of PCR products were proposed soon after its introduction, these techniques were originally developed for DNA quantification (Abbott *et al.*, 1988; Syvanen *et al.*, 1988). Isolation and exploitation of reverse transcriptase enzymes to convert messenger RNA (mRNA) into complementary DNA soon extended the method's utility to the measurement of gene expression (Becker-Andre and Hahlbrock, 1989; Wang *et al.*, 1989; Gilliland *et al.*, 1990). However, it was not until the introduction of fluorescent techniques to monitor and quantify the accumulation of PCR products in 'real time' in the mid-1990s (Riedy *et al.*, 1995; Chiang *et al.*, 1996; Gibson *et al.*, 1996; Heid *et al.*, 1996) that gene expression analysis by reverse-transcription quantitative PCR (RT-qPCR) became a key enabling technology in the genomics field. The conceptual simplicity, practical ease and high-throughput capacity, coupled with the requirement for only very small amounts of input RNA, have made the RT-qPCR assay the most widely used method for quantification of moderate gene numbers (Klein, 2002) and the 'gold standard' for validation of microarray discoveries (Canales *et al.*, 2006).

The performance of a RT-qPCR experiment consists of a number of steps. In general these are (1) isolation and characterization of RNA, (2) conversion of RNA to complementary DNA (cDNA) via a reverse transcription reaction, (3) amplification of cDNA via PCR and acquisition of real time data, and (4) normalization and analysis of the acquired data. While in theory, PCR is quite robust and predictable, in reality even minor variations in template, reaction components and thermal cycling conditions can lead to large changes in the overall amount of amplified product. In order to garner relevant gene expression results from real time RT-qPCR experiments, it is not only important to have an understanding of the basic principles and chemistries behind the technique, but also to be acutely aware of the caveats, sources of variation and limitations at each step of the procedure.

A significant advantage of real-time PCR is that no post PCR handling is required, such as gel electrophoretic analysis of products, which effectively eliminates the risk of carry-over contamination, a vitally important feature for clinical settings. As the assay can be assessed in real time, several quality controls can be integrated to monitor the efficacy of the PCR assay, such as final fluorescence, replicate uniformity, slope of the fluorescence curve, and melting curve analysis. Perhaps that most important advantage of qPCR over earlier methods for measuring RNA is its ability to quantify over an extraordinarily wide dynamic range (at least five log units). This is coupled to extreme sensitivity, allowing the detection of less than five copies of a target sequence, making it possible to analyse samples where input material is extremely limited, such as clinical biopsies or samples obtained by laser capture microdissection (Valasek and Repa, 2005). Real-time PCR assays are 1000 to 100,000-fold more sensitive than RNase protection assays (Wang and Brown, 1999), 1000-fold more sensitive than dot blot hybridizations (Malinen *et al.*, 2003), and have been shown capable to even detect a single copy of a specific transcript (Palmer *et al.*, 2003). With appropriate internal standards and calculations, mean variation coefficients are 1–2%, allowing reproducible analysis of subtle gene expression changes even at low levels of expression (Klein, 2002; Luu-The *et al.*, 2005). In addition, some real-time platforms can be automated for high-throughput applications. Furthermore, full advantage can be made of the fluorescent signal detection system in that multiple fluorophores can be monitored simultaneously. The system is very amenable to automation and is constantly being adapted to new formats which permit high-throughput, such as the recently developed 384-well format, which requires very little reagent and permits very rapid PCR cycles. With the use of four different fluorophores, up to 1536 different targets can be simultaneously analysed in a real time setting, with results being obtained in as little as 10 minutes (Hahn *et al.*, 2005).

RNA isolation and characterization

While in theory RT-qPCR differs from qPCR only by the addition of the initial conversion of RNA into a DNA template, in practice this additional step results in a much more tenuous and variable assay (Bustin and Mueller, 2005). In any experiment, the accumulation of errors begins at the initial steps. Although it is known that the accuracy of gene-expression profiling is highly dependent on mRNA quality (Nolan *et al.*, 2006a,b), very few qPCR users, even those experienced, give this area due consideration (Bustin, 2008a; Derveaux *et al.*, 2010). A study of the impact of RNA quality on the expression stability of reference genes indicated that it is inappropriate to compare degraded and intact samples (Perez-Novo *et al.*, 2005). Significant degradation of target RNA can result in a *Cq* that is artificially high, leading to an underestimation of its concentration and copy number. In fact, a difference of more than six cycles in *Cq* values for degraded versus intact RNA samples may be observed (Bustin *et al.*, 2009a). In addition to the profound effect RNA quality may have on the analytical results of a single gene, this impact can be much more pronounced when multigene signatures are being assessed and normalized to reference genes (Derveaux *et al.*, 2010). Hence, sample quality control prior to RT-qPCR measurements is an absolute necessity.

Unlike DNA, RNA is extremely fragile once removed from its cellular environment. Therefore, its purification and handling is much trickier than that of DNA. Procurement of samples and RNA extraction therefore requires adherence to strict experimental protocols for acquisition, processing, and storage to ensure reproducibility and minimize standard deviations between replicates. The choice of method used for nucleic acid extraction can be a major determinant of the final transcript quantification. Nucleic acid extraction efficiencies can vary considerably between different methods and the final RNA yield is dependent on both, the method used and the type of sample being studied (e.g. soil, water, food, blood, etc.) (Smith and Osborn, 2009). To properly compare transcript numbers from different patients or environmental samples, the same extraction procedure is recommended to be used for each sample.

A major problem when working with prokaryotic mRNA as compared with eukaryotic mRNA, is its distinct instability. Prokaryotic mRNA only has a half-life in the range of 40 seconds to 20 minutes for individual transcripts with the activity of bacterial RNAses being responsible for this rapid turnover (Kushner, 2002). Clearly, prokaryotic RNA is much harder to handle than eukaryotic RNA. This instability demands special care during isolation of prokaryotic RNA to avoid artefacts. It is possible to fail to observe the expression of certain genes simply because of degradation of the corresponding mRNA during an experiment. It must also be considered that depending on the promoter and its regulation, bacterial transcription initiation and elongation rates can be measured in seconds (Ehrenreich, 2006). Microorganisms can quickly adjust their transcription to environmental changes that might occur during harvesting of the culture and cell disruption. Therefore, extreme care during harvest, cell disruption and RNA isolation using methods that stop transcription immediately and prevent RNA breakdown is required to avoid measuring changes in transcript levels solely due to these stresses. Methods to aid this purpose can include immediate cooling of the cells upon harvest, disrupting cells using liquid nitrogen and a cooled ball mill, and dissolving disrupted cells immediately in buffer containing guanidium isothiocyanate to inhibit RNase activity (Ehrenreich, 2006). If samples must be collected over a period of time or in too large numbers to be processed immediately, they should be stored in appropriate conditions (frozen at −80°C and/or in RNA storage solution) until use and once thawed, be kept on ice whenever possible. Handling time during the RNA extraction procedure should be minimized by processing samples in small batches (Taylor, 2010). The RNA extraction procedure should also include a DNase I treatment step to remove any contaminating genomic DNA. Many commercially available kits include an on-column DNase I treatment step.

Ensuring that RNA extracted from samples is of high purity (no contaminants) and high integrity (not degraded) is one of the most critical points in the RT-qPCR work flow. Impurities can lead to inhibition of the reverse transcription reaction, yielding biased data. Use of partially degraded RNA can lead to varying and incorrect quantification results (Fleige and Pfaffl, 2006; Gingrich *et al.*, 2006). Since sample purity and integrity are not related, both should be assessed to ascertain that the RNA sample meets minimal acceptance criteria for downstream applications. Purity of the sample can be assessed spectrophotometrically by measuring the optical density at different wave lengths: 260 nm (absorption maxima of nucleic acids), 280 nm (absorption maxima of proteins), 230 nm (absorption of contaminants), and 320 nm (absorption of contaminants). The OD260/OD280 ratio, which is commonly used as an indicator for RNA purity, is optimal between 1.8 and 2.0 (Becker *et al.*, 2010).

RNA integrity can be assessed using several methods. The traditional method is visual inspection after electrophoresis on a formaldehyde agarose gel in the presence of a fluorescent dye such as ethidium bromide. The observation of two sharp bands, representing the large and small subunits of rRNA (28S and 18S for eukaryotes; 23S and 16S for prokaryotes), with intensity of the larger band approximately twice that of the small band is indicative of intact RNA. While this method is inexpensive, the interpretation of data are mostly subjective and the procedure is cumbersome, low through-put and requires significant amounts of total RNA, which may be difficult if sample availability is limited. A major improvement in RNA integrity analysis has been the introduction of microfluidics-based electrophoresis systems (Imbeaud *et al.*, 2005) such as the Experion (Bio-Rad Laboratories) or the Bioanalyzer 2100 (Agilent Technologies) which combine integrity and quantification of RNA in a single step using only nanogram or picogram quantities. In addition to generating a virtual gel, an electropherogram, and calculating the 28S/18S (or 23S/16S) ratio, the software automatically calculates and reports a quality index scale to represent the level of degradation in a sample (Denisov *et al.*, 2008; Becker *et al.*, 2010). The Agilent RNA Integrity Number (RIN) and the Bio-Rad RNA Quality Indicator (RQI) are both numbers from 1 to 10 derived by algorithms applied to the electrophoretic traces. A RQI/RIN of 1 represents fragmented and degraded RNA and a RQI/RIN of 10 represents intact and non-fragmented RNA (Schroeder *et al.*, 2006). Fleige and Pfaffl (2006) recommend an RNA quality score (RIN or RQI) higher than five as indicative of acceptable total RNA quality and higher than eight as indicative of perfect total RNA for downstream applications such as RT-qPCR. RNAs of widely dissimilar quality (e.g. RQI/RIN values of 5.0 vs. 9.5) should not be quantitatively compared. By ensuring consistency in both purity and quality across RNA samples, variability between biological replicates can be reduced. It is recommended that once a batch of RNA samples has been fully analysed, the samples should be immediately converted into much more stable cDNA to avoid degradation from multiple freeze/thaws or extended storage (Bustin *et al.*, 2010; Taylor, 2010).

Even when RNA quality is assessed using gel electrophoresis or microfluidics-based systems, users should be aware that such measurements only look at ribosomal RNA without relating the results to mRNA integrity. Hence, rRNAs yielding similar RQI numbers may contain mRNAs that differ significantly in their integrity (Bustin *et al.*, 2010). PCR based tests can be used to better indicate the status of mRNA integrity. In one such test, cDNA is synthesized from the RNA template using oligo-dT primers for eukaryotes (i.e. from the 3′ end of the RNA). Amplicons for a universally expressed gene, or panel of reference genes, located at the 3′ and 5′ ends of the transcript are amplified via PCR and the ratio between their abundance calculated (3′:5′). A ratio greater than 5 suggests that the RNA is degraded. The integrity of the measured transcript(s) is taken as representative of the integrity of all mRNAs in a given sample (Nolan *et al.*, 2006b).

It is worth mentioning that *in vivo* mRNA is subject to constant degradation by exo- and endonucleases. This is likely to result in significant natural variability of mRNA levels. This is in addition to any degradation introduced during the extraction or storage of RNA. While these comments are obvious, their implications have never been explored (Bustin, 2008b).

Every RNA sample or, in the case of large numbers, representative samples should also be tested for the presence/absence of inhibitors of the reverse transcription and PCR reactions. Co-purification of inhibitors of the RT-qPCR assay during RNA template preparation can present a serious problem to accurate and reproducible quantification of mRNA transcript levels (Cone *et al.*, 1992). Common inhibitors can include various components of body fluids and reagents encountered in clinical and forensic science (e.g. haemoglobin and urea), food constituents (e.g. organic and phenolic compounds, fats) and environmental compounds (e.g. humic acids and heavy metals) (Wilson, 1997). The presence of such inhibitors can make the comparison of RT-qPCR results from different patients/environments or different samples from the same patient/environment next to impossible as it results in different amplification efficiencies and hence different *Cq* values (Bustin and Mueller, 2005). The so-called SPUD assay, a PCR-based test, can be used for this purpose (Nolan *et al.*, 2006a). By comparing the *Cq* value of a known concentration of a spiked RNA molecule in both a control sample and in the sample of unknown quality, the extent of inhibition of the RT-qPCR reaction can be determined.

Due to the sensitivity of fluorescence-based RT-qPCR methods, it is particularly important that the RNA template is free from contaminating DNA that could contribute to the final amplification signal. While it is sometimes possible to design an assay that spans an intron such that residual DNA is not co-amplified, this strategy can be problematic. First, it makes the assay design process more cumbersome, and secondly it will not work for genes that are either single exon genes or have one or more processed pseudogenes (retropseudogene or intronless copy) in the genome (Derveaux *et al.*, 2010). The most efficient strategy is to do a proper DNase treatment, followed by a careful check for absence of DNA through qPCR analysis of a DNA target on the crude RNA (Vandesompele *et al.*, 2002b).

A last point of attention with respect to RNA samples is the increased use of sample pre-amplification protocols. Indeed, many biological specimens are valuable and often the amount of RNA extracted is limiting for large-scale genome-wide expression studies. RNA pre-amplification methods address this issue by producing micrograms of cDNA starting from a few nanograms of total RNA. Prior to using one of the different amplification protocols that are available, there is only one important criterion that has to be evaluated: fold changes in expression between two samples should be preserved before and after amplification.

Hence, in order to obtain reliable quantification results on RT-qPCR analysis, particular attention must be paid to all steps in the RNA preparation workflow. To be considered a suitable RNA template for inclusion in a RT-qPCR assay, the RNA sample should fulfil the following criteria: (1) be of high quality (i.e. unfragmented) (Fleige and Pfaffl, 2006; Becker *et al.*, 2010), (2) be free of DNA, especially if the target is an intronless gene,

(3) be free of protein and of enzymatic inhibitors (Nolan *et al.*, 2006a) and (4) be free of RNases (Bustin and Nolan, 2004).

Reverse transcription

Although seemingly a small step, the conversion of RNA to an amplifiable cDNA template is one of the main contributors to variability in RT-qPCR assays (Bustin *et al.*, 2005; Tichopod *et al.*, 2009). The contribution of the reverse transcription step to variance in gene expression measurements has been shown to be as much as 20% (Kitchen *et al.*, 2010). The lack of standardization for this crucial step is mirrored in the wide range of commercially available reverse transcriptase enzymes, and in the diversity of priming strategies and protocols used. During the reverse transcription reaction, a RNA-dependent DNA polymerase transcribes single-stranded RNA into complementary DNA (in the presence of primers, an optimized buffer system, and free dNTPs). Ideally, the reaction should result in a cDNA population that reflects the original mRNA population in terms of transcript abundance and complexity (Stangegaard *et al.*, 2006; Smith and Osborn, 2009). Also, the RT reaction should be as efficient as possible to give maximum sensitivity in the final assay. RNA quality (Mannhalter *et al.*, 2000; Bustin and Nolan, 2004), priming strategy (Brooks *et al.*, 1995; Stahlberg *et al.*, 2004a), and enzyme efficiency (Stahlberg *et al.*, 2004b; Okello *et al.*, 2010) are important parameters for obtaining high yield of good quality cDNA.

Most known RT enzymes are encoded by retroviruses (where their *in vivo* role is to copy the viral RNA genome into DNA prior to its integration into host cells). The first isolated RT enzymes were AMV (avian myeloblastosis virus) and MMLV (Moloney murine leukaemia virus). Since then, several engineered versions, derived from these original isolates, have been developed to enhance polymerase activity or decrease unwanted nuclease activities (e.g. Omniscript®, PowerScript™, SuperScript™ III, etc.) (Valasek and Repa, 2005). It has been demonstrated that different reverse transcriptases can have significantly different efficiencies of reverse transcription (as much as 100-fold differences in yields), and that these efficiencies can be target-dependent as a consequence of mRNA secondary and tertiary structures (Stahlberg *et al.*, 2004a,b). Moreover, the conversion efficiency can be significantly (more than 3-fold) lower when target templates are rare. This observation is of prime importance, since transcripts of gene regulators or modulators, such as transcription factors, silencers, and enhancers, are often present in low abundance. These rare transcripts are sometimes not detected by microarray (Cao *et al.*, 2004; Okello *et al.*, 2010). Levesque-Sergerie *et al.* (2007) evaluated the influence of non-specific or total background RNA amounts and transcript abundance on the efficiency of different commercial RTases. They found that although the RNA-to-cDNA conversion of abundant transcripts was efficient with all the systems tested in the presence of various amounts of background RNA, cDNA could not be detected with certain RT systems when low-abundance transcripts were involved. Similarly, Okello *et al.* (2010) performed RT-qPCR to assess the sensitivity and reproducibility of 11 commercially available reverse transcriptases in cDNA synthesis from low copy number RNA levels. The enzymes varied in sensitivity (from 10 to 1000 copies). Those that demonstrated the best sensitivity still showed considerable variation in reproducibility and overall amplification efficiency. The authors recommended the use of Accuscript® or Superscript™ III when dealing with low copy number RNA levels. The components of the RT reaction (enzyme and buffer system) itself can be an important source of PCR inhibitors, interfering as much as 70% with the qPCR assay (Levesque-Sergerie *et al.*, 2007). If present at high concentrations, the RT enzyme can inhibit the PCR reaction presumably due to its RNase H domain which cleaves DNA–RNA hybrids. In many commercial enzymes, this domain has been inactivated by point mutation (Hawkins *et al.*, 2003) or the enzyme engineered for reduced RNase H activity. However, Stahlberg *et al.* (2004b) found no advantage when using reverse transcriptases without RNase H activity on yields, reproducibility and sensitivity of low to intermediate expressed genes evaluated by RT-qPCR. A compilation of some available RT enzymes is provided in Table 10.4.

Table 10.4 A compilation of some commercially available reverse transcriptase enzymes

Name	Source	RNase H activity	Recommended temperature range	Dynamic range	Company
AccuScript®	MMLV	RNase H–	37°C–42°C	10 ng–5 µg	Agilent Technologies
AffinityScript™	MMLV	RNase H–	42°C–55°C	1 ng–500 ng	Agilent Technologies
ArrayScript™	MMLV	RNase H–	42°C	1 µg–5 µg	Ambion
GoScript™	MMLV	?	37°C–55°C	1 pg–5 µg	Promega Corporation
ImProm-II™	Proprietary	?	37°C–55°C	1 pg–1 µg	Promega Corporation
iScript™	MMLV	RNase H+	42°C	1 pg–1 µg	Bio-Rad Laboratories
Maxima®	MMLV	RNase H+	50°C–60°C	1 pg–5 µg	Fermentas
Omniscript®	Proprietary	RNase H+	37°C–42°C	50 ng–2 µg	Qiagen
PowerScript™	MMLV	RNase H–	42°C	1 µg–5 µg	Clontech Laboratories
ProtoScript®	MMLV	RNase H–	42°C	10 pg–1 µg	New England Biolabs
QuantiTect®	Proprietary	RNase H+	42°C	10 pg–1 µg	Qiagen
Sensiscript®	Proprietary	RNase H+	37°C–42°C	<50 ng	Qiagen
SMART™	MMLV	↓* RNase H	42°C	100 ng–5 µg	Clontech Laboratories
SuperScript™ III	MMLV	↓ RNase H	42°C–55°C	1 pg–5 µg	Invitrogen
ThermoScript™	AMV	↓ RNase H	50°C–65°C	10 pg–5 µg	Invitrogen
Thermo-X™	Proprietary	RNase H–	50°C–70°C	1 pg–5 µg	Invitrogen

↓ RNase H, decreased RNase H activity; ?, information could not be found.

The mechanism of cDNA priming has a significant effect on the outcome of any quantification experiment. In general, the reverse transcription reaction can be initiated in three different ways: using gene-specific primers, random primers or oligo-dT primers. The three methods produce diverse results in terms of cDNA yield, variety and specificity that are distinct for different mRNA targets. Since the choice of RT priming strategy can cause marked variation in calculated mRNA levels (Zhang and Byrne, 1999; Lekanne Deprez *et al.*, 2002), the implications of using any particular method should be considered carefully

cDNA synthesis using oligo-dT primers (consisting of 16–25 deoxythymidine residues) is more specific to mRNA than random priming, as it will not transcribe rRNA. Oligo-dT priming produces cDNA from the 3′ end of polyadenylated (polyA) mRNA, which often results in a 3′ bias of the synthesized products (Brooks *et al.*, 1995; Decraene *et al.*, 1999). Hence, using oligo-dT priming can limit the selection of PCR primers to a region near the polyA tail, sometimes requiring placement within the 3′ UTR (most probably not crossing an exon/exon junction). The oligo-dT priming method can also struggle to generate transcripts from mRNAs with significant secondary structure (and where the polyA tail may be protected due to fold-back structures), and may fail to reach a PCR primer-binding site if located at the extreme 5′ end of a long mRNA (even without interference from secondary structures). This may be problematic if trying to target specific splice variants differing at the 5′ end of the mRNA (Bustin *et al.*, 2005). Obviously oligo-dT priming will not work for any RNAs that lack a polyA tail (e.g. viral RNAs). For yeast, it has been demonstrated that many to all of the 24 putative and known mitochondrial mRNAs are not stabilized by a polyA tail (Dieckmann and Tzagoloff, 1985; Butow *et al.*, 1989; Manthey and McEwen, 1995; Ellis *et al.*, 2004). Since oligo-dT priming requires very high quality full length RNA, it is not a good choice for transcribing RNA that is fragmented or partially degraded (Bustin and Nolan, 2004).

Random primers (most commonly hexamers or nonamers) anneal at multiple points along the transcript giving multiple start points for the RT enzyme, and hence producing multiple cDNA transcripts of varying lengths per original target. Random priming is by definition non-specific, but yields the most cDNA and is most useful for transcripts with significant secondary structure. Although this priming method can theoretically effectively reverse transcribe the entire RNA sample, and hence produce a total cDNA pool representative of all mRNA transcripts present within the sample, the method can be problematic. It is possible that priming may be initiated within the target amplicon and therefore eliminates the transcript from detection. Also, the majority of cDNA synthesized from total RNA will be ribosomal RNA or transfer RNA-derived. This could create problems if the target of interest is present at only low levels, as it may not be primed proportionately and its subsequent amplification may not be quantitative. Indeed, it has been demonstrated that random hexamers can overestimate mRNA copy numbers by up to 19-fold compared with a sequence-specific primer (Zhang and Byrne, 1999). It has been described as the least reliable method of priming (Lekanne Deprez *et al.*, 2002). However, an advantage of random priming is that it is less likely to give a 3′ end bias in the resulting cDNA (Ginzinger, 2002). PolyA-selected RNA, like an isolated mRNA fraction or amplified RNA (aRNA), are preferably reverse transcribed with random primers (Stangegaard *et al.*, 2006). The length of random primers in the reverse transcription reaction has been demonstrated to influence cDNA yield and amount of detectable genes upon expression analysis. Stangegaard *et al.* (2006) demonstrated that the use of 15-nucleotide-long random oligonucleotides (pentadecamers) consistently yielded at least 2-fold more cDNA as random hexamers using either polyA RNA or an amplified version of messenger RNA (aRNA) as a template. The increased priming efficiency resulted in reverse transcription of >80% of the template aRNA, while random hexamers induced RT of only 40% of the template aRNA. 11-fold more genes were detected in DNA microarray experiments using the random pentadecamer-primed reactions compared with random hexamer-primed reactions. Hence the use of random pentadecamers yielded better coverage of the transcriptome and better yield, resulting in improved sensitivity in gene expression analysis using microarrays or real-time PCR. Some investigators use a combination of random and oligo-dT primers for RT-qPCR assays to reap the benefits of both strategies. Caution must be taken if using this approach, however, since it could serve to simply exacerbate the problems inherent to the individual methods (Bustin *et al.*, 2005).

Target-specific primers synthesize the most specific cDNA and are probably the most sensitive option for quantification as their use can reduce the amount of non-specific background cDNA (Lekanne Deprez *et al.*, 2002; Smith and Osborn, 2009). Bustin and Nolan (2004) demonstrated that a reaction primed by target-specific primers is likely to be linear over a wider range than a similar reaction primed by random primers. The main disadvantage of the method is that it requires separate priming reactions for each target; hence it is not possible to return to the same preparation and amplify other targets at a later stage. It may be wasteful if only limited amounts of RNA are available. Although this priming technique is usually only considered useful if one is targeting relatively few mRNA species (Iturriza-Gomara *et al.*, 1999), Stanley and Szewczuk (2005) recently reported using gene-specific primers for reliable and specific amplification of 72 genes from limiting amounts of RNA using a multiplexed tandem PCR approach. However, while it may be possible to specifically prime more than one target in a single reaction, doing so requires careful experimental design and optimization of reaction conditions if quantitative data are to accurately reflect the target mRNA levels.

A specific problem of working with prokaryotic organisms is that their mRNA lacks polyA tails. Hence random or target-specific priming must be used on total RNA. However, only approximately 4% of the total RNA is mRNA, the rest being mainly rRNA and tRNA (Talaat *et al.*, 2000). The large amount of non-target RNA produced can significantly limit the sensitivity of the downstream assay method (microarray or RT-qPCR), especially for low copy number transcripts. Numerous attempts have been made to resolve this problem, such as preparation of polyadenylated mRNA from prokaryotes (Wendisch *et al.*, 2001) or priming with a primer set with higher probability of priming the RT of mRNA than rRNA (Talaat *et al.*, 2000). Ribosomal RNA sequences in many prokaryotes are GC rich relative to the rest of the genome and are highly conserved. These properties have been exploited to design non-random hexamers (HD/DHTTT) to prime RT reactions (Gonzalez and Robb, 2007). The result was a counter-selective synthesis of cDNA corresponding to mRNA from prokaryotic total RNA extractions. Application of gene-specific primers on a genomic scale requires synthesis of multiple primers. An algorithm to predict the minimal number of non-degenerate genome-directed primers that specifically anneal to all genes in a given genome has been designed and successfully applied in bacteria (Talaat *et al.*, 2000). Another recently developed method relies on a collection of short, computationally selected oligonucleotides ('not-so-random' (NSR) primers) to obtain full-length, strand-specific representation of non-ribosomal RNA transcripts (Armour *et al.*, 2009). Selective enrichment of non-rRNA targets was achieved by computationally subtracting rRNA priming sequences from a random hexamer library. Recently Adomas *et al.* (2010) designed an algorithm to find common sequence motifs that were present within most protein-coding genes of *Saccharomyces cerevisiae* and of *Neurospora crassa*, but that were not present within their ribosomal RNA or transfer RNA genes (providing selective targeting of multiple genes and counter-selection against undesirable sequences). They demonstrated that degenerately priming these motifs with multitargeted primers improved the accuracy and completeness of transcriptomic assays (microarray profiling and next-generation RNA tag sequencing). Furthermore, priming with the multitargeted primers in addition to oligo-dT resulted in higher sensitivity, a larger number of well-measured gene transcripts and greater power to detect differences in gene expression.

In general, random and oligo-dT priming requires lower reaction temperatures than the PCR step (i.e. room temperature). However, the lower reaction temperature can contribute to a low degree of sequence specificity in the priming event. In fact, the RT reaction can be primed by other RNA molecules present in the sample or by dNTPs (Agranovsky, 1992). This problem may be overcome by the use of thermostable reverse transcriptases such as SuperScript™ III (Invitrogen). Primer hybridization relies on access to the appropriate target site in the mRNA and may vary substantially because of mRNA folding (Sohail and Southern, 2000). In general, a higher RT annealing temperature improves RT yields by reducing formation of mRNA secondary structures (Brooks *et al.*, 1995; Kuo *et al.*, 1997; Stahlberg *et al.*, 2004b).

It is important to note that in any good experimental design, replicates should be set up as early as possible. Ideally replicates would begin at the sampling step. At the latest, replicates should begin at the RT step. A common error made when performing RT-qPCR assays is to use one replicate in the RT reaction and then take duplicates or triplicates from this sample for the qPCR assay. As the main technical variance in RT-qPCR reactions belongs to the RT step and not to the PCR step, this practice is poor. Choosing triplicates in the RT step and doing single analysis in qPCR is more representative of the true variance (Stahlberg *et al.*, 2004b; Tichopad *et al.*, 2009).

Hence, when performing gene expression measurements by RT-qPCR, the following general recommendations should be followed for the reverse transcription step: (1) use the same method of RNA quantitation for all samples; (2) where possible use high-quality RNA; (3) adjust the total RNA concentrations to be the same in all samples (adding carrier RNA if necessary); (4) perform the experiment in at least duplicate, preferably triplicate, starting with the RT step; (5) use a thermotolerant RT enzyme and heat samples at 60–65°C for 5 minutes before the RT reaction to open up secondary RNA structure; and (6) always use the same enzyme, RT priming strategy and reaction conditions in experiments that are to be compared (Stahlberg *et al.*, 2004b; Mackay, 2007).

The RT-qPCR reaction

When quantifying mRNA, RT-qPCR amplifications can be performed as either a one-step or a two-step reaction. In a one-step RT-qPCR protocol both the RT step and the qPCR step are carried out consecutively in a single tube. In a two-step reaction reverse transcription and PCR amplification are done in separate tubes. There are several advantages and disadvantages associated with each method. In addition to the reduced risk of contamination and the convenience of setting up only a single reaction, one-step RT-qPCR is thought to minimize experimental variation because both enzymatic reactions occur in a single tube. However, due to the fragile nature of the RNA starting template required for this method, it may not be suitable where the same sample needs to be assayed repeatedly over time. One-step protocols are also reportedly less sensitive than two-step protocols as reaction conditions are compromised to accommodate the two different enzymes required within a single reaction (Bustin, 2002). Two-step protocols separate the RT reaction from the qPCR assay, allowing the optimal conditions to be used for each step. Furthermore, cDNA generated in the RT reaction can be used as a template for a number of different qPCR assays. Because the RT process is notorious for its highly variable reaction efficiency, using dilutions from the same cDNA template ensures that reactions from subsequent assays have the same amount of template as those assayed earlier. If random or oligo-dT primers are used in the initial RT reaction, then any number of subsequent gene-specific qPCR amplifications may be conducted using this material (Wong and Medrano, 2005; Smith and Osborn, 2009).

Primer design for the qPCR step

Both primer design and careful choice of target sequence are essential to ensure specific and efficient amplification of products. The assay can be designed using free or commercially available software, bought from specialized vendors that provide assays off the shelf or design on demand, or garnered from a search of published and validated assays in public databases (Bustin *et al.*, 2010; Derveaux *et al.*, 2010). The first step is to find the correct sequence for the gene of interest. Several databases are available, the most popular being NCBI and ENSEMBL. Specialized databases exist for bacterial and viral sequences (http://www.ebi.ac.uk/genomes/). As more gene sequences are identified and published, the selection of a target sequence (amplicon) can become very complex/confusing because each gene may have several sequence variations posted (i.e. different isoforms, splice variants, SNPs, mutations, partial sequences) or several sequence variations may exist amongst species and strains. Amplicon selection can be particularly difficult where micro-organisms are involved. The target sequence must be unique in order to identify a specific organism or an organism group (e.g. group A *Streptococcus* or *Mycobacterium* genus screen), quantitate a microbe (e.g. cytomegalovirus), or identify unique virulence genes (e.g. verotoxin genes) or genes or mutations associated with antimicrobial resistance (e.g. *mecA* gene or mutations in *rpoB* gene associated with rifampin resistance) which can occur across strains or species (Espy *et al.*, 2006). Historically, and until recently, sequence information has primarily been derived from genome or gene fragment sequences from cultured organisms and/or from clone libraries generated by PCR using primers that are themselves based on current sequence knowledge. In the case of environmental microorganisms, molecular analysis has repeatedly shown that the majority of micro-organisms (and their genes) in the environment are highly divergent from those of most cultured organisms (Smith and Osborn, 2009). In general, target (amplicon) sequences should be unique and as short as reasonably possible (between 50 and 150 bp in length). Shorter amplicons amplify more efficiently and are more tolerant of reaction conditions (Wong and Medrano, 2005). If possible, the amplicon sequence should span an exon/exon boundary to exclude amplification from contaminating residual DNA. The GC content of the amplicon should be between 50 and 60%, and it should not contain secondary structures (Bustin and Nolan, 2004; Stahlberg *et al.*, 2004b).

Primers for the qPCR reaction can have GC contents ranging from 20% to 80% (optimally 50–60%), with higher-end GC contents tending to increase the specificity of the reaction. The optimal length for single-stranded primers is 15–20 bases. Long G or C stretches in the primer should be avoided, but it is recommended that G or C are at the end of the primers (Taylor, 2010). Hybridization probed-based and SYBR Green primers can have melting temperatures (Tm) ranging from 55°C to 65°C, whereas the T_m for hydrolysis probe-based primers should be slightly higher (68–70°C). The melting temperatures of paired primers should be similar (i.e. not differ by more than 1–2°C). When designing a hydrolysis probe (i.e. TaqMan), the probe should be situated as close as possible to the forward primer without overlapping. The probe should not have a guanine nucleotide at the 5′ end or have more guanines than cytosines as guanine residues are natural quenchers. The T_m of the probe should be 8–10°C above the T_m of the primers to ensure that the probe is fully hybridized during primer extension. Non-specific priming may be minimized by selecting primers that have only one or two GCs within the last five nucleotides at the 3′ end. If using SYBR Green, the PCR primers must not form an appreciable amount of primer-dimer bands. A melting curve analysis of each product is needed to ensure that the fluorescent signal observed is from the desired PCR product (Wong and Medrano, 2005; Smith and Osborn, 2009).

qPCR primer sequences should be compared with databases to ensure they are unique and specific for the gene of interest by doing a BLAST or BiSearch query (Arányi and Tusnády, 2007). The potential presence of SNPs in the primer annealing regions, which may hamper efficient annealing of the primer or prevent amplification of the variant allele, should be inspected. Programs such as UNAFold (http://mfold.rna.albany.edu) can then be used to analyse the amplicon sequence for potential secondary structures that may prevent efficient amplification at the predicted annealing temperature (Zuker, 2003; Markham and Zuker, 2008). There are a number of programs available that integrate these tools to help design primer pairs and pick target and probe sequences (Table 10.5), and the use of such primer design software is highly recommended (Bustin *et al.*, 2010). In practice, often calculated primers do not work well in qPCR. This problem may be particularly noted when dealing with clinical or environmental samples where the primers must be able to identify with high efficiency and specificity the target amplicon in the specimen of interest (e.g. stool, soil, water, blood) (Smith and Osborn, 2009). Therefore, at least two sets of oligonucleotides that amplify in different regions of the target gene should be tested for their performance in a qPCR reaction, the amplicon length verified using gel electrophoresis (once), and the melt curve inspected (when using SYBR Green). Additionally, since the databases currently available represent only a small portion of the nucleic acid sequences for microorganisms present in complex specimen matrices, related organisms should also be tested to confirm the lack of cross-reactivity (Espy *et al.*, 2006). Finally, a standard or dilution curve should be run to estimate the PCR efficiency. The more dilution points and the wider the range, the more precisely PCR efficiency can be determined (Hellemans *et al.*, 2007). The best practice is to use a mixture of representative samples as input material for the dilution series.

Fluorescent detection chemistries

Real-time PCR is the technique of collecting data throughout the PCR process as it occurs, thus combining amplification and detection into a single step. This can be achieved using a variety of different fluorescent chemistries that correlate PCR product concentration to fluorescence intensity. Reactions are characterized by the point in time (or PCR cycle) where the target amplification is first detected. This value is usually referred to as cycle threshold (*Ct*) or crossing point (*Cp*), the time at which fluorescence intensity is greater than background fluorescence. More recently, the term *Cq* has been recommended as the standard term to use for this value according to RDML guidelines (Lefever *et al.*, 2009). Consequently, the greater the quantity of target nucleic acid in the starting material, the sooner a significant increase in fluorescent signal will appear (Bustin and Mueller, 2005; Wong and Medrano, 2005).

Currently, there are four main fluorescent detection chemistries in general use for real-time PCR: (1) probe sequences that fluoresce upon hydrolysis (TaqMan; Applied Biosystems) or hybridization (LightCycler; Roche); (2) fluorescent hairpin probes (Molecular beacons, Scorpion probes; Sigma-Aldrich); (3) fluorophores attached to primers (Lux primers; Invitrogen, Plexor primers; Promega) and (4) intercalating DNA binding dyes (SYBR Green; Molecular Probes).

All of these chemistries allow detection of PCR products via the generation (or reduction) of a fluorescent signal. DNA binding dyes, such

Table 10.5 A selection of PCR primer design software

Program	Company/author	Reference/URL	Availability
AlleleID	Premier Biosoft	http://www.premierbiosoft.com/bacterial-identification/index.html	WD
AmplifX	Jean Roche Institute	http://ifrjr.nord.univ-mrs.fr/AmplifX	WD
Amplify v3.1	University of Wisconsin	http://engels.genetics.wisc.edu/amplify/index.html	WD
AutoPrime	Wrobel *et al.* (2004)	http://www.autoprime.de/AutoPrimeWeb	WS
BatchPrimer3	You *et al.* (2008b)	http://probes.pw.usda.gov/cgi-bin/batchprimer3/batchprimer3.cgi	WS
BeaconDesigner	Premier Biosoft Int.	http://www.premierbiosoft.com/molecular_beacons/index.html	WD
BiSearch	Arányi *et al.* (2006)	http://bisearch.enzim.hu/	WS
EasyExonPrimer	Wu and Munroe (2006)	http://129.43.22.27/~primer/EasyExonPrimer.html	WS
ExonPrimer	Helmholtz Centre, Munich	http://ihg2.helmholtz-muenchen.de/ihg/ExonPrimer.html	WS
GeneFisher2	Bielefeld University Bioinformatics Server	http://bibiserv.techfak.uni-bielefeld.de/genefisher2/	WS
GeneWorks	GeneWorks Inc.	http://www.geneworks.com.au/	WS
GenoFrag	Ben Zakour *et al.* (2004)	http://www.irisa.fr/symbiose/lavenier/genofrag.html	FA
GSTPrime	Leister and Varotto (2007)	http://www.ncbi.nlm.nih.gov/pubmed/17951794	FA
Hyden	Linhart and Shamir (2002)	http://acgt.cs.tau.ac.il/hyden/HYDEN.htm	WD
iCodeHop	University of Washington	https://icodehop.cphi.washington.edu/i-codehop-context/Welcome	WS
In-silico PCR	UCSC	http://genome.ucsc.edu/cgi-bin/hgPcr	WS
JCVI Primer Designer	J. Craig Venter Institute	http://sourceforge.net/projects/primerdesigner/	WD
MacVector	MacVector Inc.	http://www.macvector.com/	WD
MEDUSA	Podowski and Sonnhammer (2001)	http://www.cgr.ki.se/cgr/MEDUSA	NA
MethPrimer	Li and Dahiya (2002)	http://www.urogene.org//methprimer/index1.html	WS
Mpprimer	Shen *et al.* (2010)	http://biocompute.bmi.ac.cn/MPprimer/	WS, WD
MultiPLX	Kaplinski and Remm (2007)	http ://bioinfo.ut.ee/multiplx/	WS
MultiPrimer	Fernandes and Skiena (2007)	http ://www.ncbi.nlm.nih.gov/pubmed/17951802	FA
muPlex	Rachlin *et al.* (2005)	http ://genomics14.bu.edu :8080/MuPlex/MuPlex.html	NA
MutaPrimer	Premier Biosoft	http://www.premierbiosoft.com/mutagenesis/index.html	WD
NetPrimer	Premier Biosoft	http ://www.premierbiosoft.com/netprimer/	WS
PathoGene	University of Texas	http://pathogene.vbi.vt.edu/	WS
PerlPrimer	Marshal (2004)	http://perlprimer.sourceforge.net/	WD
PIRA PCR	Ke *et al.* (2001)	http://cedar.genetics.soton.ac.uk/public_html/primer2.html	WS
PRIDE	Haas *et al.* (1998)	http://pride.molgen.mpg.de/pride.html	WS
PriFi	Fredslund *et al.* (2005)	http://cgi-www.daimi.au.dk/cgi-chili/PriFi/main?configurex=101andconfigurey=30	WS

Table 10.5 Continued

Program	Company/author	Reference/URL	Availability
Primaclade	Gadberry *et al.* (2005)	http://www.umsl.edu/services/kellogg/primaclade.html	WS
Prime	GCG	http://ranger.med.nyu.edu/gcg/prime.html	NA
PrimerCE	Cao *et al.* (2010)	http://tch.hebau.edu.cn/shengm/download/down.html	NA
PrimerD	Washington University	http://mblab.wustl.edu/software/primerd/	NA
PrimerStation	Yamada *et al.* (2006)	http://ps.cb.k.u-tokyo.ac.jp/index.html	WS
PrimerX	Carlo Lapid	http://www.bioinformatics.org/primerx/	WS
Primer3	Whitehead_Institute and Howard Hughes Medical Institute	http://primer3.sourceforge.net/	WD
Primer-BLAST	NCBI	http://www.ncbi.nlm.nih.gov/tools/primer-blast/index.cgi?LINK_LOC=NcbiHomeAd	WS
Primer Design Assistant	Chen *et al.* (2003)	http://dbb.nhri.org.tw/primer/	WS
Primer3 Plus	Untergasser *et al.* (2007)	http://www.bioinformatics.nl/cgi-bin/primer3plus/primer3plus.cgi	WS
Primer Quest	Integrated DNA Technologies	http://www.idtdna.com/Scitools/Applications/Primerquest/Default.aspx	WS
PrimerPlex	Premier Biosoft Int.	http://www.premierbiosoft.com/primerplex/index.html	WD
Primer Premier	Premier Biosoft Int.	http://www.scubeindia.com/premier_biosoft/primer_premier.html	WD
Primo Pro	Chang Bioscience, Inc.	http://www.changbioscience.com/primo/primo.html	WS
QuantPrime	Arvidsson *et al.* (2008)	http://www.quantprime.de/	WS
RAPD-primer generator	Schäfer and Wöstemeyer (1992)	http://www2.uni-jena.de/biologie/mikrobio/tipps/rapd.html	FA
Real-time PCR (TaqMan) Primer Design	GenScript USA Inc.	https://www.genscript.com/ssl-bin/app/primer	WS
RealTimeDesign Software	Biosearch Technologies, Inc.	http://www.biosearchtech.com/realtimedesign	WD
SNPBox	Weckx *et al.* (2007)	http://www.snpbox.org/	WS
The PCR Suite	van Baren and Heutink (2004)	http://pcrsuite.cse.ucsc.edu/	WS
Vector NTI Software	Invitrogen	http://www.invitrogen.com/site/us/en/home/LINNEA-Online-Guides/LINNEA-Communities/Vector-NTI-Community/vector-nti-software.html	WD
Visual OMP	DNA Software, Inc.	http://www.dnasoftware.com/VisualOMP/tabid/108/Default.aspx	WD
WebPrimer	Stanford University	http://www.yeastgenome.org/cgi-bin/web-primer	WS
Xpression Primer	Premier Biosoft Int.	http://www.premierbiosoft.com/expression/index.html	WD

NA, not available; FA, from authors; WD, web download; WS, web server.

as SYBR Green, fluoresce upon light excitation when intercalated into double stranded DNA (Wittwer *et al.*, 1997; Morrison *et al.*, 1998). These are cheap, easily used with any pair of primers and for any target, and amplification products can be verified by the use of melt curves. They can lack specificity (the dye will bind to all dsDNAs formed during the reaction i.e. non-specific products and primer-dimers) and fluorescence varies with amplicon length. In general, they are one

Cq or so more sensitive than probe-based assays. Their main drawback is that the no-template controls often produce signals around *Cq* values of 36 or more, although melting curves can often distinguish genuine amplification from non-specific noise (Ririe *et al.*, 1997). A characteristic peak at the melting temperature (T_m) of the amplicon will distinguish it from amplification artefacts.

Hydrolysis (TaqMan) probes depend on the 5′-nuclease activity of the DNA polymerase used for PCR to hydrolyse an oligonucleotide that is hybridized to the target amplicon. These so-called hydrolysis probes utilize covalently bound fluorescent dye and quencher dye at the 5′ and 3′ ends, respectively. Various fluorescent reporter dyes (e.g. FAM, TET, HEX, VIC) can be used in combination with various quenchers (e.g. TAMRA, DABCYL). When the probe is intact, the proximity of the reporter and quencher dyes prevents fluorescence emission. During PCR amplification, the probe anneals to the target, and Taq polymerase cleaves the probe allowing an increase in fluorescence emission. The increase in fluorescence intensity is directly proportional to the amount of amplicon produced. TaqMan minor groove-binding probes have more recently been developed. In this chemistry, the standard TAMRA quencher at the 3′ end is replaced by a non-fluorescent quencher and a minor groove-binder molecule is also incorporated at the 3′ terminus. The latter stabilizes the probe–target complex by folding into the minor groove of the dsDNA. Additionally, the T_m of the probes is increased, allowing the use of very short oligo-probes (14 nucleotides in length) and providing more accurate allelic discrimination (de Kok *et al.*, 2002; van Hoeyveld *et al.*, 2004), making these probes ideal for detecting single nucleotide polymorphisms.

Dual hybridization probes, such as those used in the LightCycler system, use one probe carrying a donor fluorophore at its 3′ end and another probe labelled with an acceptor fluorophore at its 5′ end. Both probes hybridize to their target sequence in a head-to-tail arrangement during the annealing step. This brings the two dyes in close proximity, allowing fluorescence emission. This reaction has a high degree of specificity as a fluorescent signal is only detected when two independent probes hybridize to their correct target sequence (Aliyu *et al.*, 2004).

Molecular beacons also contain covalently bound fluorescent and quenching dyes at either end of a single-stranded DNA probe. However, they are also designed to adopt a hairpin or stem-and-loop structure while free in solution to bring the dye and quencher in close proximity (Tyagi and Kramer, 1996; Abravaya *et al.*, 2003). The loop portion of the molecule is complementary to the target nucleic acid molecule, and the stem is formed by the annealing of complementary arm sequences on the end of the probe sequence. When the probe sequence in the loop hybridizes to its target sequence during the annealing step, a conformational change occurs that forces the stem apart, allowing for separation of the fluorophore from the quencher and an increase in fluorescence emission.

Similar to Molecular beacons, scorpions adopt a hairpin configuration with a 5′ fluorophore and 3′ quencher. The specific probe sequence is held within the hairpin loop, which is attached to the 5′ terminus of a PCR primer sequence by a non-amplifiable monomer (termed the PCR stopper). This chemical modification prevents PCR from copying the stem-loop sequence of the scorpion primer. During PCR, scorpion primers are extended to form amplicon. In the annealing phase, the specific probe sequence in the scorpion tail curls back to hybridize to the complementary target sequence in the amplicon, thus opening up the hairpin loop and allowing signal emission (Whitcombe *et al.*, 1999; Thelwell *et al.*, 2000). Duplex scorpions are a modification where the fluorophore and quencher dyes are separated onto different and complementary oligonucleotides. The greater separation between reporter and quencher allows for greater signal intensity when the probe is bound to its target (Solinas *et al.*, 2001).

With probe-based methods (TaqMan, Molecular beacons, Scorpions) non-specific amplification due to mispriming or primer-dimer formation does not generate signal, as specific hybridization between probe and template is necessary for fluorescence emission. Also, probes can be labelled with different and distinguishable fluorescent reporter dyes, thus allowing specific detection and quantification of amplicons produced by several primer pairs in a single PCR reaction (multiplex qPCR). The main disadvantages of these chemistries are cost, complexity and occasional fragility of probe synthesis. There are also potential problems associated with the fact that probe-based assays do not report primer-dimers that can interfere with the efficiency of the amplification reaction.

Fluorophores attached to primers (e.g. Lux, Plexor) are relatively inexpensive and amplification products can be verified by melt curves. Light upon extension (LUX) primers are self-quenching single-fluorophore labelled primers. Rather than using a quencher fluorophore, the secondary structure of the 3′ end of the primer reduces initial fluorescence to a minimum (Nazarenko *et al.*, 2002). Specificity depends on the primers and specific (usually company-specific) design software needs to be used for optimal performance.

New fluorescent chemistries are continually being developed with a focus on increasing sensitivity (by increasing the signal-to-noise ratio) and specificity, enhancing multiplexing capabilities, and reducing cost (Valasek and Repa, 2005).

Normalization

Any gene expression quantification result is composed of two main sources of variation: technical variation (whether inherent or experimentally induced) and the true biological variation that underlies the goal of the investigation. The purpose of any normalization strategy is to remove the technical variation as much as possible, ending up with changes that are only due to true biological events. Several genes (*GAPDH*, *β-actin*, *HPRT*), ribosomal RNA, and total cellular RNA have all been proposed as means to normalize results, however each method has reported disadvantages (Bustin, 2000; Goidin *et al.*, 2001; Selvey *et al.*, 2001; Solanas *et al.*, 2001; Dheda *et al.*, 2004; Huggett *et al.*, 2005; Gutierrez *et al.*, 2008;). Reference genes are by far the most widely used method for normalization of RT-qPCR data. For each experimental sample the value of both the target and the reference gene are extrapolated from the respective standard curve equation, or relatively quantified against a calibrator sample. The target value is then divided by the endogenous reference value to obtain a normalized target value (Vandesompele *et al.*, 2002a). The identification of a valid reference for data normalization is a crucial issue in qPCR experimental design. The underlying assumptions when using reference genes to normalize data are that they are expressed at constant levels across the samples, that their expression does not vary in response to experimental manipulation and that they undergo all steps of the qPCR with the same kinetics as the gene(s) of interest (Bustin, 2000; Bustin and Nolan, 2004). Because expression of the target gene is measured relative to the reference gene, it is essential that these assumptions hold. It has become clear, however, that no single reference gene can be used for all studies and that the choice of the control gene depends on the sample type and the experimental conditions (Schmittgen and Zakrajsek, 2000; Gorzelniak *et al.*, 2001; Deindl *et al.*, 2002; Tricarico *et al.*, 2002; Schmid *et al.*, 2003; Aerts *et al.*, 2004). Vandesompele and colleagues (2002a) showed that normalization based on a single non-validated reference gene led to erroneous quantifications of up to 3-fold in 25% of cases and 6.4-fold in 10% of cases, with sporadic cases showing errors greater than 20-fold. Hence the utility of a reference gene must be validated experimentally for the particular sample type and using the specific experiment in question. Unless fully validated single reference genes are used, normalization should be performed against multiple reference genes to reduce the degree of quantification error (Bustin *et al.*, 2010).

A method employing multiple reference genes has emerged to minimize problems associated with using single, inadequately validated reference genes for data normalization (Vandesompele *et al.*, 2002a; Pfaffl *et al.*, 2004). In this method, multiple reference genes (10–13) in the samples of interest are assayed and their relative gene expression levels determined. The genes are selected such that they represent different biological pathways and expression abundance levels. Microarray results may be exploited to identify potential normalization candidates. Medium throughput RT-qPCR methods, such as TaqMan Low-Density Arrays that allow for the assay of up to 384 genes on a single microfluidics card, may also be beneficial for this purpose (Abruzzo *et al.*, 2005). Many algorithms have been reported to evaluate expression stability and suitability of potential reference genes. The geNorm method (Vandesompele *et al.*, 2002a) was the first and has become the standard (http://medgen.ugent.be/genorm). BestKeeper (Pfaffl *et al.*, 2004) is a similar program (http://www.gene-quantification.info). Both are Microsoft Excel applets that estimate gene stability through numerous pair-wise comparisons of the relative expression data. The software is able to indicate which reference genes to use and how many reference genes are optimally required to remove most of the technical variation. Typically, 3–5 genes are required for accurate normalization (Vandesompele *et al.*, 2002a; Hellemans *et al.*, 2007). Once a suitable panel of reference genes is selected for the experiment, a normalization factor is calculated from the geometric mean of their expression levels. While the use of multiple housekeeping genes may be a labour-intensive method, it is also the most conservative method of data normalization (Wong and Medrano, 2005).

Although normalization is recognized as an essential component of proper data analysis (Vandesompele *et al.*, 2002a), many investigators still continue to perform the task in an inappropriate manner. A high proportion of papers still report expression patterns of target genes normalized against a single, invalidated reference gene (Bustin, 2008b, 2010; Becker *et al.*, 2010).

Analysis

The qPCR reaction can be broken into four major phases: the linear ground phase, early exponential phase, log-linear (also known as exponential) phase, and plateau phase (Tichopad *et al.*, 2003). During the linear ground phase (usually the first 10–15 cycles), PCR is just beginning, and fluorescence emission at each cycle has not yet risen above background. Baseline fluorescence is calculated at this time. At the early exponential phase, the amount of fluorescence has reached a threshold where it is significantly higher (usually 10 times the standard deviation of the baseline) than background levels (Heid *et al.*, 1996). The cycle at which this occurs is known as the cycle threshold (*Ct*) or crossing point (*Cp*). To avoid confusion, *Cq* has recently been recommended as the standard term to use for this value according to RDML guidelines (Lefever *et al.*, 2009). The *Cq* is used in the calculation of gene expression levels. During the log-linear phase, PCR reaches its optimal amplification period with the PCR product doubling after every cycle in ideal reaction conditions. Finally, the plateau stage is reached when reaction components become limited and the fluorescence intensity is no longer useful for data calculation (Bustin, 2000; Wong and Medrano, 2005).

Quantification of the target mRNA transcript(s) may be described in either absolute or relative terms. Absolute quantification uses serially diluted standards of known amounts (e.g. copy number, nanomole) to generate a standard curve. The standard curve produces a linear relationship between *Cq* and initial amounts of total RNA or cDNA, allowing the determination of the concentration of unknowns based on their *Cq* values (Heid *et al.*, 1996). This method assumes all standards and samples have approximately equal amplification efficiencies. Although DNA standards can be produced by cloning the target sequence into a plasmid, purifying a conventional PCR product, or directly synthesizing the target sequence (Wong and Medrano, 2005), an RNA-based standard is preferable for RT-qPCR to control for reverse transcription efficiency (Giulietti *et al.*, 2001). A simple protocol for constructing a complementary (cRNA) standard for RT-qPCR can be found in Fronhoffs *et al.* (2002). Commercially available universal reference RNAs may also be used as standards (Pfaffl and Hageleit, 2001). Absolute quantification is most obviously useful for quantifying viral load in body fluids or environmental samples. However, care must be taken in interpretation of results. Although termed 'absolute', in reality this method represents quantification of the target transcript in comparison against a constructed standard curve, rather than as an absolute measurement of the number of target genes present within an environmental or biological sample. Any number of factors involved in the construction of the standard curve including the initial quantification of the standard curve template, serial dilution of the template and the algorithmic determination of the *Cq* value (Love *et al.*, 2006) contribute to the final quantification. Moreover, Smith *et al.* (2006)

have previously demonstrated that even highly reproducible standard curves may result in statistically significant differences in gene transcript numbers for the same template when quantified within different RT-qPCR assays due to the log nature of the curve, whereby minor differences in *Cq* values and standard curves can result in large differences in copy numbers. An alternative computation method for absolute quantification has been proposed that does not require the use of standard curves (Wilkening and Bader, 2004).

Relative quantification expresses changes in sample gene expression compared with both a calibrator sample (e.g. total RNA prepared from culture) and to an endogenous reference gene (Livak and Schmittgen, 2001; Pfaffl, 2001). In such experiments, the calibrator and the reference gene(s) are processed in parallel with the experimental samples and the target genes. Because the calibrator and samples are processed under identical thermal cycling conditions, normalization to the calibrator adjusts for minor differences in those conditions between runs. Normalization of the target genes to the reference gene(s) can adjust for differences in RNA quantity and quality across samples. Using this method, the calibrator is usually designated as 1-fold, and all sample quantities reported as an n-fold difference relative to the calibrator.

Several methods exist for calculation of relative quantification using *Cq* values. The comparative *Ct* (or $2^{-\Delta\Delta Ct}$) method has been the most commonly used mathematical model (Livak and Schmittgen, 2001). This model assumes that the amplification efficiency is 100% (i.e. PCR product amount doubles during every cycle) for both the target gene and the reference gene (Gibson *et al.*, 1996). However, since many PCR reactions do not have perfect amplification efficiencies, such calculations used without employing an appropriate correction factor may lead to erroneous quantification of transcript amounts (Livak and Schmitttgen, 2001). Particularly in later PCR cycles (i.e. 26 or greater), as little as 5% difference in amplification efficiencies can result in a 2-fold difference of PCR product amount (Freeman *et al.*, 1999). A review of various methods that can be used to estimate qPCR assay efficiency may be found in Cikos *et al.* (2007).

The Pfaffl model for relative quantification (Pfaffl, 2001) is a variation of the comparative *Ct* method that combines gene quantification and normalization into a single calculation. This model incorporates the amplification efficiencies of the target and reference genes to correct for differences between the two assays, and hence improves accuracy of the results. The relative expression software tool (REST), which runs in Microsoft Excel, automates data analysis using this model (Pfaffl *et al.*, 2002). A limitation of this model is that only one reference gene can be inserted in the equation for normalization. The qBase quantification model (Hellemans *et al.*, 2007) is a modified method allowing for the use of efficiency corrected multiple reference genes. Additional updates and variations continue to be introduced for the comparative *Ct* (*Cq*) model (Schmittgen and Livak, 2008; Yuan *et al.*, 2008).

Inappropriate experimental designs, improper analyses, subjective interpretation of RT-qPCR data and variability of results depending on the choice of analysis algorithms all combine to compromise the interpretation and confident application of quantitative mRNA data (Bustin and Mueller, 2005). Thus, the use of appropriate statistical methodologies for experimental design, data handling and data processing is an essential element to obtaining relevant and reliable RT-qPCR results. Independent biological replicates are a necessity to minimize the impact of biological variability on end results. It is key to remember that the use of technical replicates at the qPCR level has minimal impact on the precision of the estimated *Cq* value (Bengtsson *et al.*, 2008; Tichopad *et al.*, 2009; Kitchen *et al.*, 2010). In general, the most effective means of increasing the power and validity of qPCR experiments is to increase the number of independent biological replicates randomly selected from within each sample population. The number required depends on the statistical test to be used for data analysis and on the power one wants to achieve (e.g. for confidence interval analysis, at least three replicates are needed). It is recommended that a biostatistician be consulted to ensure design and data analysis methods are appropriate to answer the question at hand (Derveaux *et al.*, 2010). The experimental lay-out must also be considered. Bustin *et al.*

(2010) recommend the use of a sample maximization strategy whereby as many samples as possible are assayed in the same run. This lay-out is in preference to a gene maximization strategy that assays as many genes as possible in the same run. The sample maximization method helps to minimize any technical, run-to-run variation between different samples for comparison of gene expression levels. Measuring the difference in *Cq* or the normalized relative quantity between the calibrator sample across different runs can also be used to calculate a correction or calibration factor to remove run-to-run differences (Derveaux *et al.*, 2010). Hence, variance can be minimized by effective, informed experiment design and sampling plans that use replicates where they have the greatest benefit. Kitchen *et al.* (2010) developed powerNest (http://www.powernest.net) to create optimal designs for large qPCR experiments through exploiting the results of a pilot study employing a small number of technical replicates at various stages of sample preparation. The software calculates biological and technical variance components from the small pilot study and can deliver cost-optimal, variance-minimizing experiment designs that can enhance the statistical resolution of the larger scale assay.

There are a large number of statistical tools that can be used to process RT-qPCR results, and since there are no guidelines or universally accepted standards for data management, handling and interpretation, the use of different statistical tools by different researchers can add to variability and discordance in observed RT-qPCR results (Pfaffl *et al.*, 2009; Bustin *et al.*, 2010). Bustin and colleagues highly recommend the use of specialized software to manage and analyse data (Bustin, 2008a, 2010). Some of the available programs for this purpose include PREXCEL-Q (Gallup and Ackermann, 2008), Q-Gene (Muller *et al.*, 2002), qPCR-DAMS (Jin *et al.*, 2006), qBase and qBase plus (Hellemans *et al.*, 2007) and qpcR (Ritz and Spiess, 2008). However, a major limitation for data exchange and analysis by external software is that the qPCR instrument software used and the third party data analysis software often do not speak the same language. To address this issue, the international Real-time PCR Data Markup Language (RDML) consortium was founded (http://www.rdml.org) (Lefever *et al.*, 2009) with the goal to develop an XML-based RDML standard to enable straightforward exchange of qPCR data and related information.

The ubiquity of qPCR and lack of strict guidelines for design, performance and reporting of experiments has resulted in a proliferation of protocols that vary at every stage of the workflow and are a major source of discordance between published conclusions. Bustin and colleagues have expressed concerns for some time regarding the importance of establishing a framework for producing consistent, reproducible, high quality data from real time PCR experiments given the current status of qPCR as a benchmark or 'gold standard' technology (Bustin, 2008a,b, 2010; Bustin *et al.*, 2009a,b). Towards this goal, the minimum information for publication of quantitative real-time PCR experiments (MIQE) guidelines were recently published (Bustin *et al.*, 2009b), outlining a standardized approach to each step of the RT-qPCR workflow and providing a detailed checklist of information regarding every aspect of the experiment (sample acquisition, assay design and performance and validation and data analysis) that should be reported by researchers when publishing their results. These criteria have recently been simplified in a concise, eight step experimental approach to qPCR (Taylor *et al.*, 2010).

'Validation' of microarray results using RT-qPCR

Although the implementation of microarray technology has revolutionized gene expression analysis, the high volumes of data produced by these experiments present significant challenges with regards to their generation, analyses, and in particular, interpretation (Slonim, 2002). The uncertainties associated with validity and reliability of results have been highlighted by studies demonstrating lack of agreement in results across different microarray platforms and even using the same platform (Grunenfelder and Winzeler, 2002; Kuo *et al.*, 2002; Bammler *et al.*, 2005; Irizarry *et al.*, 2005; Zhu *et al.*, 2005; Draghici *et al.*, 2006; Shields, 2006; Walker and Hughes, 2008). Additionally, limitations currently inherent to microarrays can make it difficult to obtain reliable quantitative data or adequate statistical power to definitively identify differentially expressed genes (Valasek and Repa, 2005). Although a consensus currently exists that validation of microarray results using independent measurement techniques is a desirable facet of a complete experiment, the choice of a validation methodology remains a contentious issue (Rockett and Hellman, 2004). RT-qPCR is the method of validation that has been used in the majority of published microarray studies, presumably because of its rapid, sensitive, specific, medium throughput nature and the requirement for minimal amounts of input RNA material (Klein, 2002).

There are many potential explanations for discrepancies in microarray experiments. As for RT-qPCR, a microarray experiment is a multistep process including RNA extraction and purification, fluorescent dye labelling/incorporation, hybridization, image analysis, statistical analysis and data interpretation. At each step of the protocol, an abundance of choices must be made, including platform type and manufacturer, types and composition of probes, experimental design, imaging instrumentation and settings, labelling and hybridization protocols, data preprocessing methods and data analysis methods (Hardiman, 2004; Yauk *et al.*, 2004; Irizarry *et al.*, 2005; Larkin *et al.*, 2005; Draghici *et al.*, 2006; Wang *et al.*, 2006; Ding *et al.*, 2007; Walker and Hughes, 2008). The complex nature of microarray technology alone can contribute to discordance in results. There is no single microarray technology, but rather a collection of different platforms including Affymetrix GeneChips, PCR-product-based cDNA arrays and long oligomer arrays that are manufactured in-house or by Agilent, and newer platforms such as the Illumina Bead array (Fan *et al.*, 2004) or the Universal Hexamer Array from Agilix (Roth *et al.*, 2004).

In many instances, microarray experiments are prone to the same biological and technical sources of variation as RT-qPCR assays. Error in microarrays can be random (e.g. lint covering a spot on the array) or systematic. Systematic errors can be reproducible and due to numerous known or unknown variables (e.g. specific gene sequence, dye effects, slide production defects) (Loeb and Beaumont, 2009). The goal of any sound microarray experiment design is to minimize variables so any changes observed in gene expression are due only to the variable of interest. A fundamental approach is to use a combination of biological and technical replicates that include 'dye-flipping' or control–control hybridizations (using a common reference) to reduce errors. A sufficient number of replicate measurements must be performed to discriminate truly predictive results from spurious ones. Additionally, these replicates must be of the proper type. Most of the observed expression variation in microarrays is due to biological variability (Chen *et al.*, 2004; Yauk *et al.*, 2004; van Hijum *et al.*, 2005; Zakharkin *et al.*, 2005; Breitling, 2006). It is therefore best to perform as many biological replicates as possible to enable detection of those expression changes that are reproducible and relevant.

As in RT-qPCR, a critical component of any microarray experiment is the quality of input material. RNA must be handled, processed and characterized with great care prior to its use. An added challenge for microarray experiments is the requirement for large amounts of starting RNA material. Although various techniques can be used for mRNA amplification when samples are limited (Nygaard and Hovig, 2006), an understanding of their potential to introduce bias is required for correct interpretation of results (Chuaqui *et al.*, 2002).

Perhaps the most significant reason as to why experimental validations are necessary is because microarray data are inherently noisy. By virtue of their sheer numbers alone, the large numbers of gene expression measurements obtained in a typical microarray experiment can often yield significant numbers of false positive and false negative results. Such artefacts can arise due to numerous experimental and technical reasons which can include spurious signals caused by microarray probes cross-hybridizing to related transcripts of similar sequence, mistakes in the assignment of probe identities (of particular concern when dealing with cDNA microarrays, as many cDNA clone libraries invariable contain a certain degree of mistaken assignments), artefacts induced by the sample preparation technique (e.g. preferential enrichment of certain transcripts during RNA amplification procedures, degradation of RNA during the extraction process), and artefacts caused by the hybridization procedure itself (e.g. biased hybridization due to the fluorophore or conditions used) (Draghici *et al.*, 2006; Ding *et al.*, 2007; Walker and Hughes, 2008). An inherent limitation of DNA microarrays is that not all of the thousands of genes included on the array can be expected to generate a perfectly specific and linear signal. A single hybridization condition is applied to tens of thousands of distinct nucleic acid hybridization reactions, resulting in suboptimal conditions for some reactions. Some degree of non-specific and cross-hybridization for a substantial number of probes is unavoidable (Ding *et al.*, 2007). It has been shown previously that a large proportion of microarray probes produce significant cross-hybridization signals (Wu *et al.*, 2005; Zhang *et al.*, 2005) for both oligonucleotide and cDNA microarrays. Even a limited stretch of sequence complementarity might be sufficient to enable binding between two unrelated sequences. Evaluating the overall impact of cross-hybridizations on the accuracy of microarray measurements can be very difficult. For example, in Affymetrix arrays, the effect of a single cross-hybridizing probe can be down-weighted by the rest of the probe set (10 other probes in HU-133 chips). Furthermore, the impact of cross-hybridization strongly depends on the relative concentration and the relative affinities of the correct target and the cross-hybridizing target(s). The latter must be present in sufficient quantities to interfere with specific signals (Draghici *et al.*, 2006). Although careful attention to probe design can minimize cross-hybridizations and enhance specificity, the sequence dependence and thermodynamics of DNA–DNA and DNA–RNA hybridizations are not always predictable (Mir and Southern, 1999; Mei *et al.*, 2003; Zhang *et al.*, 2003; Wu *et al.*, 2005). Splice variants constitute another dimension that can introduce difficulties in the microarray probe design. A given short oligonucleotide probe may be targeted at either a constitutive exon (present in all splice variants) or at an exon specific for certain splice variants. cDNA microarrays usually have a unique long probe with which the abundance of several splice variants can be measured. These design differences might explain some of the discrepancies often observed between cDNA and short oligonucleotide microarrays (Draghici *et al.*, 2006). Folding of the target transcripts (Wu *et al.*, 2005) can also contribute to the variation between different probes targeting the same region of a given transcript.

Tools and specialized statistics abound for use in interpreting the mountains of data collected from microarray experiments and turning this data into relevant biological insights. Prior to analysis of the microarray experiment, the data usually need to be normalized to remove obvious systematic differences between samples. Many methods have been proposed to perform this task, including, for two-colour arrays, simple quantile normalization (Bolstad *et al.*, 2003), variance stabilizing normalization (Durbin and Rocke, 2004) and started log transformation (Rocke and Durbin, 2003). For one-colour arrays, MAS 5.0 (Shedden *et al.*, 2005; Irizarry *et al.*, 2006; Qin *et al.*, 2006), Robust Multichip Analysis (RMA) (Irizarry *et al.*, 2003a,b), its modification GC-RMA (Wu *et al.*, 2004), and dChip (Li and Wong, 2001; Wu *et al.*, 2006) are among the normalization methods employed.

Detecting genes that show reproducible differences in mRNA abundance between sample classes is probably the most basic and important step of microarray analysis. A huge variety of methods have been used for this task to discriminate between changes that are spurious and those that are reproducible. These methods include SAM (Tusher *et al.*, 2001), Rank products (Breitling *et al.*, 2004; Breitling and Herzyk, 2005), the Empirical Bayes modification of the t-test (Baldi and Long, 2001; Efron *et al.*, 2001) and other Bayesian approaches (Townsend and Hartl, 2002; Smyth, 2004), and analysis of variance – ANOVA (Kerr *et al.*, 2001). The end result of all statistical tests is a *P*-value for each gene, which basically describes how likely it would be to observe a particular differential expression by chance. This estimate is based on the observed variability and effect size in replicated independent measurements (Breitling, 2006).

To interpret the resulting microarray data analyses in a systematic way, some 'dimensionality reduction' is required to extract and emphasize factors or phenomena that dominate the observed data. There are two main ways of doing this, both with advantages and disadvantages. The first, cluster analysis (and relatives such as PCA) (Eisen *et al.*, 1998; Datta and Datta, 2003; Girolami and Breitling, 2004), is data driven, while the second, annotation analysis (Khatri and Draghici, 2005), is knowledge driven. Cluster analysis may miss subtle patterns that might be obvious if previous biological knowledge is considered. Conversely, annotation analysis may be overly restrictive or biased by its focus on genes and processes that have already been reported. Several automated methods have been devised to perform the task of annotation analysis and remove the effects of interpreter bias, but their basic principle is the same. Given a list of genes and the annotation of the genome, they all calculate if a certain functional category is statistically over-represented in the selected group of genes, compared with the rest of the genome. The most popular software tools used for this purpose are Onto-Express (Khatri *et al.*, 2002), GOstat (Beissbarth and Speed, 2004), GoMiner (Zeeberg *et al.*, 2003), EASE (Hosack *et al.*, 2003), and Gene Ontology (Ashburner *et al.*, 2000), but many others are available.

The abundance of statistical methods available for data pre-processing and analysis, and the lack of standardization of such techniques contribute significantly to the variability of results observed amongst microarray experiments. Several studies have looked at the multitude of statistical methods employed to identify differentially expressed genes to address their performance, point out inefficiencies and introduce improved algorithms or improve upon existing algorithms (Zervakis *et al.*, 2009). Breitling (2006) provides a comprehensive review of statistical and annotation methods used to interpret microarray data and the advantages and disadvantages of several of these techniques.

A number of comparative evaluation studies and the accumulated experience from many thousands of microarray studies have led to the emergence of a set of guidelines and informal standards to aid in organization, analysis and annotation of differential expression information (Breitling, 2006). The Minimum Information About Microarray Experiments (MIAME) guidelines describe the reporting standards required when recording gene expression data to ensure an experiment's easy interpretation and potential to be independently verified (Brazma *et al.*, 2001). Public databases to store results from DNA microarray experiments have been developed which conform to these standards (Stoeckert *et al.*, 2002). Additionally, scientists from the US Food and Drug Administration (FDA) established the MicroArray Quality Control (MAQC) consortium to evaluate the performance of several microarray platforms as well as three quantitative gene expression assays (Patterson *et al.*, 2006; Shi *et al.*, 2006). Reports from the MAQC project, which profiled two standardized reference RNA samples, contain important findings on the performance of different expression measurement technologies and give insights into the level of cross-platform comparability amongst different technologies and advice on analysis methods that can improve agreement in results. Additionally, the project sought to generate a thoroughly characterized reference data set against which new modifications in the existing microarray platforms and other expression measurement technologies could be compared and validated. This goal was accomplished by introducing two commercially available high-quality human reference RNA samples that can be used as tools for calibration and quality control as well as for performance assessment and validation of assays. The two RNA samples used in the MAQC project were the Stratagene Universal Human Reference RNA (comprising RNA from 10 different cell lines) and the Ambion Human Brain Reference RNA. Similarly, the External RNA Controls Consortium (ERCC) strives to provide external RNA controls for microbiological applications (ERCC, 2005).

When using RT-qPCR to validate microarray data, it is important to define what is meant by validation in this context. There are currently no generally accepted guidelines or recommendations concerning the definition of validation, the level of validation for which to strive, or methods to use to perform validation. Is RT-qPCR simply to be used as a complementary method to verify and better quantify expression levels of genes of interest derived from the microarray data, or is the goal of performing RT-qPCR measurements to demonstrate validity of the microarray experiment on a global level? Are the validations to be performed in the same laboratory that generated the original microarray data using the same or similar biological samples used for the microarray experiment, or are independently generated biological samples to be measured by researchers in another laboratory? In any case, it must be taken into account that quantitative PCR, while considered a 'gold standard' is not without its own limitations (as described in a previous section) and it has often been questioned as the optimal technology through which to confirm or confute microarray results (Allison *et al.*, 2006; Ding *et al.*, 2007). Several alternative approaches for validation have been proposed including the use of two combined microarray platforms to perform the experiment as a built-in cross-comparison (Bosotti *et al.*, 2007), and in fact the necessity to always validate microarray data using other quantification methods with their own inherent pitfalls has been questioned (Rockett and Hellman, 2004).

Since it is not feasible to validate every gene expression measurement from a microarray experiment, the question arises as to how to select which transcripts to examine by RT-qPCR. Following initial preprocessing of the microarray data a reduced number of genes is often identified that exhibit differential regulation. For many experiments, the numbers of genes within this subset will usually still be too large for follow-up on every member. Often, the selection of 'which genes to validate' from this group is done in an ad-hoc manner and can be based on the absolute expression level of the gene (abundantly expressed genes are typically easier to measure than rare transcripts), the prior availability of detection reagents (primers and probes for PCR), or involvement of the gene in a particular biological process of interest to the researcher. Using such criteria for gene selection, however, makes poor use of the plethora of expression data generated by the original microarray experiment. Many important regulatory genes such as transcription factors are not abundantly expressed, although such genes may be highly relevant to the biological question at hand. Although it might be more convenient to follow-up with genes that are already well known, the characterization of novel transcripts may ultimately provide a greater overall insight. Another common strategy for gene selection is to identify those transcripts exhibiting the largest fold changes (FCs) or the most statistically significant differential expression, based on the idea that large effects are more probable to be valid (Chuaqui *et al.*, 2002). Using such a non-random method for gene selection, however, severely limits what can be said about the remaining genes that were not selected for validation. If applying RT-qPCR as a global validation approach, the results should provide an index of the quality of the fold-change estimates regarding all differentially expressed genes on the microarray. Selecting only top-ranked differentially expressed genes for validation can lead to underestimating the level of agreement between microarray and RT-qPCR validation FC values. The use of random or stratified sampling procedures to select genes for follow-up for global validation purposes has been proposed (Miron *et al.*, 2006). Using random sampling procedures results in the selection of genes representative of a wide range of intensity levels and fold changes, thereby providing an index of reliability across the dynamic range of the array platform for the experimental conditions used.

A question also arises whether agreement between microarray data and RT-qPCR data need be only qualitatively shown (i.e. a similar direction, either up or down, in differential expression is observed) or also quantitatively assessed. If quantitative validation is required, which measurement should be validated? Fold change (FC), rather than raw expression, has been proposed as the appropriate measure for comparing results across gene quantification platforms (Irizarry *et al.*, 2005; Larkin *et al.*, 2005). Because different gene expression platforms utilize different technologies, quantitation, and normalization methods, the absolute signal values for each platform tend to be somewhat arbitrary and not suitable for correlation analysis across different platforms (Wang *et al.*, 2006). Various technology-specific artefacts (e.g. probe-specific biases across microarray platforms, amplification bias in PCR) can compromise direct comparisons between gene expression measurements. Fold change metrics tend to cancel out systematic platform biases in absolute signal values. In addition, this is the most biologically meaningful metric (Wang *et al.*, 2006). Finally, a decision must be made as to what statistical evidence is indicative of global validation. Miron *et al.* (2006) argue that a one-to-one correspondence between microarray and RT-qPCR FC estimates is the gold standard for validation. Furthermore, this group suggests that the concordance correlation coefficient (CCC), combining accuracy and precision coefficients in one index (Lin, 1989; King and Chinchilli, 2001; Lin *et al.*, 2002) is a better indicator of the agreement between two sets of FC measurements across quantification platforms compared with regression analysis employing the Pearson *r* correlation coefficient.

Although RT-qPCR is commonly used to confirm gene expression results obtained from microarray analysis, microarray and qPCR data often result in disagreement. Presently, correlations of qPCR and microarray data are seldom presented in the literature and, where they are reported, cover a broad range (e.g. from 0.48 to 0.94) (Beckman *et al.*, 2004; Etienne *et al.*, 2004; Dallas *et al.*, 2005; Canales *et al.*, 2006; Morey *et al.*, 2006; Wang *et al.*, 2006; Arikawa *et al.*, 2008). These correlations are rarely presented with statistical analyses and non-agreeing data are rarely explained (Bosotti *et al.*, 2007). Furthermore, few authors define the criteria by which acceptable validation of microarray results is determined. The variety of platforms that exist for both microarray and qPCR measurements, and the inherent pitfalls of both quantification technologies contribute to the complexity of data analysis and interpretation and cause difficulties in pinpointing potential sources of discordance. Since both biological and technical variability can have a great impact on microarray and qPCR results (Chuaqui *et al.*, 2002; Wurmbach *et al.*, 2003), care must be taken in the experimental design. The quality of RNA is essential to the accuracy of results, as is the efficiency of reverse transcription and suitability of priming method (Freeman *et al.*, 1999). The effects of dye biases such as differences in efficiencies of incorporation during labelling procedures (Yang *et al.*, 2002) and non-specific and/or cross hybridizations of labelled targets to array probes (Chuaqui *et al.*, 2002) are sources of error unique to microarray procedures. Amplification biases (Chuaqui *et al.*, 2002), exponential amplification of errors (Freeman *et al.*, 1999), mispriming or the formation of primer-dimers (Bustin, 2002), and the changing efficiency of qPCR at later cycles (Freeman *et al.*, 1999; Bustin, 2002) are some of the qPCR-related sources of error. These sources of error can all compound to cause erroneous estimates of agreement between microarray and qPCR.

The choice of microarray and RT-qPCR preprocessing methods (e.g. background correction, normalization, transformation, calibration) can effect agreement between the methods (Harr and Schlotterer, 2006; Cikos *et al.*, 2007; Devonshire *et al.*, 2010). In general, data normalization methods differ between microarray and qPCR; the former usually employing a global normalization strategy, and the latter generally utilizing a set of reference genes (Morey *et al.*, 2006). Hence, selection and appropriate application of normalization criteria has been proposed to play a potential role in determining the extent of data agreement between the methods. However, Dallas *et al.* (2005) found that the methods employed to normalize data did not have a significant impact on platform agreement. The group assessed correlations between expression scores obtained for 48 human genes using oligonucleotide microarrays and the expression levels for the same genes measured by RT-qPCR. The microarray data was processed using two different algorithms for normalization of the data: robust multiarray analysis (RMA) and the MAS 5.0 algorithm. They found that the degree of correlation between microarray and RT-qPCR results was very similar irrespective of the normalization procedure employed. Furthermore, their results indicated that when identical transcripts were targeted by the two methods, correlations between RT-qPCR and microarray data were generally strong ($r=0.89$). However, poor correlations were observed for 13–16% of genes assessed. The authors postulated differences between platform probe sets, the existence of alternative cross-hybridizing transcripts differentially recognized by the probe sets, gene specific variation related to the different hybridization kinetics associated with the two technologies, and misleading results associated with errors in GenBank sequence data and/or probe set annotations as possible contributors to the lack of agreement for these genes.

Lower correlations between microarray and qPCR results have consistently been reported for genes exhibiting small degrees of change (< 2-fold), as compared with genes showing greater than 2-fold change based on the microarray results (Rajeevan *et al.*, 2001a,b; Wurmbach *et al.*, 2003; Etienne *et al.*, 2004; Wang *et al.*, 2006). Wurmbach *et al.* (2003) reported 100% validation of results for genes exhibiting at least 1.6-fold change; however, they defined validation as directional confirmation only and large discrepancies in the amount of change were not addressed. Dallas *et al.* (2005) reported decreased correlations for genes expressing less than a 1.5-fold change in microarray analysis. It is probable that this trend relates to the fact that small variations in mRNA levels (< 2-fold) can be accurately detected by RT-qPCR, while the smaller dynamic range of microarrays means that the same changes may not be accurately reflected, especially for genes expressed at low levels. Abruzzo *et al.* (2005) compared normalized RT-qPCR values for a panel of 96 genes with results obtained from GeneChip oligonucleotide microarray (Affymetrix) experiments on the same samples. They found the dynamic range of the RT-qPCR measurements to be much wider than the dynamic range of the GeneChip microarrays. The dynamic range of the RT-qPCR was approximately seven orders of magnitude, whereas the dynamic range of the U133A GeneChip has been reported to be about three orders of magnitude based on spike-in experiments (Irizarry *et al.*, 2003b; Zhang *et al.*, 2003). Thus, genes whose mean expression level is low, based on RT-qPCR, are less likely to be measured accurately on the U133A GeneChip, where they will be obscured by relatively large noise levels. When the analysis was restricted to higher intensity genes whose expression varied significantly between samples, the results were well correlated between the two technologies (Abruzzo *et al.*, 2005). Hence, sensitivity and valid linear ranges of detection for microarray and qPCR methods differ substantially and can affect the accuracy of mRNA detection and estimates of fold change differences at different abundance levels. Fold change results determined by RT-qPCR are consistently significantly greater than fold changes assessed for the same genes by microarray analyses (Yuen *et al.*, 2002; Dallas *et al.*, 2005). Fold changes determined by microarray may be compressed by as much as 10-fold compared with those determined by qPCR (Conway and Schoolnik, 2003). In addition, the degree of compression can vary depending on the microarray platform used and methods applied for normalization and data processing (Wang *et al.*, 2006). As a further example, a 2004 study examining gene expression profiles in *Arabidopsis* (Czechowski *et al.* 2004) reported good agreement between oligonucleotide microarray and RT-qPCR data when ratios of gene expression were compared for highly expressed genes. However, the correlations were generally poor for genes expressed at low levels. Etienne *et al.* (2004) hypothesized that in addition to genes with low expression levels, those with very high expression levels, or a greater percentage of absent calls, may show lower levels of correlation between microarray expression scores and RT-qPCR data. The dynamic range for most of the commercial and home brewed microarray platforms usually falls into 3–4 orders of magnitude (Tan *et al.*, 2003; Shippy *et al.*, 2004), while TaqMan based real-time PCR can achieve 6–8 orders of magnitude dynamic range (Yang *et al.*, 2004; Zhao *et al.*, 2005). Microarray reportedly has valid detection dynamics for high mRNA levels in the range from 10 to 7000 pg, whereas RT-qPCR has a better detection range at lower levels from 100 fg to 1000 pg (Liu and Slininger, 2007). Hence, it is not surprising that the poorer correlations of expression data across microarray and RT-qPCR platforms have been observed for high and particularly, low abundance genes (Yuen *et al.*, 2002; Czechowski *et al.*, 2004; Etienne *et al.*, 2004; Pappas *et al.*, 2004; Dallas *et al.*, 2005). Similarly, Morey *et al.* (2006) found a lack of concurrence in direction of change between methods for genes exhibiting low levels of change (< 1.4-fold). A better correlation between the methods (0.7, Spearman's Rho) was observed among genes showing up-regulation by microarray than the correlation (0.356, Spearman's Rho) observed among down-regulated genes. Overall, 72% of down-regulated genes, by microarray, exhibited less than 1.4-fold change whereas only 60% of up-regulated genes exhibited less than 1.4-fold change. A similar trend of higher correlations among up-regulated genes was observed by Beckman *et al.* (2004), who proposed that this effect might be due to the increased variability observed in low-intensity array spots. However, Morey *et al.* (2006) found no apparent trend in the effects of average array spot intensity on correlation of data (but, this may have been due to the feature analysis software used by the authors which identified and excluded spots likely to introduce errors due to signal strength, high background, and/or poor spot morphology, etc.). Morey *et al.* (2006) suggest that lower correlations between array and qPCR for down-regulated genes may be due to the effects of greater variability associated with decreased reaction efficiencies found in qPCR measurements at later cycles, where genes with low expression levels respond. They observed significantly lower correlations at early and late cycle thresholds, especially in samples with $Cq < 17$ or $Cq \leq 31$. The low correlations observed at early Cq values (i.e. highly expressed genes or markedly up-regulated genes) may be attributed to the effects of large intensity ratios due to large differences in expression between the treatments compared (Beckman *et al.*, 2004). Hence in many cases, discordance between array and qPCR results may be at least partially attributed to using data beyond the valid detection limits of the different platforms and emphasizes that for quantitative data confirmations, only the overlapping range of detection between these methods should be considered as valid.

Microarray spot *P*-values appear to have a considerable effect on the correlations between array and qPCR results according to Morey *et al.* (2006). In their investigation, genes yielding highly significant array results generated high correlations regardless of spot intensity or Cq values. Among the genes assayed by qPCR, those that were labelled as 'signature' (a composite $P \leq 0.01$) had a correlation of 0.847 (Spearman's rho), whereas genes that were not called 'signature' only exhibited a correlation of 0.435 (Spearman's rho). The calculation of composite *P*-value includes measurements of signal strength, background levels, spot morphology, and fold change from replicate arrays. Therefore, the smaller the *p*-value reported, the more confidence in the accuracy of the microarray results, as errors such as experimental noise and non-specific or cross hybridization, are discounted (Chuaqui *et al.*, 2002). Hence the stringency applied for filtering of array data using signal to noise ratios and the coefficient of variation from replicate arrays can yield a final data set of high quality and increased accuracy, and thus, increased correlations with qPCR results. Their results demonstrate the importance of microarray data quality, not only fold change, on the agreement of results.

Different strategies in probe designs for microarrays (usually 3′ biased and targeting a composite of transcripts) and RT-qPCR assays (usually without *a priori* bias and targeting a single or subset of transcripts) may also account for a percentage of the discordance observed between microarrays and real-time PCR results (Wang *et al.*, 2006). Differential RNA splicing on alternative transcripts and polymorphisms can influence results depending on probe/primer design and target (Dallas *et al.*, 2005; Ding *et al.*, 2007). Etienne *et al.* (2004) found that increased distance between the location of the PCR primers and microarray probes on a given gene decreased the correlation between the two methods. Similarly, Arikawa *et al.* (2008) and Canales and co-workers (2006) concluded that a primary cause in the discordance of results they observed was due to differences in the regions of genes being probed. Thus, differences in probe/primer designs may result in quantification of different transcript populations (e.g. product of alternative splicing, polymorphism, or degradation) by microarrays versus qPCR assays. Arikawa and colleagues (2008) also suggested that a cause for discrepancy may be due to inaccuracy of sequence information provided by the NCBI RefSeq database. The study highlighted the need to periodically check sequences of primers and probes against the most updated version of the RefSeq database to ensure assay accuracy. In addition, differences in probe/primer hybridization/annealing conditions between qPCR and microarray measures may influence results. In RT-qPCR design, assay conditions may be optimized for a specific set of primers and nucleotide sequences, whereas in microarray experiments thousands of probes are used under the same hybridization conditions which may not be optimal for any one particular gene/probe set.

In efforts to improve consistency across gene expression measurement platforms, several groups support the use of external quality controls (Bammler *et al.*, 2005; ERCC, 2005; Irizarry *et al.*, 2005; Liu and Slininger, 2007). Regulatory bodies such as the FDA and EU commission have highlighted the need to standardize results from 'omics' platforms using reference materials (RMs). For microarray experiments, the application of exogenous nucleic acids to calibrate quantitative expression and measure differential expression ratios has been widely practiced (Badiee *et al.*, 2003; van de Peppel *et al.*, 2003), but largely with platform-specific controls. To allow for direct comparison of gene expression data from different microarray platforms, universal microarray reference materials have been developed (Cronin *et al.*, 2004; Anderson and Foy, 2005; Kawasaki, 2006; Tong *et al.*, 2006). For human gene studies, two commercial sources of universal RNA microarray controls are available (Stratagene and Ambion) as previously discussed. Several RNA quality controls exist for RT-qPCR applications, particularly for clinical and industrial applications (Baker *et al.*, 2005; ERCC, 2005). *In vitro* produced RNA spike-ins are independent of the biological process and can act as controls for both the RT and qPCR steps (Huggett *et al.*, 2005; Gilsbach *et al.*, 2006). Few researchers, however, exploit the existence of such RNA controls as tools by which to standardize and normalize gene expression results across microarray and qPCR platforms. Liu and colleagues recently developed a set of six universal external RNA quality controls for microbial mRNA expression analysis that they applied to both DNA oligo microarray and real-time RT-qPCR assays (Liu and Slininger, 2007). Appropriate application of the universal controls, providing the same sequences to be assayed in both platforms, resulted in a better agreement in expression data. Hence, the use of generic RNA standards can aid in validation of research findings from diverse platforms such as microarrays and RT-qPCR by providing a means to compare and correct for inter-platform variations (Devonshire *et al.*, 2010).

Previous work on microarray data validation using RT-qPCR

The use of DNA microarrays for detection, identification and quantification of microbial populations in a diversity of environments, such as soil (Reich *et al.*, 2009), water (Peplies *et al.*, 2004; Urakawa *et al.*, 2007; Martins *et al.*, 2010), compost (Franke-Whittle *et al.*, 2005, 2009), sludge (Loy *et al.*, 2005), air filtrates (Wilson *et al.*, 2002), milk (Giannino *et al.*, 2009), human intestine (Wang *et al.*, 2002b), human sputum (Fukushima *et al.*, 2003) and clinical samples (Wang *et al.*, 2002a) is still a major microbiology challenge. Despite their major advantage of high multiplexing experimental capacity, some limitations exist in the improvement of microarray technology and results interpretation, such as the complexity and the time required to design and validate discriminatory genetic regions that help with the species and subspecies differentiation (Everett *et al.*, 2010).

A series of alternative techniques are used to validate microarray results and reduce the false discovery rate, such as Northern blot hybridization (Taniguchi *et al.*, 2001), enzymatic assays, reporter gene systems, RNase protection assays, primer extension and real-time quantitative PCR (RT-qPCR), the later being increasingly more popular in the research community (Rajeevan *et al.*, 2001a,b). The RT-qPCR technique is highly sensitive and it is typically used to verify and quantify a selection of DNA regions representing differentially expressed genes, genetic markers or species and subspecies signatures. This technique has been so far successfully applied on validation of data produced with cDNA and gene microarrays.

RT-qPCR applied on cDNA microarrays

Quinn *et al.* (2006) investigated the expression profiling of 2300 ESTs from the Coccolithophorid *E. huxleyi* cultures grown under calcifying and non-calcifying conditions. The study identified 127 significantly up- or down-regulated transcripts corresponding to proteins associated with bio-mineralization and the results were successfully validated via real-time RT-qPCR.

Manos *et al.* (2008) studied biofilm-forming characteristics of clonal *Pseudomonas aeruginosa* cystic fibrosis isolates from the sputum of eight chronically infected patients using gene expression microarray analyses based on PAO1 chips (Affymetrix). Their observation that the proportion of genes showing differential expression during biofilm growth for all isolates was ~0.3%, which is much lower than in other studies (1–20%), suggesting that no particular set of differentially expressed genes in the clonal strain AES-1 could definitively be ascribed a role in its infectivity, thus increasing the likelihood that infectivity is multifactorial. The validation of their results was performed with quantitative SYBR Green RT-PCR using eight genes (*exoT, pvdA, lasB, bdhA, aprA, prtB, PA0986* and *P. aeruginosa PAK* strain *flaA*) with previously reported high differential expression, association with virulence and known function.

Pandolfi *et al.* (2010) studied the gene expression profile of the plant pathogen *Fusarium graminearum* under the antagonistic effect of *Pantoea agglomerans,* using a cDNA microarray spotted with 1014 transcripts from an *F. graminearum* cDNA library. Using *GAPDH* as a reference endogenous gene, they selected and quantified a set of 29 differentially expressed genes (19 up-regulated, 10 down-regulated) by performing RT-qPCR and calculating relative changes with the aid of the $2^{-\Delta\Delta Ct}$ method (Livak and Schmittgen, 2001).

Simultaneous application of RT-qPCR and microarrays

Tracz *et al.* (2007) developed two novel assays for the molecular typing of pathogenic *Vibrio* species. The Luminex liquid microsphere assay involved a PCR procedure that can amplify the *ftsZ* locus from all clinically relevant *Vibrio* spp. and utilizes species-specific probes targeting the major *Vibrio* pathogens. In the same study, Light Upon eXtension (LUX) real-time PCR assays were developed to target these species-specific polymorphisms, and were successful in rapidly differentiating the major pathogenic *Vibrio* species: *V. cholerae, V. parahaemolyticus* and *V. vulnificus.*

Khodakov *et al.* (2008) describe a novel microarray-based approach for the simultaneous identification and quantification of human immunodeficiency virus type 1 (HIV-1) and hepatitis B and C viruses (HBV and HCV) in donor plasma specimens. The method is based on multiplex real-time RT-PCR performed within the microarray hydrogel pads and the reported specificity and sensitivity was 100% on 132 blood specimens with a 95% detection limit per assay.

RT-qPCR applied on gene microarrays

Wilson *et al.* (2006) have demonstrated the feasibility of using a dual-genome microarray to simultaneously study the expression profile of the pea aphid, *Acyrthosiphon pisum* under heat shock and its obligate bacterial symbiont, *Buchnera aphidicola.* Using RT-qPCR for validation and two control genes (*argG* and *cysG*), they found good congruence between methods. In a similar study, Wilcox *et al.* (2003) studied gene expression under heat shock in *Buchnera* derived from the aphid *Schizaphis graminum* and identified a set of nine genes that were significantly up-regulated. They have also applied RT-qPCR to validate the microarray results on three select instances: two genes exhibiting significantly increased gene expression with heat stress; one exhibiting marked decreased expression; one gene for which expression changed little between treatments.

Lee *et al.* (2006) developed and evaluated DNA microarray techniques for bacterial pathogen identification via direct hybridization of wastewater microbial genomic DNA and hybridization of PCR amplified 23S rRNA gene. Using 10 functional genes as detection targets, their microarray attained a sensitivity of 1 μg of *E. coli* genomic DNA and was further enhanced approximately by 6 orders of magnitude when the target 23S rRNA gene sequences were PCR amplified with a novel universal primer set and allowed hybridization to 24 species-specific oligonucleotide probes. Using quantitative RT-PCR technology to validate the results, they concluded that in this particular experiment, direct detection of genomic DNA using a microarray was less sensitive than desired for monitoring of wastewater microorganisms.

You *et al.* (2008) established an enteropathogen gene microarray technique that includes 34 virulence genes from eight target bacteria species (*Escherichia coli*, *V. cholerae*, *V. parahaemolyticus*, *Salmonella enterica*, *Campylobacter jejuni*, *Shigellae*, *Yersinia enterocolitica*, and *Listeria monocytogenes*) and successfully applied this technique to identify the pathogens in stool specimens from patients with diarrhoea. With an overall sensitivity of 58 copies/μl validated with the aid of RT-PCR runs on a LineGene Thermal cycler (BIOER Technology Co., Ltd, Hangzhou, China), their study suggests that microarrays can also be used for characterization of viral and bacterial infections present in food products and environmental and clinical specimens, and thus can be a useful platform for identification and characterization of infectious pathogens.

RT-qPCR applied on microarray AA-aRNA leftovers

Recent work by Jeanty *et al.* (2010) introduced an optimized protocol for microarray data validation by quantitative RT-PCR when limited amounts of amplified amino allyl labelled RNA (AA-aRNA) are available. While their study focused on identification of novel biomarkers of graft rejection by the transcriptomic analysis of blood cells from brain-dead organ donors, their optimized validation protocol can be also applied on RNA material originating from microbial samples.

Discussion

From a technical standpoint, gene expression microarrays and RT-qPCR would seem to be extremely complementary techniques. On the other hand, when gene expression microarrays first became mainstream it was common to think of them as 'fishing expeditions'; a general overview of global gene expression patterns, but lacking precision and accuracy. Conversely RT-qPCR was thought to be an extremely precise technique, but permitting the analysis of only a few genes at one time. From that perspective, it is obvious why RT-qPCR became the method of choice for the validation of gene expression microarray data. In fact, RT-qPCR is still considered by many to be a gold standard for microarray results validation, although some studies suggest that it may be useful as complementation rather than validation (Yang and Speed, 2002; Allison *et al.*, 2006; Gaj *et al.*, 2008). Interestingly, there is now a tendency to reduce the size of microarrays and create small groups of genes whose expression profiles are indicative of disease, or the presence of a pathogen. Further, a vast improvement in microarray fabrication, experimental designs and analysis algorithms means gene expression arrays are now capable of high levels of precision and accuracy. At the same time, the capacity for multigene analysis has greatly improved in RT-qPCR systems; a number of fluorophores permit at least some level multiplex analysis, 96- and 384-well systems are now commonplace, and more specialized systems permit the analysis of thousands of genes simultaneously (BioTrove). The distinction between the technologies becomes even less evident in cases where quantification is unimportant; both techniques are extremely powerful when simply determining the presence or absence of an organism within a sample. It is thus of utmost importance to choose a technology that fits the purpose of your experimentation, rather that adapting your experimentation to accommodate the advantages of a specific technology.

Fitting an assay to a specific purpose

If RT-qPCR and microarray platforms are complementary technologies, how should one choose between them? It is a question of fitting the technology to its purpose.

From a diagnostic perspective where the goal is simply to ascertain the presence or absence of a single organism, RT-qPCR may be the only sensible choice. For example the World Organization for Animal Health (OIE) maintains a database of animal pathogens and list validated tests for their detection. Many of these tests rely on cell culture or histopathology, but increasingly they include molecular methods as well. For example the OIE manual lists three molecular methods for *Aphanomyces invadans*, the causative agent of the Epizootic Ulcerative Syndrome in fish (OIE Manual). In this case, PCR is the obvious choice for detection as (a) it readily identifies the pathogen to the species level, (b) a single PCR reaction is still far less expensive than a single microarray reaction and (c) the large amounts of ancillary data captured by a microarray experiment are neither important nor desirable in this type of diagnostic context. Conversely, there are situations where multiple microbial entities should be identified simultaneously. For example, Paliy *et al.* (2009) developed a microarray system capable of identifying 775 different bacterial species that are found in human stool. The system was successfully utilized to demonstrate the presence of over 225 bacterial species in stool samples from children, and 200 bacterial species from adult stool samples. This microarray system was based on unique sequences found in the 16S rDNA gene of the bacteria. The system was quite sensitive, capable of detecting as little as 1 pg from a specific bacterial species. This corresponds to only 0.00025% of the total DNA from the sample. To achieve this level of sensitivity, the rDNA gene of all species in the sample was amplified using a few rounds of PCR with universal bacterial rDNA primers. The authors acknowledge that this increased sensitivity came at the cost of lower specificity. The lower specificity, which resulted from unequal amplification of the various rDNA genes in the PCR reaction could be mitigated by pooling a few PCR reactions together prior to hybridizing the amplicons to the microarray chip. Again, if the relative abundance of the individual species is of lesser importance, biased PCR amplification is a non-issue. On the other hand, if quantification of the individual species is important, more work will need to be done in order to validate this method.

Zhou *et al.* (2003) described the construction of a functional gene array (FGA) for microbial community analysis at the species-strain level. This array was constructed using oligonucleotides designed to hybridize to functional enzyme genes, involved in the biogeochemical cycling processes of bacteria. The use of 50-mer oligonucleotides, rather than longer (200–1000 nt) PCR products made for an array that was much easier to construct. Further, additions or substitutions to the array are also far easier to manage. The initial tested array contained just over 1000 probes targeted to nitrogen cycling and sulfite reduction genes. As is often the case, the much increased specificity resulting from the use of oligonucleotide probes was accompanied by an at least 10-fold decrease in sensitivity. As total genomic DNA is not usually a limiting factor in this type of study, the decrease in sensitivity is not considered to be a major issue.

Clearly, the first consideration for choosing qPCR (RT-qPCR) over DNA microarrays, or vice-versa, is to define the purpose of the assay, as this will guide the experimentation in all subsequent steps. Choosing a test that is fit to purpose sounds obvious but can only be accomplished if the purpose of the experiment is fully defined prior to beginning the experimentation. Despite both the popularity of smaller arrays, and the increased capacity of qPCR instrumentation, the number of desired simultaneous analyses remains an obvious criterion for choosing one technology over another. Above we mentioned that qPCR was the obvious choice for the detection of a single species. Conversely, when hundreds or thousands of species must be determined simultaneously, the obvious choice is some form of array. The choice is not so obvious with moderate numbers, and there are no 'right' numbers indicating when to switch from one technology to the other.

Assay validation

Assay validation means different things to different people. To an inexperienced user, an assay validation may simply be the design of PCR primers, the running of a PCR reaction, and the showing of a single amplicon on an agarose gel. For commercial entities, validation may be an ongoing process comprising of stage after stage of trials. In short, however, 'Validation is a process that determines the fitness of an assay, which has been properly developed, optimized and standardized, for an intended purpose' (OIE manual). As validation is directly linked to the purpose of an assay, it is clear that the level of validation required for your assay will affect the choice of technology.

Full scale validation of an assay for the presence of a pathogen that could affect international trade relations, for example, is considered to have four stages, covering 12 criteria, including the determination of analytical sensitivity and specificity of the assay, diagnostic sensitivity and specificity of the assay, as well as assay robustness

(OIE manual). This is followed by an ongoing monitoring of the validation criteria. This level of validation is extremely work-intensive even for the detection of a single pathogen in a single qPCR reaction. To maintain this level of validation for even a small array for the detection of only dozens of distinct organisms may prove impossible. On the other hand, as stated above, diagnostic tests that could affect international trade are usually focused towards a specific pathogen. For the analysis of microbial communities, the advantage of housing hundreds of probes on a single array far outweighs the shortcomings in the validation process. Simple analytical validations of various array technologies have been demonstrated on multiple occasions (Ye, 2001; Palmer, 2006; Zhang, 2007; Franke-Whittle, 2009), and are more than sufficient for most experiments examining microbial communities or phylogenies. This is not to say arrays need not be validated, rather they must be designed and validated to fit their intended purpose.

General considerations

It is still very common to validate the results obtained with the use of various microarray platforms by using RT-qPCR. Despite this tendency, it seems to be quite an inappropriate use of both platforms. It is our hope that many will realize that the choice of technology should be made by validating the use of a technology to its intended purpose. In this sense, the 'validation' of microarray results by qPCR is an oxymoron. Suppose you have determined that a properly designed and constructed array system is the best fit for your purpose, and you have validated this thought process with experimentation. There is then no context in which the use of RT-qPCR can be thought of as a further validation for your array results. At best, the RT-qPCR results can be considered redundant information. Should the results of the array and RT-qPCR be discordant, under what circumstances should the RT-qPCR results, which have not been validated for your purpose, be considered 'right'? It seems more and more likely that both techniques will continue to have their own (dis)advantages and specificities, and that both technologies will develop as complementary techniques rather than one being used to validate the other (Yang and Speed, 2002; Allison *et al.*, 2006; Gaj *et al.*, 2008).

Future trends

The two main methods for quantitative detection of amplified products with RT-qPCR include the use of (i) double-stranded DNA-binding agents, such as SYBR Green I and (ii) FRET fluorescent probes. While the high sensitivities that can be achieved with double-stranded DNA-binding agents are well known, their lack of specificity creates difficulties for discrimination between specific and non-specific products generated during the PCR process, raising the potential of obtaining a false-positive result (Tan, 2004). A possible improvement of such technique could reside in augmenting the specificity of the oligonucleotide primers used to amplify template sequences using in the first stage computational approaches that better capture the physic-chemical aspects of the hybridization process. The introduction of FRET fluorescent probes alleviate a second disadvantage of the dsDNA-binding agents, namely their inability to be quantified in the same reaction. The FRET fluorescent probes also allow multilabelling approaches where different distinguishable reporter dyes are used with the only downside that experimental costs will increase.

A potential successful trend that can dramatically improve the multiplex limitations of the current RT-qPCR technology as suggested by Loy *et al.* (2006) is the immobilization of large numbers of amplification primers on solid substrates. Such an approach combines the multiplexing advantage of microarrays with the increased sensitivity of RT-qPCR, thus opening new horizons in the field. Also, technological improvements for RT-qPCR where reaction volumes become smaller while throughput increases, guarantee the future success and usability of this technique primarily for microarray data validation in the years to come (VanGuilder *et al.*, 2008).

While RT-qPCR is considered by many researchers to be a 'golden standard' for microarray data validation, some studies suggest that it may be more useful as a complementary approach rather than a validation strategy (Yang and Speed, 2002; Allison *et al.*, 2006; Gaj *et al.*, 2008). Recent studies (Streit *et al.*, 2009) also suggest that by combining microarrays with RT-qPCR technologies, the identification of clinically relevant diagnostic and prognostic biomarkers and the discovery of novel therapeutic targets become more efficient and reliable procedures.

A new trend in microarray development for microbial studies focuses on reducing chip size and the number of features deployed on the chip, given that minimal advantage is obtained with mismatch probes and small numbers of well chosen probes produce similar or better results than exhaustive selections (Bahl *et al.*, 2010). Such an approach will bring close together, quality-wise, the microarray and the RT-qPCR technologies.

Last but not the least, the utilization of microarrays as diagnostic tools is an area with great perspectives and scientific and technological breakthroughs are expected to smooth the technical transfer from purely research microarray based studies to medical applications. Since the introduction of the first Food and Drug Administration (FDA) approved microarray chip in 2005, namely the Roche AmpliChip Cytochrome P450 test for use on the Affymetrix GeneChip Microarray Instrumentation System to help doctors personalize treatments for their patients, less than a dozen microarrays have been given the green light to be used in clinical environments in different parts of the world. Another commercial RNA-based diagnostic microarray (ParoCheck) developed by Greiner Bio One, Frickenhausen, Germany and approved for usage in Europe is able to identify 20 oral bacterial species and used for semi-quantitative detection of periodontitis-associated pathogens in clinical samples. One year later, in January 2006, due to problems involving performance, safety and lack of standards reasons, FDA has stopped authorizations for any microarrays to be sold as analyte specific reagents (ASRs), thus leaving only one viable option for the development and commercialization of clinical diagnostic assays, namely conducting internal validations of assays (home-brew tests) using general-purpose reagents. This route can be pursued by using one of the very few available commercial platforms provided by CombiMatrix (the CustomArray and Microarray Synthesizer). Therefore, the need for high-quality chip designs and proper validation mechanisms are crucial for the use of microarrays in clinical settings.

Web resources

Akonni Biosystems – TruArray Tests: http://www.akonni.com/products/mdx-tests.html

NimbleGen Gene Expression – Prokayotic Designs: http://www.nimblegen.com/products/exp/prokaryotic.html

Oxford Gene Technology – microbial array products: http://www.ogt.co.uk/prokaryotic_designs.html

Veredus Laboratories – chips: http://www.vereduslabs.com/products.html

World Organization for Animal Health (OIE): http://www.oie.int

References

Abbott, M.A., Poiesz, B.J., Byrne, B.C., Kwok, S., Sninsky, J.J., and Ehrlich, G.D. (1988). Enzymatic gene amplification: qualitative and quantitative methods for detecting proviral DNA amplified *in vitro*. J. Infect. Dis. *158*, 1158–1169.

Abravaya, K., Huff, J., Marshall, R., Merchant, B., Mullen, C., Schneider, G., and Robinson, J. (2003). Molecular beacons as diagnostic tools: technology and applications. Clin. Chem. Lab. Med. *41*, 468–474.

Abruzzo, L.V., Lee, K.Y., Fuller, A., Silverman, A., Keating, M.J., Medeiros, L.J., and Coombes, K.R. (2005). Validation of oligonucleotide microarray data using microfluidic low-density arrays: a new statistical method to normalize real-time RT-PCR data. Biotechniques *38*, 785–792.

Adomas, A.B., Lopez-Giraldez, F., Clark, T.A., Wang, Z., and Townsend, J.P. (2010). Multi-targeted priming for genome-wide gene expression assays. BMC Genomics *11*, 477.

Aerts, J.L., Gonzales, M.I., and Topalian, S.L. (2004). Selection of appropriate control genes to assess expression of tumor antigens using real-time RT-PCR. Biotechniques *36*, 84–86, 88, 90–91.

Agranovsky, A.A. (1992). Exogenous primer-independent cDNA synthesis with commercial reverse transcriptase preparations on plant virus RNA templates. Anal. Biochem. *203*, 163–165.

Aliyu, S.H., Aliyu, M.H., Salihu, H.M., Parmar, S., Jalal, H., and Curran, M.D. (2004). Rapid detection and quantitation of hepatitis B virus DNA by real-time PCR using a new fluorescent (FRET) detection system. J. Clin. Virol. *30*, 191–195.

Allison, D.B., Cui, X., Page, G.P., and Sabripour, M. (2006). Microarray data analysis: from disarray to consolidation and consensus. Nat. Rev. Genet. *7*, 55–65.

Alwine, J.C., Kemp, D.J., and Stark, G.R. (1977). Method for detection of specific RNAs in agarose gels by transfer to diazobenzyloxymethyl-paper and hybridization with DNA probes. Proc. Natl. Acad. Sci. U.S.A. *74*, 5350–5354.

Andersen, M.T., and Foy, C.A. (2005). The development of microarray standards. Anal. Bioanal. Chem. *381*, 87–89.

Anjum, M.F., Tucker, J.D., Sprigings, K.A., Woodward, M.J., and Ehricht, R. (2006). Use of miniaturized protein arrays for *Escherichia coli* O serotyping. Clin. Vaccine Immunol. *13*, 561–567.

Aranyi, T., and Tusnady, G.E. (2007). BiSearch: ePCR tool for native or bisulfite-treated genomic template. Meth. Mol. Biol. *402*, 385–402.

Aranyi, T., Varadi, A., Simon, I., and Tusnady, G.E. (2006). The BiSearch web server. BMC Bioinformatics *7*, 431.

Arikawa, E., Sun, Y., Wang, J., Zhou, Q., Ning, B., Dial, S.L., Guo, L., and Yang, J. (2008). Cross-platform comparison of SYBR Green real-time PCR with TaqMan PCR, microarrays and other gene expression measurement technologies evaluated in the MicroArray Quality Control (MAQC) study. BMC Genomics *9*, 328.

Armour, C.D., Castle, J.C., Chen, R., Babak, T., Loerch, P., Jackson, S., Shah, J.K., Dey, J., Rohl, C.A., Johnson, J.M., *et al.* (2009). Digital transcriptome profiling using selective hexamer priming for cDNA synthesis. Nat. Methods *6*, 647–649.

Arvidsson, S., Kwasniewski, M., Riano-Pachon, D.M., and Mueller-Roeber, B. (2008). QuantPrime – a flexible tool for reliable high-throughput primer design for quantitative PCR. BMC Bioinformatics *9*, 465.

Ashburner, M., Ball, C.A., Blake, J.A., Botstein, D., Butler, H., Cherry, J.M., Davis, A.P., Dolinski, K., Dwight, S.S., Eppig, J.T., *et al.* (2000). Gene ontology: tool for the unification of biology. The Gene Ontology Consortium. Nat. Genet. *25*, 25–29.

Avarre, J.C., de Lajudie, P., and Bena, G. (2007). Hybridization of genomic DNA to microarrays: a challenge for the analysis of environmental samples. J. Microbiol. Methods *69*, 242–248.

Badiee, A., Eiken, H.G., Steen, V.M., and Lovlie, R. (2003). Evaluation of five different cDNA labeling methods for microarrays using spike controls. BMC Biotechnol. *3*, 23.

Bahl, A., Davis, P.H., Behnke, M., Dzierszinski, F., Jagalur, M., Chen, F., Shanmugam, D., White, M.W., Kulp, D., and Roos, D.S. (2010). A novel multifunctional oligonucleotide microarray for *Toxoplasma gondii*. BMC Genomics *11*, 603.

Baker, S.C., Bauer, S.R., Beyer, R.P., Brenton, J.D., Bromley, B., Burrill, J., Causton, H., Conley, M.P., Elespuru, R., Fero, M., *et al.* (2005). The External RNA Controls Consortium: a progress report. Nat. Methods *2*, 731–734.

Baldi, P., and Long, A.D. (2001). A Bayesian framework for the analysis of microarray expression data: regularized t-test and statistical inferences of gene changes. Bioinformatics *17*, 509–519.

Bammler, T., Beyer, R.P., Bhattacharya, S., Boorman, G.A., Boyles, A., Bradford, B.U., Bumgarner, R.E., Bushel, P.R., Chaturvedi, K., Choi, D., *et al.* (2005). Standardizing global gene expression analysis between laboratories and across platforms. Nat. Methods *2*, 351–356.

Bao, Y.P., Huber, M., Wei, T.F., Marla, S.S., Storhoff, J.J., and Muller, U.R. (2005). SNP identification in unamplified human genomic DNA with gold nanoparticle probes. Nucleic Acids Res. *33*, e15.

van Baren, M.J., and Heutink, P. (2004). The PCR suite. Bioinformatics *20*, 591–593.

Bavykin, S.G., Akowski, J.P., Zakhariev, V.M., Barsky, V.E., Perov, A.N., and Mirzabekov, A.D. (2001). Portable system for microbial sample preparation and oligonucleotide microarray analysis. Appl. Environ. Microbiol. *67*, 922–928.

Becker, C., Hammerle-Fickinger, A., Riedmaier, I., and Pfaffl, M.W. (2010). mRNA and microRNA quality control for RT-qPCR analysis. Methods *50*, 237–243.

Becker-Andre, M., and Hahlbrock, K. (1989). Absolute mRNA quantification using the polymerase chain reaction (PCR). A novel approach by a PCR aided transcript titration assay (PATTY). Nucleic Acids Res. *17*, 9437–9446.

Beckman, K.B., Lee, K.Y., Golden, T., and Melov, S. (2004). Gene expression profiling in mitochondrial disease: assessment of microarray accuracy by high-throughput Q-PCR. Mitochondrion *4*, 453–470.

Behr, M.A., Wilson, M.A., Gill, W.P., Salamon, H., Schoolnik, G.K., Rane, S., and Small, P.M. (1999). Comparative genomics of BCG vaccines by whole-genome DNA microarray. Science *284*, 1520–1523.

Beissbarth, T., and Speed, T.P. (2004). GOstat: find statistically overrepresented Gene Ontologies within a group of genes. Bioinformatics *20*, 1464–1465.

Belcher, C.E., Drenkow, J., Kehoe, B., Gingeras, T.R., McNamara, N., Lemjabbar, H., Basbaum, C., and Relman, D.A. (2000). The transcriptional responses of respiratory epithelial cells to Bordetella pertussis reveal host defensive and pathogen counter-defensive strategies. Proc. Natl. Acad. Sci. U.S.A. *97*, 13847–13852.

Belosludtsev, Y.Y., Bowerman, D., Weil, R., Marthandan, N., Balog, R., Luebke, K., Lawson, J., Johnston, S.A., Lyons, C.R., Obrien, K., *et al.* (2004). Organism identification using a genome sequence-independent universal microarray probe set. Biotechniques *37*, 654–658, 660.

Ben Zakour, N., Gautier, M., Andonov, R., Lavenier, D., Cochet, M.F., Veber, P., Sorokin, A., and Le Loir, Y. (2004). GenoFrag: software to design primers optimized for whole genome scanning by long-range PCR amplification. Nucleic Acids Res. *32*, 17–24.

Bengtsson, M., Hemberg, M., Rorsman, P., and Stahlberg, A. (2008). Quantification of mRNA in single cells and modelling of RT-qPCR induced noise. BMC Mol. Biol. *9*, 63.

Blazewicz, J., and Kasprzak, M. (2006). Computational complexity of isothermic DNA sequencing by hybridization. Discrete Applied Mathematics *154*, 718–729.

Bodrossy, L., and Sessitsch, A. (2004). Oligonucleotide microarrays in microbial diagnostics. Curr. Opin. Microbiol. *7*, 245–254.

Bolstad, B.M., Irizarry, R.A., Astrand, M., and Speed, T.P. (2003). A comparison of normalization methods for high density oligonucleotide array data based on variance and bias. Bioinformatics *19*, 185–193.

Bosotti, R., Locatelli, G., Healy, S., Scacheri, E., Sartori, L., Mercurio, C., Calogero, R., and Isacchi, A. (2007). Cross platform microarray analysis for robust identification of differentially expressed genes. BMC Bioinformatics 8 Suppl *1*, S5.

Bozdech, Z., Zhu, J., Joachimiak, M.P., Cohen, F.E., Pulliam, B., and DeRisi, J.L. (2003). Expression profiling of the schizont and trophozoite stages of Plasmodium falciparum with a long-oligonucleotide microarray. Genome Biol. *4*, R9.

Brazma, A., Hingamp, P., Quackenbush, J., Sherlock, G., Spellman, P., Stoeckert, C., Aach, J., Ansorge, W., Ball, C.A., Causton, H.C., *et al.* (2001). Minimum information about a microarray experiment (MIAME)-toward standards for microarray data. Nat. Genet. *29*, 365–371.

Breitling, R. (2006). Biological microarray interpretation: the rules of engagement. Biochim. Biophys. Acta *1759*, 319–327.

Breitling, R., Armengaud, P., Amtmann, A., and Herzyk, P. (2004). Rank products: a simple, yet powerful, new method to detect differentially regulated genes in replicated microarray experiments. FEBS Lett. *573*, 83–92.

Breitling, R., and Herzyk, P. (2005). Rank-based methods as a non-parametric alternative of the T-statistic for the analysis of biological microarray data. J. Bioinform. Comput. Biol. *3*, 1171–1189.

Brooks, E.M., Sheflin, L.G., and Spaulding, S.W. (1995). Secondary structure in the 3′ UTR of EGF and the choice of reverse transcriptases affect the detection of message diversity by RT-PCR. Biotechniques *19*, 806–812, 814–805.

Brunstein, J., and Thomas, E. (2006). Direct screening of clinical specimens for multiple respiratory pathogens using the Genaco Respiratory Panels 1 and 2. Diagn. Mol. Pathol. *15*, 169–173.

Bustin, S.A. (2000). Absolute quantification of mRNA using real-time reverse transcription polymerase chain reaction assays. J. Mol. Endocrinol. *25*, 169–193.

Bustin, S.A. (2002). Quantification of mRNA using real-time reverse transcription PCR (RT-PCR): trends and problems. J. Mol. Endocrinol. *29*, 23–39.

Bustin, S.A. (2008a). Molecular medicine, gene-expression profiling and molecular diagnostics: putting the cart before the horse. Biomark. Med. *2*, 201–207.

Bustin, S.A. (2008b). Real-time polymerase chain reaction – towards a more reliable, accurate and relevant assay. Eur. Pharma. Rev. *6*, 19–27.

Bustin, S.A. (2010). Why the need for qPCR publication guidelines? – The case for MIQE. Methods *50*, 217–226.

Bustin, S.A., and Mueller, R. (2005). Real-time reverse transcription PCR (qRT-PCR) and its potential use in clinical diagnosis. Clin. Sci. (Lond) *109*, 365–379.

Bustin, S.A., and Nolan, T. (2004). Pitfalls of quantitative real-time reverse-transcription polymerase chain reaction. J. Biomol. Tech. *15*, 155–166.

Bustin, S.A., Benes, V., Nolan, T., and Pfaffl, M.W. (2005). Quantitative real-time RT-PCR – a perspective. J. Mol. Endocrinol. *34*, 597–601.

Bustin, S.A., Vandesompele, J., and Pfaffl, M. (2009a). Standardization of qPCR and RT-qPCR. New guidelines seek to promote accurate interpretation of data and reliable results. Gen. Eng. Biotechnol. News *9*, 14.

Bustin, S.A., Benes, V., Garson, J.A., Hellemans, J., Huggett, J., Kubista, M., Mueller, R., Nolan, T., Pfaffl, M.W., Shipley, G.L., *et al.* (2009b). The MIQE guidelines: minimum information for publication of quantitative real-time PCR experiments. Clin. Chem. *55*, 611–622.

Bustin, S.A., Beaulieu, J.F., Huggett, J., Jaggi, R., Kibenge, F.S., Olsvik, P.A., Penning, L.C., and Toegel, S. (2010). MIQE precis: Practical implementation of minimum standard guidelines for fluorescence-based quantitative real-time PCR experiments. BMC Mol. Biol. *11*, 74.

Butow, R.A., Zhu, H., Perlman, P., and Conrad-Webb, H. (1989). The role of a conserved dodecamer sequence in yeast mitochondrial gene expression. Genome *31*, 757–760.

Call, D.R., Borucki, M.K., and Loge, F.J. (2003). Detection of bacterial pathogens in environmental samples using DNA microarrays. J. Microbiol. Methods *53*, 235–243.

Canales, R.D., Luo, Y., Willey, J.C., Austermiller, B., Barbacioru, C.C., Boysen, C., Hunkapiller, K., Jensen, R.V., Knight, C.R., Lee, K.Y., *et al.* (2006). Evaluation of DNA microarray results with quantitative gene expression platforms. Nat. Biotechnol. *24*, 1115–1122.

Cannon, C.H., Kua, C.S., Lobenhofer, E.K., and Hurban, P. (2006). Capturing genomic signatures of DNA sequence variation using a standard anonymous microarray platform. Nucleic Acids Res. *34*, e121.

Cao, W., Epstein, C., Liu, H., DeLoughery, C., Ge, N., Lin, J., Diao, R., Cao, H., Long, F., Zhang, X., *et al.* (2004). Comparing gene discovery from Affymetrix GeneChip microarrays and Clontech PCR-select cDNA subtraction: a case study. BMC Genomics *5*, 26.

Cao, Y., Sun, J., Zhu, J., Li, L., and Liu, G. (2010). PrimerCE: designing primers for cloning and gene expression. Mol. Biotechnol. *46*, 113–117.

Chandler, D.P., Alferov, O., Chernov, B., Daly, D.S., Golova, J., Perov, A., Protic, M., Robison, R., Schipma, M., White, A., *et al.* (2006). Diagnostic oligonucleotide microarray fingerprinting of *Bacillus* isolates. J. Clin. Microbiol. *44*, 244–250.

Charbonnier, Y., Gettler, B., Francois, P., Bento, M., Renzoni, A., Vaudaux, P., Schlegel, W., and Schrenzel, J. (2005). A generic approach for the design of whole-genome oligoarrays, validated for genomotyping, deletion mapping and gene expression analysis on *Staphylococcus aureus*. BMC Genomics *6*, 95.

Chen, H., and Sharp, B. (2002). Oliz, a suite of Perl scripts that assist in the design of microarrays using 50mer oligonucleotides from the 3′ untranslated region. BMC Bioinformatics 3, 27.

Chen, J.J., Delongchamp, R.R., Tsai, C.A., Hsueh, H.M., Sistare, F., Thompson, K.L., Desai, V.G., and Fuscoe, J.C. (2004). Analysis of variance components in gene expression data. Bioinformatics *20*, 1436–1446.

Chen, S.H., Lin, C.Y., Cho, C.S., Lo, C.Z., and Hsiung, C.A. (2003). Primer Design Assistant (PDA): A web-based primer design tool. Nucleic Acids Res. *31*, 3751–3754.

Chen, S.H., Lo, C.Z., Su, S.Y., Kuo, B.H., Hsiung, C.A., and Lin, C.Y. (2010). UPS 2.0: unique probe selector for probe design and oligonucleotide microarrays at the pangenomic/genomic level. BMC Genomics 11 (Suppl 4), S6.

Chiang, P.W., Song, W.J., Wu, K.Y., Korenberg, J.R., Fogel, E.J., Van Keuren, M.L., Lashkari, D., and Kurnit, D.M. (1996). Use of a fluorescent-PCR reaction to detect genomic sequence copy number and transcriptional abundance. Genome Res. *6*, 1013–1026.

Chizhikov, V., Rasooly, A., Chumakov, K., and Levy, D.D. (2001). Microarray analysis of microbial virulence factors. Appl. Environ. Microbiol. *67*, 3258–3263.

Chou, H.H., Hsia, A.P., Mooney, D.L., and Schnable, P.S. (2004). Picky: oligo microarray design for large genomes. Bioinformatics *20*, 2893–2902.

Chuaqui, R.F., Bonner, R.F., Best, C.J., Gillespie, J.W., Flaig, M.J., Hewitt, S.M., Phillips, J.L., Krizman, D.B., Tangrea, M.A., Ahram, M., *et al.* (2002). Post-analysis follow-up and validation of microarray experiments. Nat. Genet. 32 (Suppl), 509–514.

Chung, W.H., Rhee, S.K., Wan, X.F., Bae, J.W., Quan, Z.X., and Park, Y.H. (2005). Design of long oligonucleotide probes for functional gene detection in a microbial community. Bioinformatics *21*, 4092–4100.

Cikos, S., Bukovska, A., and Koppel, J. (2007). Relative quantification of mRNA: comparison of methods currently used for real-time PCR data analysis. BMC Mol. Biol. *8*, 113.

Cohen, P., Bouaboula, M., Bellis, M., Baron, V., Jbilo, O., Poinot-Chazel, C., Galiegue, S., Hadibi, E.H., and Casellas, P. (2000). Monitoring cellular responses to *Listeria monocytogenes* with oligonucleotide arrays. J. Biol. Chem. *275*, 11181–11190.

Colombo, A.P., Boches, S.K., Cotton, S.L., Goodson, J.M., Kent, R., Haffajee, A.D., Socransky, S.S., Hasturk, H., Van Dyke, T.E., Dewhirst, F., *et al.* (2009). Comparisons of subgingival microbial profiles of refractory periodontitis, severe periodontitis, and periodontal health using the human oral microbe identification microarray. J. Periodontol. *80*, 1421–1432.

Cone, R.W., Hobson, A.C., and Huang, M.L. (1992). Coamplified positive control detects inhibition of polymerase chain reactions. J. Clin. Microbiol. *30*, 3185–3189.

Consortium, E.R.C. (2005). Proposed methods for testing and selecting the ERCC external RNA controls. BMC Genomics *6*, 150.

Conway, T., and Schoolnik, G.K. (2003). Microarray expression profiling: capturing a genome-wide portrait of the transcriptome. Mol. Microbiol. *47*, 879–889.

Coombes, B.K., and Mahony, J.B. (2001). cDNA array analysis of altered gene expression in human endothelial cells in response to Chlamydia pneumoniae infection. Infect. Immun. *69*, 1420–1427.

Couzinet, S., Jay, C., Barras, C., Vachon, R., Vernet, G., Ninet, B., Jan, I., Minazio, M.A., Francois, P., Lew, D., *et al.* (2005). High-density DNA probe arrays for identification of staphylococci to the species level. J. Microbiol. Methods *61*, 201–208.

Cronin, M., Ghosh, K., Sistare, F., Quackenbush, J., Vilker, V., and O'Connell, C. (2004). Universal RNA reference materials for gene expression. Clin. Chem. *50*, 1464–1471.

Cummings, C.A., and Relman, D.A. (2000). Using DNA microarrays to study host–microbe interactions. Emerg. Infect. Dis. *6*, 513–525.

Cutler, D.J., Zwick, M.E., Carrasquillo, M.M., Yohn, C.T., Tobin, K.P., Kashuk, C., Mathews, D.J., Shah, N.A., Eichler, E.E., Warrington, J.A., *et al.* (2001). High-throughput variation detection and genotyping using microarrays. Genome Res. *11*, 1913–1925.

Czechowski, T., Bari, R.P., Stitt, M., Scheible, W.R., and Udvardi, M.K. (2004). Real-time RT-PCR profiling of over 1400 *Arabidopsis* transcription factors: unprecedented sensitivity reveals novel root- and shoot-specific genes. Plant J. *38*, 366–379.

Dallas, P.B., Gottardo, N.G., Firth, M.J., Beesley, A.H., Hoffmann, K., Terry, P.A., Freitas, J.R., Boag, J.M., Cummings, A.J., and Kees, U.R. (2005). Gene expression levels assessed by oligonucleotide microarray analysis and quantitative real-time RT-PCR – how well do they correlate? BMC Genomics *6*, 59.

Datta, S. (2003). Comparisons and validation of statistical clustering techniques for microarray gene expression data. Bioinformatics *19*, 459–466.

Decraene, C., Reguigne-Arnould, I., Auffray, C., and Pietu, G. (1999). Reverse transcription in the presence of dideoxynucleotides to increase the sensitivity of expression monitoring with cDNA arrays. Biotechniques *27*, 962–966.

Deindl, E., Boengler, K., van Royen, N., and Schaper, W. (2002). Differential expression of GAPDH and beta3-actin in growing collateral arteries. Mol. Cell Biochem. *236*, 139–146.

Denisov, V., Strong, W., Walder, M., Gingrich, J., and Wintz, H. (2008). Development and validation of RQI: an RNA quality indicator for the Experion automated electrophoresis system. Bio-Rad. Bull. 5761.

Derveaux, S., Vandesompele, J., and Hellemans, J. (2010). How to do successful gene expression analysis using real-time PCR. Methods *50*, 227–230.

DeSantis, T.Z., Stone, C.E., Murray, S.R., Moberg, J.P., and Andersen, G.L. (2005). Rapid quantification and taxonomic classification of environmental DNA from both prokaryotic and eukaryotic origins using a microarray. FEMS Microbiol. Lett. *245*, 271–278.

Devonshire, A.S., Elaswarapu, R., and Foy, C.A. (2010). Evaluation of external RNA controls for the standardisation of gene expression biomarker measurements. BMC Genomics *11*, 662.

Dheda, K., Huggett, J.F., Bustin, S.A., Johnson, M.A., Rook, G., and Zumla, A. (2004). Validation of housekeeping genes for normalizing RNA expression in real-time PCR. Biotechniques *37*, 112–114, 116, 118–119.

Dieckmann, C.L., and Tzagoloff, A. (1985). Assembly of the mitochondrial membrane system. CBP6, a yeast nuclear gene necessary for synthesis of cytochrome b. J. Biol. Chem. *260*, 1513–1520.

Ding, Y., Xu, L., Jovanovic, B.D., Helenowski, I.B., Kelly, D.L., Catalona, W.J., Yang, X.J., Pins, M., and Bergan, R.C. (2007). The methodology used to measure differential gene expression affects the outcome. J. Biomol. Tech. *18*, 321–330.

Doran, M., Raicu, D.S., Furst, J.D., Settimi, R., Schipma, M., and Chandler, D.P. (2007). Oligonucleotide microarray identification of *Bacillus anthracis* strains using support vector machines. Bioinformatics *23*, 487–492.

Draghici, S., Khatri, P., Eklund, A.C., and Szallasi, Z. (2006). Reliability and reproducibility issues in DNA microarray measurements. Trends Genet. *22*, 101–109.

Dufour, Y.S., Wesenberg, G.E., Tritt, A.J., Glasner, J.D., Perna, N.T., Mitchell, J.C., and Donohue, T.J. (2010). chipD: a web tool to design oligonucleotide probes for high-density tiling arrays. Nucleic Acids Res. *38*, W321–325.

Durbin, B.P., and Rocke, D.M. (2004). Variance-stabilizing transformations for two-color microarrays. Bioinformatics *20*, 660–667.

Efron, B., Tibshirani, R., Storey, J.D., and Tusher, V. (2002). Empirical Bayes analysis of a microarray experiment. J. Am. Stat. Assoc. *456*, 1151–1160.

Ehrenreich, A. (2006). DNA microarray technology for the microbiologist: an overview. Appl. Microbiol. Biotechnol. *73*, 255–273.

Eisen, M.B., Spellman, P.T., Brown, P.O., and Botstein, D. (1998). Cluster analysis and display of genome-wide expression patterns. Proc. Natl. Acad. Sci. U.S.A. *95*, 14863–14868.

Elad, T., Lee, J.H., Belkin, S., and Gu, M.B. (2008). Microbial whole-cell arrays. Microb. Biotechnol. *1*, 137–148.

Ellis, T.P., Helfenbein, K.G., Tzagoloff, A., and Dieckmann, C.L. (2004). Aep3p stabilizes the mitochondrial bicistronic mRNA encoding subunits 6 and 8 of the H+-translocating ATP synthase of *Saccharomyces cerevisiae*. J. Biol. Chem. *279*, 15728–15733.

Emrich, S.J., Lowe, M., and Delcher, A.L. (2003). PROBEmer: A web-based software tool for selecting optimal DNA oligos. Nucleic Acids Res. *31*, 3746–3750.

Espy, M.J., Uhl, J.R., Sloan, L.M., Buckwalter, S.P., Jones, M.F., Vetter, E.A., Yao, J.D., Wengenack, N.L., Rosenblatt, J.E., Cockerill, F.R., 3rd, *et al.* (2006). Real-time PCR in clinical microbiology: applications for routine laboratory testing. Clin. Microbiol. Rev. *19*, 165–256.

Fan, J.B., Yeakley, J.M., Bibikova, M., Chudin, E., Wickham, E., Chen, J., Doucet, D., Rigault, P., Zhang, B., Shen, R., *et al.* (2004). A versatile assay for high-throughput gene expression profiling on universal array matrices. Genome Res. *14*, 878–885.

Fawcett, P., Eichenberger, P., Losick, R., and Youngman, P. (2000). The transcriptional profile of early to middle sporulation in *Bacillus subtilis*. Proc. Natl. Acad. Sci. U.S.A. *97*, 8063–8068.

Fay, J.C., McCullough, H.L., Sniegowski, P.D., and Eisen, M.B. (2004). Population genetic variation in gene expression is associated with phenotypic variation in *Saccharomyces cerevisiae*. Genome Biol. *5*, R26.

Feng, S., and Tillier, E.R. (2007). A fast and flexible approach to oligonucleotide probe design for genomes and gene families. Bioinformatics *23*, 1195–1202.

Fernandes, R., and Skiena, S. (2007). MultiPrimer: a system for microarray PCR primer design. Meth. Mol. Biol. *402*, 305–314.

Fleige, S., and Pfaffl, M.W. (2006). RNA integrity and the effect on the real-time qRT-PCR performance. Mol. Aspects Med. *27*, 126–139.

Franke-Whittle, I.H., Klammer, S.H., and Insam, H. (2005). Design and application of an oligonucleotide microarray for the investigation of compost microbial communities. J. Microbiol. Methods *62*, 37–56.

Franke-Whittle, I.H., Knapp, B.A., Fuchs, J., Kaufmann, R., and Insam, H. (2009). Application of COMPOCHIP microarray to investigate the bacterial communities of different composts. Microb. Ecol. *57*, 510–521.

Fredslund, J., Schauser, L., Madsen, L.H., Sandal, N., and Stougaard, J. (2005). PriFi: using a multiple alignment of related sequences to find primers for amplification of homologs. Nucleic Acids Res. *33*, W516–520.

Freeman, W.M., Walker, S.J., and Vrana, K.E. (1999). Quantitative RT-PCR: pitfalls and potential. Biotechniques *26*, 112–122, 124–125.

Fronhoffs, S., Totzke, G., Stier, S., Wernert, N., Rothe, M., Bruning, T., Koch, B., Sachinidis, A., Vetter, H., and Ko, Y. (2002). A method for the rapid construction of cRNA standard curves in quantitative real-time reverse transcription polymerase chain reaction. Mol. Cell Probes *16*, 99–110.

Fu, L.M., and Tai, S.C. (2009). The differential gene expression pattern of *Mycobacterium tuberculosis* in response to Capreomycin and PA-824 versus first-line TB drugs reveals stress- and PE/PPE-related drug targets. Int. J. Microbiol. *2009*, 879621.

Fukushima, M., Kakinuma, K., Hayashi, H., Nagai, H., Ito, K., and Kawaguchi, R. (2003). Detection and identification of *Mycobacterium* species isolates by DNA microarray. J. Clin. Microbiol. *41*, 2605–2615.

Gadberry, M.D., Malcomber, S.T., Doust, A.N., and Kellogg, E.A. (2005). Primaclade – a flexible tool to find conserved PCR primers across multiple species. Bioinformatics *21*, 1263–1264.

Gaj, S., Eijssen, L., Mensink, R.P., and Evelo, C.T. (2008). Validating nutrient-related gene expression changes from microarrays using RT(2) PCR-arrays. Genes Nutr. *3*, 153–157.

Gallup, J.M., and Ackermann, M.R. (2008). The 'PREXCEL-Q Method' for qPCR. Int. J. Biomed. Sci. *4*, 273–293.

Gerber, A.P., Herschlag, D., and Brown, P.O. (2004). Extensive association of functionally and cytotopically related mRNAs with Puf family RNA-binding proteins in yeast. PLoS Biol. *2*, E79.

Giancarlo, R., Scaturro, D., and Utro, F. (2008). Computational cluster validation for microarray data analysis: experimental assessment of Clest, Consensus Clustering, Figure of Merit, Gap Statistics and Model Explorer. BMC Bioinformatics *9*, 462.

Giannino, M.L., Aliprandi, M., Feligini, M., Vanoni, L., Brasca, M., and Fracchetti, F. (2009). A DNA array based assay for the characterization of microbial community in raw milk. J. Microbiol. Methods *78*, 181–188.

Gibson, U.E., Heid, C.A., and Williams, P.M. (1996). A novel method for real time quantitative RT-PCR. Genome Res. *6*, 995–1001.

Gill, R.T., DeLisa, M.P., Valdes, J.J., and Bentley, W.E. (2001). Genomic analysis of high-cell-density recombinant *Escherichia coli* fermentation and "cell conditioning" for improved recombinant protein yield. Biotechnol. Bioeng. *72*, 85–95.

Gilliland, G., Perrin, S., Blanchard, K., and Bunn, H.F. (1990). Analysis of cytokine mRNA and DNA: detection and quantitation by competitive polymerase chain reaction. Proc. Natl. Acad. Sci. U.S.A. *87*, 2725–2729.

Gilsbach, R., Kouta, M., Bonisch, H., and Bruss, M. (2006). Comparison of *in vitro* and *in vivo* reference genes for internal standardization of real-time PCR data. Biotechniques *40*, 173–177.

Gingrich, J., Rubio, T., and Karlak, C. (2006). Effect of RNA degradation on the data quality in quantitative PCR and microarray experiments. Bio-Rad Bull. 5452A.

Ginzinger, D.G. (2002). Gene quantification using real-time quantitative PCR: an emerging technology hits the mainstream. Exp. Hematol. *30*, 503–512.

Girolami, M., and Breitling, R. (2004). Biologically valid linear factor models of gene expression. Bioinformatics *20*, 3021–3033.

Giulietti, A., Overbergh, L., Valckx, D., Decallonne, B., Bouillon, R., and Mathieu, C. (2001). An overview of real-time quantitative PCR: applications to quantify cytokine gene expression. Methods *25*, 386–401.

Goidin, D., Mamessier, A., Staquet, M.J., Schmitt, D., and Berthier-Vergnes, O. (2001). Ribosomal 18S RNA prevails over glyceraldehyde-3-phosphate dehydrogenase and beta-actin genes as internal standard for quantitative comparison of mRNA levels in invasive and noninvasive human melanoma cell subpopulations. Anal. Biochem. *295*, 17–21.

Gomez, P., Hackett, T.L., Moore, M.M., Knight, D.A., and Tebbutt, S.J. (2010). Functional genomics of human bronchial epithelial cells directly interacting with conidia of Aspergillus fumigatus. BMC Genomics *11*, 358.

Gonzalez, J.M., and Robb, F.T. (2007). Counterselection of prokaryotic ribosomal RNA during reverse transcription using non-random hexameric oligonucleotides. J. Microbiol. Methods *71*, 288–291.

Gordon, P.M., and Sensen, C.W. (2004). Osprey: a comprehensive tool employing novel methods for the design of oligonucleotides for DNA sequencing and microarrays. Nucleic Acids Res. *32*, e133.

Gorzelniak, K., Janke, J., Engeli, S., and Sharma, A.M. (2001). Validation of endogenous controls for gene expression studies in human adipocytes and preadipocytes. Horm. Metab. Res. *33*, 625–627.

Gowda, M., Jantasuriyarat, C., Dean, R.A., and Wang, G.L. (2004). Robust-LongSAGE (RL-SAGE): a substantially improved LongSAGE method for gene discovery and transcriptome analysis. Plant Physiol. *134*, 890–897.

Greninger, A.L., Chen, E.C., Sittler, T., Scheinerman, A., Roubinian, N., Yu, G., Kim, E., Pillai, D.R., Guyard, C., Mazzulli, T., *et al.* (2010). A metagenomic analysis of pandemic influenza A (2009 H1N1) infection in patients from North America. PLoS ONE *5*, e13381.

Grunenfelder, B., and Winzeler, E.A. (2002). Treasures and traps in genome-wide data sets: case examples from yeast. Nat. Rev. Genet. *3*, 653–661.

Guschin, D.Y., Mobarry, B.K., Proudnikov, D., Stahl, D.A., Rittmann, B.E., and Mirzabekov, A.D. (1997). Oligonucleotide microchips as genosensors for determinative and environmental studies in microbiology. Appl. Environ. Microbiol. *63*, 2397–2402.

Gutierrez, L., Mauriat, M., Guenin, S., Pelloux, J., Lefebvre, J.F., Louvet, R., Rusterucci, C., Moritz, T., Guerineau, F., Bellini, C., *et al.* (2008). The lack of a systematic validation of reference genes: a serious pitfall undervalued in reverse transcription-polymerase chain reaction (RT-PCR) analysis in plants. Plant Biotechnol. J. *6*, 609–618.

Haas, S., Vingron, M., Poustka, A., and Wiemann, S. (1998). Primer design for large scale sequencing. Nucleic Acids Res. *26*, 3006–3012.

Haas, S.A., Hild, M., Wright, A.P., Hain, T., Talibi, D., and Vingron, M. (2003). Genome-scale design of PCR primers and long oligomers for DNA microarrays. Nucleic Acids Res. *31*, 5576–5581.

Hahn, S., Mergenthaler, S., Zimmerman, B., and Holzgreve, W. (2005). Nucleic acid based biosensors: the desires of the user. Bioelectrochemistry *67*, 151–154.

Hakenbeck, R., Balmelle, N., Weber, B., Gardes, C., Keck, W., and de Saizieu, A. (2001). Mosaic genes and mosaic chromosomes: intra- and interspecies genomic variation of *Streptococcus pneumoniae*. Infect. Immun. *69*, 2477–2486.

Hardiman, G. (2004). Microarray platforms – comparisons and contrasts. Pharmacogenomics *5*, 487–502.

Harr, B., and Schlotterer, C. (2006). Comparison of algorithms for the analysis of Affymetrix microarray data as evaluated by co-expression of genes in known operons. Nucleic Acids Res. *34*, e8.

Hawkins, P.R., Jin, P., and Fu, G.K. (2003). Full-length cDNA synthesis for long-distance RT-PCR of large mRNA transcripts. Biotechniques *34*, 768–770, 772–773.

Hecker, M., and Engelmann, S. (2000). Proteomics, DNA arrays and the analysis of still unknown regulons and unknown proteins of *Bacillus subtilis* and pathogenic gram-positive bacteria. Int. J. Med. Microbiol. *290*, 123–134.

Heid, C.A., Stevens, J., Livak, K.J., and Williams, P.M. (1996). Real time quantitative PCR. Genome Res. *6*, 986–994.

Hellemans, J., Mortier, G., De Paepe, A., Speleman, F., and Vandesompele, J. (2007). qBase relative quantification framework and software for management and automated analysis of real-time quantitative PCR data. Genome Biol. *8*, R19.

van Hijum, S.A., de Jong, A., Baerends, R.J., Karsens, H.A., Kramer, N.E., Larsen, R., den Hengst, C.D., Albers, C.J., Kok, J., and Kuipers, O.P. (2005). A generally applicable validation scheme for the assessment of factors involved in reproducibility and quality of DNA-microarray data. BMC Genomics *6*, 77.

Hornshoj, H., Stengaard, H., Panitz, F., and Bendixen, C. (2004). SEPON, a Selection and Evaluation Pipeline for OligoNucleotides based on ESTs with a non-target Tm algorithm for reducing cross-hybridization in microarray gene expression experiments. Bioinformatics *20*, 428–429.

Hosack, D.A., Dennis, G., Jr., Sherman, B.T., Lane, H.C., and Lempicki, R.A. (2003). Identifying biological themes within lists of genes with EASE. Genome Biol. *4*, R70.

Hovik, H., and Chen, T. (2010). Dynamic probe selection for studying microbial transcriptome with high-density genomic tiling microarrays. BMC Bioinformatics *11*, 82.

Huggett, J., Dheda, K., Bustin, S., and Zumla, A. (2005). Real-time RT-PCR normalisation; strategies and considerations. Genes Immun. *6*, 279–284.

Hughes, S., Arneson, N., Done, S., and Squire, J. (2005). The use of whole genome amplification in the study of human disease. Prog. Biophys. Mol. Biol. *88*, 173–189.

Ichikawa, J.K., Norris, A., Bangera, M.G., Geiss, G.K., van 't Wout, A.B., Bumgarner, R.E., and Lory, S. (2000). Interaction of pseudomonas aeruginosa with epithelial cells: identification of differentially regulated genes by expression microarray analysis of human cDNAs. Proc. Natl. Acad. Sci. U.S.A. *97*, 9659–9664.

Imbeaud, S., Graudens, E., Boulanger, V., Barlet, X., Zaborski, P., Eveno, E., Mueller, O., Schroeder, A., and Auffray, C. (2005). Towards standardization of RNA quality assessment using user-independent classifiers of microcapillary electrophoresis traces. Nucleic Acids Res. *33*, e56.

Irizarry, R.A., Bolstad, B.M., Collin, F., Cope, L.M., Hobbs, B., and Speed, T.P. (2003a). Summaries of Affymetrix GeneChip probe level data. Nucleic Acids Res. *31*, e15.

Irizarry, R.A., Hobbs, B., Collin, F., Beazer-Barclay, Y.D., Antonellis, K.J., Scherf, U., and Speed, T.P. (2003b). Exploration, normalization, and summaries of high density oligonucleotide array probe level data. Biostatistics *4*, 249–264.

Irizarry, R.A., Warren, D., Spencer, F., Kim, I.F., Biswal, S., Frank, B.C., Gabrielson, E., Garcia, J.G., Geoghegan, J., Germino, G., *et al.* (2005). Multiple-laboratory comparison of microarray platforms. Nat. Methods *2*, 345–350.

Irizarry, R.A., Wu, Z., and Jaffee, H.A. (2006). Comparison of Affymetrix GeneChip expression measures. Bioinformatics *22*, 789–794.

Israel, D.A., Salama, N., Arnold, C.N., Moss, S.F., Ando, T., Wirth, H.P., Tham, K.T., Camorlinga, M., Blaser, M.J., Falkow, S., *et al.* (2001). Helicobacter pylori strain-specific differences in genetic content, identified by microarray, influence host inflammatory responses. J. Clin. Invest. *107*, 611–620.

Iturriza-Gomara, M., Green, J., Brown, D.W., Desselberger, U., and Gray, J.J. (1999). Comparison of specific and random priming in the reverse transcriptase polymerase chain reaction for genotyping group A rotaviruses. J. Virol. Methods *78*, 93–103.

Iyer, V.R., Horak, C.E., Scafe, C.S., Botstein, D., Snyder, M., and Brown, P.O. (2001). Genomic binding sites of the yeast cell-cycle transcription factors SBF and MBF. Nature *409*, 533–538.

Jacobson, R., and Wright, P. (2009). Principles and methods of validation of diagnostic assays for infectious diseases. In Manual of diagnostic tests for aquatic animals, Epizooties, O.I.E., ed. (World Organization for Animal Health, Paris).

Jin, N., He, K., and Liu, L. (2006). qPCR-DAMS: a database tool to analyze, manage, and store both relative and absolute quantitative real-time PCR data. Physiol. Genomics *25*, 525–527.

Jourdren, L., Duclos, A., Brion, C., Portnoy, T., Mathis, H., Margeot, A., and Le Crom, S. (2010). Teolenn: an efficient and customizable workflow to design high-quality probes for microarray experiments. Nucleic Acids Res. *38*, e117.

Kaderali, L., and Schliep, A. (2002). Selecting signature oligonucleotides to identify organisms using DNA arrays. Bioinformatics *18*, 1340–1349.

Kaplinski, L., Andreson, R., Puurand, T., and Remm, M. (2005). MultiPLX: automatic grouping and evaluation of PCR primers. Bioinformatics *21*, 1701–1702.

Kastenmayer, J.P., Ni, L., Chu, A., Kitchen, L.E., Au, W.C., Yang, H., Carter, C.D., Wheeler, D., Davis, R.W., Boeke, J.D., *et al.* (2006). Functional genomics of genes with small open reading frames (sORFs) in *S. cerevisiae*. Genome Res. *16*, 365–373.

Kato-Maeda, M., Gao, Q., and Small, P.M. (2001). Microarray analysis of pathogens and their interaction with hosts. Cell Microbiol. *3*, 713–719.

Kawasaki, E.S. (2006). The end of the microarray Tower of Babel: will universal standards lead the way? J. Biomol. Tech. *17*, 200–206.

Ke, X., Collins, A., and Ye, S. (2001). PIRA PCR designer for restriction analysis of single nucleotide polymorphisms. Bioinformatics *17*, 838–839.

Kerr, M.K., Martin, M., and Churchill, G.A. (2000). Analysis of variance for gene expression microarray data. J. Comput. Biol. *7*, 819–837.

Kevil, C.G., Walsh, L., Laroux, F.S., Kalogeris, T., Grisham, M.B., and Alexander, J.S. (1997). An improved, rapid Northern protocol. Biochem. Biophys. Res. Commun. *238*, 277–279.

Khatri, P., and Draghici, S. (2005). Ontological analysis of gene expression data: current tools, limitations, and open problems. Bioinformatics *21*, 3587–3595.

Khatri, P., Draghici, S., Ostermeier, G.C., and Krawetz, S.A. (2002). Profiling gene expression using onto-express. Genomics *79*, 266–270.

Khodakov, D.A., Zakharova, N.V., Gryadunov, D.A., Filatov, F.P., Zasedatelev, A.S., and Mikhailovich, V.M. (2008). An oligonucleotide microarray for multiplex real-time PCR identification of HIV-1, HBV, and HCV. Biotechniques *44*, 241–246, 248.

King, T.S., and Chinchilli, V.M. (2001). A generalized concordance correlation coefficient for continuous and categorical data. Stat. Med. *20*, 2131–2147.

Kitchen, R.R., Kubista, M., and Tichopad, A. (2010). Statistical aspects of quantitative real-time PCR experiment design. Methods *50*, 231–236.

Klein, D. (2002). Quantification using real-time PCR technology: applications and limitations. Trends Mol. Med. *8*, 257–260.

de Kok, J.B., Wiegerinck, E.T., Giesendorf, B.A., and Swinkels, D.W. (2002). Rapid genotyping of single nucleotide polymorphisms using novel minor groove binding DNA oligonucleotides (MGB probes). Hum. Mutat. *19*, 554–559.

Kort, R., Keijser, B.J., Caspers, M.P., Schuren, F.H., and Montijn, R. (2008). Transcriptional activity around bacterial cell death reveals molecular biomarkers for cell viability. BMC Genomics *9*, 590.

Kricka, L.J., Master, S.R., Joos, T.O., and Fortina, P. (2006). Current perspectives in protein array technology. Ann. Clin. Biochem. *43*, 457–467.

Kuo, K.W., Leung, M.F., and Leung, W.C. (1997). Intrinsic secondary structure of human TNFR-I mRNA influences the determination of gene expression by RT-PCR. Mol. Cell. Biochem. *177*, 1–6.

Kuo, W.P., Jenssen, T.K., Butte, A.J., Ohno-Machado, L., and Kohane, I.S. (2002). Analysis of matched mRNA measurements from two different microarray technologies. Bioinformatics *18*, 405–412.

Kushner, S.R. (2002). mRNA decay in *Escherichia coli* comes of age. J. Bacteriol. *184*, 4658–4665; discussion 4657.

Larkin, J.E., Frank, B.C., Gavras, H., Sultana, R., and Quackenbush, J. (2005). Independence and reproducibility across microarray platforms. Nat. Methods *2*, 337–344.

Laub, M.T., McAdams, H.H., Feldblyum, T., Fraser, C.M., and Shapiro, L. (2000). Global analysis of the genetic network controlling a bacterial cell cycle. Science *290*, 2144–2148.

Lee, D.Y., Shannon, K., and Beaudette, L.A. (2006). Detection of bacterial pathogens in municipal wastewater using an oligonucleotide microarray and real-time quantitative PCR. J. Microbiol. Methods *65*, 453–467.

Lefever, S., Hellemans, J., Pattyn, F., Przybylski, D.R., Taylor, C., Geurts, R., Untergasser, A., and Vandesompele, J. (2009). RDML: structured language and reporting guidelines for real-time quantitative PCR data. Nucleic Acids Res. *37*, 2065–2069.

Leister, D., and Varotto, C. (2007). GST-PRIME: an algorithm for genome-wide primer design. Meth. Mol. Biol. *402*, 141–158.

Lekanne Deprez, R.H., Fijnvandraat, A.C., Ruijter, J.M., and Moorman, A.F. (2002). Sensitivity and accuracy of quantitative real-time polymerase chain reaction using SYBR Green I depends on cDNA synthesis conditions. Anal. Biochem. *307*, 63–69.

Leski, T.A., Lin, B., Malanoski, A.P., Wang, Z., Long, N.C., Meador, C.E., Barrows, B., Ibrahim, S., Hardick, J.P., Aitichou, M., *et al.* (2009). Testing and validation of high density resequencing microarray for broad range biothreat agents detection. PLoS ONE *4*, e6569.

Levesque-Sergerie, J.P., Duquette, M., Thibault, C., Delbecchi, L., and Bissonnette, N. (2007). Detection limits of several commercial reverse transcriptase enzymes: impact on the low- and high-abundance transcript levels assessed by quantitative RT-PCR. BMC Mol. Biol. *8*, 93.

Li, C., and Wong, W.H. (2001). Model-based analysis of oligonucleotide arrays: expression index computation and outlier detection. Proc. Natl. Acad. Sci. U.S.A. *98*, 31–36.

Li, F., and Stormo, G.D. (2001). Selection of optimal DNA oligos for gene expression arrays. Bioinformatics *17*, 1067–1076.

Li, H., McCormac, M.A., Estes, R.W., Sefers, S.E., Dare, R.K., Chappell, J.D., Erdman, D.D., Wright, P.F., and Tang, Y.W. (2007). Simultaneous detection and high-throughput identification of a panel of RNA viruses causing respiratory tract infections. J. Clin. Microbiol. *45*, 2105–2109.

Li, L.C., and Dahiya, R. (2002). MethPrimer: designing primers for methylation PCRs. Bioinformatics *18*, 1427–1431.

Li, W., and Ying, X. (2006). Mprobe 2.0: computer-aided probe design for oligonucleotide microarray. Appl. Bioinformatics *5*, 181–186.

Li, X., He, Z., and Zhou, J. (2005). Selection of optimal oligonucleotide probes for microarrays using multiple criteria, global alignment and parameter estimation. Nucleic Acids Res. *33*, 6114–6123.

Lin, B., Wang, Z., Vora, G.J., Thornton, J.A., Schnur, J.M., Thach, D.C., Blaney, K.M., Ligler, A.G., Malanoski, A.P., Santiago, J., *et al.* (2006). Broad-spectrum respiratory tract pathogen identification using resequencing DNA microarrays. Genome Res. *16*, 527–535.

Lin, L., Hedayat, A., Sinha, B., and Yang, M. (2002). Statistical methods in assessing agreement: Models, issues, and tools. J. Am. Stat. Assoc. *97*, 257–270.

Lin, L.I. (1989). A concordance correlation coefficient to evaluate reproducibility. Biometrics *45*, 255–268.

Linhart, C., and Shamir, R. (2002). The degenerate primer design problem. Bioinformatics *18* Suppl *1*, S172–181.

Liu, Z.L., and Slininger, P.J. (2007). Universal external RNA controls for microbial gene expression analysis using microarray and qRT-PCR. J. Microbiol. Methods *68*, 486–496.

Livak, K.J., and Schmittgen, T.D. (2001). Analysis of relative gene expression data using real-time quantitative PCR and the 2(-Delta Delta C(T)) Method. Methods *25*, 402–408.

Lloyd, A.L., Rasko, D.A., and Mobley, H.L. (2007). Defining genomic islands and uropathogen-specific genes in uropathogenic *Escherichia coli*. J. Bacteriol. *189*, 3532–3546.

Loeb, J., and Beaumont, T. (2009). What goes in is what comes out: how to design and implement a successful microarray experiment. In Bioinformatics for Systems Biology, Krawetz, S.A., ed. (Human Press, New York), pp. 209–225.

Long, W.H., Xiao, H.S., Gu, X.M., Zhang, Q.H., Yang, H.J., Zhao, G.P., and Liu, J.H. (2004). A universal microarray for detection of SARS coronavirus. J. Virol. Methods *121*, 57–63.

Love, J.L., Scholes, P., Gilpin, B., Savill, M., Lin, S., and Samuel, L. (2006). Evaluation of uncertainty in quantitative real-time PCR. J. Microbiol. Methods *67*, 349–356.

Loy, A., and Bodrossy, L. (2006). Highly parallel microbial diagnostics using oligonucleotide microarrays. Clin. Chim. Acta. *363*, 106–119.

Loy, A., Schulz, C., Lucker, S., Schopfer-Wendels, A., Stoecker, K., Baranyi, C., Lehner, A., and Wagner, M. (2005). 16S rRNA gene-based oligonucleotide microarray for environmental monitoring of the betaproteobacterial order "Rhodocyclales". Appl. Environ. Microbiol. *71*, 1373–1386.

Ludwig, W., Strunk, O., Westram, R., Richter, L., Meier, H., Yadhukumar, Buchner, A., Lai, T., Steppi, S., Jobb, G., *et al.* (2004). ARB: a software environment for sequence data. Nucleic Acids Res. *32*, 1363–1371.

Luu-The, V., Paquet, N., Calvo, E., and Cumps, J. (2005). Improved real-time RT-PCR method for high-throughput measurements using second derivative calculation and double correction. Biotechniques *38*, 287–293.

Mackay, J. (2007). Introduction to kinetic (real-time) PCR. Meth. Mol. Biol. *353*, 167–176.

Maderazo, A.B., Belk, J.P., He, F., and Jacobson, A. (2003). Nonsense-containing mRNAs that accumulate in the absence of a functional nonsense-mediated mRNA decay pathway are destabilized rapidly upon its restitution. Mol. Cell Biol. *23*, 842–851.

Majtan, T., Bukovska, G., and Timko, J. (2004). DNA microarrays – techniques and applications in microbial systems. Folia Microbiol. (Praha) *49*, 635–664.

Malinen, E., Kassinen, A., Rinttila, T., and Palva, A. (2003). Comparison of real-time PCR with SYBR Green I or 5′-nuclease assays and dot–blot hybridization with rDNA-targeted oligonucleotide probes in quantification of selected faecal bacteria. Microbiology *149*, 269–277.

Mannhalter, C., Koizar, D., and Mitterbauer, G. (2000). Evaluation of RNA isolation methods and reference genes for RT-PCR analyses of rare target RNA. Clin. Chem. Lab Med *38*, 171–177.

Manthey, G.M., and McEwen, J.E. (1995). The product of the nuclear gene PET309 is required for translation of mature mRNA and stability or production of intron-containing RNAs derived from the mitochondrial COX1 locus of *Saccharomyces cerevisiae*. EMBO J. *14*, 4031–4043.

Markham, N.R., and Zuker, M. (2008). UNAFold: software for nucleic acid folding and hybridization. Meth. Mol. Biol. *453*, 3–31.

Marshall, O.J. (2004). PerlPrimer: cross-platform, graphical primer design for standard, bisulphite and real-time PCR. Bioinformatics *20*, 2471–2472.

Martins, N., Vale, F., and Vieira, H. (2010). Design and preliminary results of a DNA chip for the detection of microorganisms in water samples. Physica Status Solidi (c) *7*, 2751–2754.

Matsumura, H., Reuter, M., Kruger, D.H., Winter, P., Kahl, G., and Terauchi, R. (2008). SuperSAGE. Meth. Mol. Biol. *387*, 55–70.

Mei, R., Hubbell, E., Bekiranov, S., Mittmann, M., Christians, F.C., Shen, M.M., Lu, G., Fang, J., Liu, W.M., Ryder, T., *et al.* (2003). Probe selection for high-density oligonucleotide arrays. Proc. Natl. Acad. Sci. U.S.A. *100*, 11237–11242.

Melton, D.A., Krieg, P.A., Rebagliati, M.R., Maniatis, T., Zinn, K., and Green, M.R. (1984). Efficient *in vitro* synthesis of biologically active RNA and RNA hybridization probes from plasmids containing a bacteriophage SP6 promoter. Nucleic Acids Res. *12*, 7035–7056.

Militon, C., Rimour, S., Missaoui, M., Biderre, C., Barra, V., Hill, D., Mone, A., Gagne, G., Meier, H., Peyretaillade, E., *et al.* (2007). PhylArray: phylogenetic probe design algorithm for microarray. Bioinformatics *23*, 2550–2557.

Miller, M.B., and Tang, Y.W. (2009). Basic concepts of microarrays and potential applications in clinical microbiology. Clin. Microbiol. Rev. *22*, 611–633.

Mir, K.U., and Southern, E.M. (1999). Determining the influence of structure on hybridization using oligonucleotide arrays. Nat. Biotechnol. *17*, 788–792.

Mira, N.P., Becker, J.D., and Sa-Correia, I. (2010). Genomic expression program involving the Haa1p-regulon in *Saccharomyces cerevisiae* response to acetic acid. OMICS *14*, 587–601.

Miron, M., Woody, O.Z., Marcil, A., Murie, C., Sladek, R., and Nadon, R. (2006). A methodology for global validation of microarray experiments. BMC Bioinformatics *7*, 333.

Morey, J.S., Ryan, J.C., and Van Dolah, F.M. (2006). Microarray validation: factors influencing correlation between oligonucleotide microarrays and real-time PCR. Biol. Proced. Online *8*, 175–193.

Morrison, T.B., Weis, J.J., and Wittwer, C.T. (1998). Quantification of low-copy transcripts by continuous SYBR Green I monitoring during amplification. Biotechniques *24*, 954–958, 960, 962.

Moy, P., Qutob, D., Chapman, B.P., Atkinson, I., and Gijzen, M. (2004). Patterns of gene expression upon infection of soybean plants by Phytophthora sojae. Mol. Plant Microbe Interact *17*, 1051–1062.

Mrowka, R., Schuchhardt, J., and Gille, C. (2002). Oligodb – interactive design of oligo DNA for transcription profiling of human genes. Bioinformatics *18*, 1686–1687.

Muller, P.Y., Janovjak, H., Miserez, A.R., and Dobbie, Z. (2002). Processing of gene expression data generated by quantitative real-time RT-PCR. Biotechniques *32*, 1372–1374, 1376, 1378–1379.

Muthaiyan, A., Silverman, J.A., Jayaswal, R.K., and Wilkinson, B.J. (2008). Transcriptional profiling reveals that daptomycin induces the *Staphylococcus aureus* cell wall stress stimulon and genes responsive to membrane depolarization. Antimicrob. Agents Chemother. *52*, 980–990.

Navin, N., Grubor, V., Hicks, J., Leibu, E., Thomas, E., Troge, J., Riggs, M., Lundin, P., Maner, S., Sebat, J., *et al.* (2006). PROBER: oligonucleotide FISH probe design software. Bioinformatics *22*, 2437–2438.

Nazarenko, I., Lowe, B., Darfler, M., Ikonomi, P., Schuster, D., and Rashtchian, A. (2002). Multiplex quantitative PCR using self-quenched primers labeled with a single fluorophore. Nucleic Acids Res. *30*, e37.

Ng, W.L., Kazmierczak, K.M., Robertson, G.T., Gilmour, R., and Winkler, M.E. (2003). Transcriptional regulation and signature patterns revealed by microarray analyses of *Streptococcus pneumoniae* R6 challenged with sublethal concentrations of translation inhibitors. J. Bacteriol. *185*, 359–370.

Nielsen, H.B., and Knudsen, S. (2002). Avoiding cross hybridization by choosing nonredundant targets on cDNA arrays. Bioinformatics *18*, 321–322.

Nielsen, H.B., Wernersson, R., and Knudsen, S. (2003). Design of oligonucleotides for microarrays and perspectives for design of multi-transcriptome arrays. Nucleic Acids Res. *31*, 3491–3496.

Nolan, T., Hands, R.E., and Bustin, S.A. (2006). Quantification of mRNA using real-time RT-PCR. Nat. Protoc. *1*, 1559–1582.

Nolan, T., Hands, R.E., Ogunkolade, W., and Bustin, S.A. (2006). SPUD: a quantitative PCR assay for the detection of inhibitors in nucleic acid preparations. Anal. Biochem. *351*, 308–310.

Nordberg, E.K. (2005). YODA: selecting signature oligonucleotides. Bioinformatics *21*, 1365–1370.

Nygaard, V., and Hovig, E. (2006). Options available for profiling small samples: a review of sample amplification technology when combined with microarray profiling. Nucleic Acids Res. *34*, 996–1014.

Oh, M.K., and Liao, J.C. (2000a). DNA microarray detection of metabolic responses to protein overproduction in *Escherichia coli*. Metab. Eng. *2*, 201–209.

Oh, M.K., and Liao, J.C. (2000b). Gene expression profiling by DNA microarrays and metabolic fluxes in *Escherichia coli*. Biotechnol. Prog. *16*, 278–286.

Okello, J.B., Rodriguez, L., Poinar, D., Bos, K., Okwi, A.L., Bimenya, G.S., Sewankambo, N.K., Henry, K.R., Kuch, M., and Poinar, H.N. (2010). Quantitative assessment of the sensitivity of various commercial reverse transcriptases based on armored HIV RNA. PLoS ONE *5*, e13931.

Palmer, C., Bik, E.M., Eisen, M.B., Eckburg, P.B., Sana, T.R., Wolber, P.K., Relman, D.A., and Brown, P.O. (2006). Rapid quantitative profiling of complex microbial populations. Nucleic Acids Res. *34*, e5.

Palmer, S., Wiegand, A.P., Maldarelli, F., Bazmi, H., Mican, J.M., Polis, M., Dewar, R.L., Planta, A., Liu, S., Metcalf, J.A., *et al.* (2003). New real-time reverse transcriptase-initiated PCR assay with single-copy sensitivity for human immunodeficiency virus type 1 RNA in plasma. J. Clin. Microbiol. *41*, 4531–4536.

Pappas, C.T., Sram, J., Moskvin, O.V., Ivanov, P.S., Mackenzie, R.C., Choudhary, M., Land, M.L., Larimer, F.W., Kaplan, S., and Gomelsky, M. (2004). Construction and validation of the *Rhodobacter sphaeroides* 2.4.1 DNA microarray: transcriptome flexibility at diverse growth modes. J. Bacteriol. *186*, 4748–4758.

Pasquer, F., Pelludat, C., Duffy, B., and Frey, J.E. (2010). Broad spectrum microarray for fingerprint-based bacterial species identification. BMC Biotechnol. *10*, 13.

Patel, V.C., Mondal, K., Shetty, A.C., Horner, V.L., Bedoyan, J.K., Martin, D., Caspary, T., Cutler, D.J., and Zwick, M.E. (2010). Microarray oligonucleotide probe designer (MOPeD): A web service. Open Access Bioinformatics *2*, 145–155.

Patterson, T.A., Lobenhofer, E.K., Fulmer-Smentek, S.B., Collins, P.J., Chu, T.M., Bao, W., Fang, H., Kawasaki, E.S., Hager, J., Tikhonova, I.R., *et al.* (2006). Performance comparison of one-color and two-color platforms within the MicroArray Quality Control (MAQC) project. Nat. Biotechnol. *24*, 1140–1150.

Peplies, J., Lau, S.C., Pernthaler, J., Amann, R., and Glockner, F.O. (2004). Application and validation of DNA microarrays for the 16S rRNA-based analysis of marine bacterioplankton. Environ. Microbiol. *6*, 638–645.

van de Peppel, J., Kemmeren, P., van Bakel, H., Radonjic, M., van Leenen, D., and Holstege, F.C. (2003). Monitoring global messenger RNA changes in externally controlled microarray experiments. EMBO Rep. *4*, 387–393.

Perez-Novo, C.A., Claeys, C., Speleman, F., Van Cauwenberge, P., Bachert, C., and Vandesompele, J. (2005). Impact of RNA quality on reference gene expression stability. Biotechniques *39*, 52, 54, 56.

Petersen, D.W., and Kawasaki, E.S. (2007). Manufacturing of microarrays. Adv Exp Med Biol *593*, 1–11.

Peterson, S., Cline, R., Tettelin, H., Sharov, V., and Morrison, D. (2000). Gene expression analysis of the *Streptococcus pneumonia* competence regulons by use of DNA microarrays. J. Bacteriol. *182*, 6192–6202.

Pfaffl, M., and Hageleit, M. (2001). Validities of mRNA quantification using recombinant RNA and recombinant DNA external calibration curves in real-time RT-PCR. Biotechnol. Lett. *23*, 275–282.

Pfaffl, M., Vandesompele, J., and Kubista, M. (2009). Data analysis software. In Real-Time PCR: Current Technology and Applications, Logan, J., Edwards, K., and Saunders, N., eds (Caister Academic Press, Norfolk), pp. 65–84.

Pfaffl, M.W. (2001). A new mathematical model for relative quantification in real-time RT-PCR. Nucleic Acids Res. *29*, e45.

Pfaffl, M.W., Horgan, G.W., and Dempfle, L. (2002). Relative expression software tool (REST) for group-wise comparison and statistical analysis of relative expression results in real-time PCR. Nucleic Acids Res. *30*, e36.

Pfaffl, M.W., Tichopad, A., Prgomet, C., and Neuvians, T.P. (2004). Determination of stable housekeeping genes, differentially regulated target genes and sample integrity: BestKeeper – Excel-based tool using pair-wise correlations. Biotechnol. Lett. *26*, 509–515.

Podowski, R.M., and Sonnhammer, E.L. (2001). MEDUSA: large scale automatic selection and visual assessment of PCR primer pairs. Bioinformatics *17*, 656–657.

Poretsky, R.S., Bano, N., Buchan, A., LeCleir, G., Kleikemper, J., Pickering, M., Pate, W.M., Moran, M.A., and Hollibaugh, J.T. (2005). Analysis of microbial gene transcripts in environmental samples. Appl. Environ. Microbiol. *71*, 4121–4126.

Postic, G., Frapy, E., Dupuis, M., Dubail, I., Livny, J., Charbit, A., and Meibom, K.L. (2010). Identification of small RNAs in *Francisella tularensis*. BMC Genomics *11*, 625.

Preza, D., Olsen, I., Willumsen, T., Boches, S.K., Cotton, S.L., Grinde, B., and Paster, B.J. (2009a). Microarray analysis of the microflora of root caries in elderly. Eur J. Clin. Microbiol. Infect. Dis. *28*, 509–517.

Preza, D., Olsen, I., Willumsen, T., Grinde, B., and Paster, B.J. (2009b). Diversity and site-specificity of the oral microflora in the elderly. Eur. J. Clin. Microbiol. Infect. Dis. 28, 1033–1040.

Qin, L.X., Beyer, R.P., Hudson, F.N., Linford, N.J., Morris, D.E., and Kerr, K.F. (2006). Evaluation of methods for oligonucleotide array data via quantitative real-time PCR. BMC Bioinformatics 7, 23.

Quinn, P., Bowers, R., Zhang, X., Wahlund, T., Fanelli, M., Olszova, D., and Read, B. (2006). cDNA microarrays as a tool for identification of biomineralization proteins in the coccolithophorid *Emiliania huxleyi* (Haptophyta). Appl. Environ. Microbiol. 72, 5512–5526.

Rachlin, J., Ding, C., Cantor, C., and Kasif, S. (2005). MuPlex: multi-objective multiplex PCR assay design. Nucleic Acids Res. *33*, W544–547.

Raddatz, G., Dehio, M., Meyer, T.F., and Dehio, C. (2001). PrimeArray: genome-scale primer design for DNA-microarray construction. Bioinformatics *17*, 98–99.

Raghunathan, A., Ferguson, H.R., Jr., Bornarth, C.J., Song, W., Driscoll, M., and Lasken, R.S. (2005). Genomic DNA amplification from a single bacterium. Appl. Environ. Microbiol. *71*, 3342–3347.

Rahmann, S. (2003). Fast large scale oligonucleotide selection using the longest common factor approach. J. Bioinform. Comput. Biol. *1*, 343–361.

Rajeevan, M.S., Ranamukhaarachchi, D.G., Vernon, S.D., and Unger, E.R. (2001a). Use of real-time quantitative PCR to validate the results of cDNA array and differential display PCR technologies. Methods 25, 443–451.

Rajeevan, M.S., Vernon, S.D., Taysavang, N., and Unger, E.R. (2001b). Validation of array-based gene expression profiles by real-time (kinetic) RT-PCR. J. Mol. Diagn. 3, 26–31.

Raphael, B.H., Joseph, L.A., McCroskey, L.M., Luquez, C., and Maslanka, S.E. (2010). Detection and differentiation of *Clostridium botulinum* type A strains using a focused DNA microarray. Mol. Cell Probes *24*, 146–153.

Raymond, F., Carbonneau, J., Boucher, N., Robitaille, L., Boisvert, S., Wu, W.K., De Serres, G., Boivin, G., and Corbeil, J. (2009). Comparison of automated microarray detection with real-time PCR assays for detection of respiratory viruses in specimens obtained from children. J. Clin. Microbiol. 47, 743–750.

Reich, M., Kohler, A., Martin, F., and Buee, M. (2009). Development and validation of an oligonucleotide microarray to characterise ectomycorrhizal fungal communities. BMC Microbiol. 9, 241.

Ren, B., Robert, F., Wyrick, J.J., Aparicio, O., Jennings, E.G., Simon, I., Zeitlinger, J., Schreiber, J., Hannett, N., Kanin, E., *et al.* (2000). Genome-wide location and function of DNA binding proteins. Science *290*, 2306–2309.

Reymond, J.L., and Babiak, P. (2007). Screening systems. Adv. Biochem. Eng. Biotechnol. *105*, 31–58.

Reymond, N., Charles, H., Duret, L., Calevro, F., Beslon, G., and Fayard, J.M. (2004). ROSO: optimizing oligonucleotide probes for microarrays. Bioinformatics *20*, 271–273.

Rhee, S., Liu, X., Wu, L., Chong, S., Wan, X., and Zhou, J. (2004). Detection of genes involved in biodegradation and biotransformation in microbial communities by using 50-mer oligonucleotide microarrays. Appl. Environ. Microbiol. *70*, 4303–4317.

Ribeiro-Guimaraes, M., and Pessolani, M. (2007). Comparative genomics of mycobacterial proteases. Microb. Pathog. *43*, 173–178.

Riedy, M.C., Timm, E.A., Jr., and Stewart, C.C. (1995). Quantitative RT-PCR for measuring gene expression. Biotechniques 18, 70–74, 76.

Riehle, M.M., Bennett, A.F., and Long, A.D. (2001). Genetic architecture of thermal adaptation in *Escherichia coli*. Proc. Natl. Acad. Sci. U.S.A. *98*, 525–530.

Rimini, R., Jansson, B., Feger, G., Roberts, T.C., de Francesco, M., Gozzi, A., Faggioni, F., Domenici, E., Wallace, D.M., Frandsen, N., *et al.* (2000). Global analysis of transcription kinetics during competence development in *Streptococcus pneumoniae* using high density DNA arrays. Mol. Microbiol. *36*, 1279–1292.

Rimour, S., Hill, D., Militon, C., and Peyret, P. (2005). GoArrays: highly dynamic and efficient microarray probe design. Bioinformatics *21*, 1094–1103.

Ririe, K.M., Rasmussen, R.P., and Wittwer, C.T. (1997). Product differentiation by analysis of DNA melting curves during the polymerase chain reaction. Anal. Biochem. *245*, 154–160.

Ritz, C., and Spiess, A.N. (2008). qpcR: an R package for sigmoidal model selection in quantitative real-time polymerase chain reaction analysis. Bioinformatics *24*, 1549–1551.

Rocke, D.M., and Durbin, B. (2003). Approximate variance-stabilizing transformations for gene-expression microarray data. Bioinformatics *19*, 966–972.

Rockett, J.C., and Hellmann, G.M. (2004). Confirming microarray data – is it really necessary? Genomics *83*, 541–549.

Roth, M.E., Feng, L., McConnell, K.J., Schaffer, P.J., Guerra, C.E., Affourtit, J.P., Piper, K.R., Guccione, L., Hariharan, J., Ford, M.J., *et al.* (2004). Expression profiling using a hexamer-based universal microarray. Nat. Biotechnol. *22*, 418–426.

Rouchka, E.C., Khalyfa, A., and Cooper, N.G. (2005). MPrime: efficient large scale multiple primer and oligonucleotide design for customized gene microarrays. BMC Bioinformatics *6*, 175.

Rouillard, J.M., Zuker, M., and Gulari, E. (2003). OligoArray 2.0: design of oligonucleotide probes for DNA microarrays using a thermodynamic approach. Nucleic Acids Res. *31*, 3057–3062.

Royce, T.E., Rozowsky, J.S., and Gerstein, M.B. (2007). Toward a universal microarray: prediction of gene expression through nearest-neighbor probe sequence identification. Nucleic Acids Res. *35*, e99.

Saha, S., Sparks, A.B., Rago, C., Akmaev, V., Wang, C.J., Vogelstein, B., Kinzler, K.W., and Velculescu, V.E. (2002). Using the transcriptome to annotate the genome. Nat. Biotechnol. *20*, 508–512.

Saiki, R.K., Scharf, S., Faloona, F., Mullis, K.B., Horn, G.T., Erlich, H.A., and Arnheim, N. (1985). Enzymatic amplification of beta-globin genomic sequences and restriction site analysis for diagnosis of sickle cell anemia. Science *230*, 1350–1354.

de Saizieu, A., Gardes, C., Flint, N., Wagner, C., Kamber, M., Mitchell, T.J., Keck, W., Amrein, K.E., and Lange, R. (2000). Microarray-based identification of a novel *Streptococcus pneumoniae* regulon controlled by an autoinduced peptide. J. Bacteriol. *182*, 4696–4703.

Salama, N., Guillemin, K., McDaniel, T.K., Sherlock, G., Tompkins, L., and Falkow, S. (2000). A whole-genome microarray reveals genetic diversity among Helicobacter pylori strains. Proc. Natl. Acad. Sci. U.S.A. *97*, 14668–14673.

SantaLucia, J., Jr., and Hicks, D. (2004). The thermodynamics of DNA structural motifs. Annu. Rev. Biophys. Biomol. Struct. *33*, 415–440.

Schäfer, C., and Wöstemeyer, J. (1992). Random primer dependent PCR differentiates aggressive from non-aggressive isolates of the oilseed rape pathogen Phoma lingam (Leptosphaeria maculans). J. Phytopathol. *136*, 124–136.

Schmid, H., Cohen, C.D., Henger, A., Irrgang, S., Schlondorff, D., and Kretzler, M. (2003). Validation of endogenous controls for gene expression analysis in microdissected human renal biopsies. Kidney Int. *64*, 356–360.

Schmittgen, T.D., and Livak, K.J. (2008). Analyzing real-time PCR data by the comparative C(T) method. Nat. Protoc. *3*, 1101–1108.

Schmittgen, T.D., and Zakrajsek, B.A. (2000). Effect of experimental treatment on housekeeping gene expression: validation by real-time, quantitative RT-PCR. J. Biochem. Biophys. Methods *46*, 69–81.

Schmoock, G., Ehricht, R., Melzer, F., Rassbach, A., Scholz, H.C., Neubauer, H., Sachse, K., Mota, R.A., Saqib, M., and Elschner, M. (2009). DNA microarray-based detection and identification of *Burkholderia mallei*, *Burkholderia pseudomallei* and *Burkholderia* spp. Mol. Cell Probes *23*, 178–187.

Schoolnik, G.K. (2002). Microarray analysis of bacterial pathogenicity. Adv. Microb. Physiol. *46*, 1–45.

Schretter, C., and Milinkovitch, M.C. (2006). OligoFaktory: a visual tool for interactive oligonucleotide design. Bioinformatics *22*, 115–116.

Schröder, S., Weber, J., and Paul, H. (2001). 50 nucleotide long probes on microarrays enable high signal intensity and high specificity. Genomic Discovery, MWG AG Biotech Application note 014.

Schroeder, A., Mueller, O., Stocker, S., Salowsky, R., Leiber, M., Gassmann, M., Lightfoot, S., Menzel, W., Granzow, M., and Ragg, T. (2006). The RIN: an RNA integrity number for assigning integrity values to RNA measurements. BMC Mol. Biol. *7*, 3.

Selinger, D.W., Cheung, K.J., Mei, R., Johansson, E.M., Richmond, C.S., Blattner, F.R., Lockhart, D.J., and Church, G.M. (2000). RNA expression analysis using a 30 base pair resolution *Escherichia coli* genome array. Nat. Biotechnol. *18*, 1262–1268.

Selvey, S., Thompson, E.W., Matthaei, K., Lea, R.A., Irving, M.G., and Griffiths, L.R. (2001). Beta-actin – an unsuitable internal control for RT-PCR. Mol. Cell Probes *15*, 307–311.

Seo, Y.S., Srinivasan, U., Oh, K.Y., Shin, J.H., Chae, J.D., Kim, M.Y., Yang, J.H., Yoon, H.R., Miller, B., DeBusscher, J., *et al.* (2010). Changing molecular epidemiology of group B streptococcus in Korea. J. Korean Med. Sci. *25*, 817–823.

Serra, M.J., and Turner, D.H. (1995). Predicting thermodynamic properties of RNA. Methods Enzymol. *259*, 242–261.

Shaw, K.J., Miller, N., Liu, X., Lerner, D., Wan, J., Bittner, A., and Morrow, B.J. (2003). Comparison of the changes in global gene expression of *Escherichia coli* induced by four bactericidal agents. J. Mol. Microbiol. Biotechnol. *5*, 105–122.

Shedden, K., Chen, W., Kuick, R., Ghosh, D., Macdonald, J., Cho, K.R., Giordano, T.J., Gruber, S.B., Fearon, E.R., Taylor, J.M., *et al.* (2005). Comparison of seven methods for producing Affymetrix expression scores based on False Discovery Rates in disease profiling data. BMC Bioinformatics *6*, 26.

Shen, Z., Qu, W., Wang, W., Lu, Y., Wu, Y., Li, Z., Hang, X., Wang, X., Zhao, D., and Zhang, C. (2010). MPprimer: a program for reliable multiplex PCR primer design. BMC Bioinformatics *11*, 143.

Shi, L., Reid, L.H., Jones, W.D., Shippy, R., Warrington, J.A., Baker, S.C., Collins, P.J., de Longueville, F., Kawasaki, E.S., Lee, K.Y., *et al.* (2006). The MicroArray Quality Control (MAQC) project shows inter- and intraplatform reproducibility of gene expression measurements. Nat. Biotechnol. *24*, 1151–1161.

Shields, R. (2006). MIAME, we have a problem. Trends Genet. *22*, 65–66.

Shippy, R., Sendera, T.J., Lockner, R., Palaniappan, C., Kaysser-Kranich, T., Watts, G., and Alsobrook, J. (2004). Performance evaluation of commercial short-oligonucleotide microarrays and the impact of noise in making cross-platform correlations. BMC Genomics 5, 61.

Smith, C.J., and Osborn, A.M. (2009). Advantages and limitations of quantitative PCR (Q-PCR)-based approaches in microbial ecology. FEMS Microbiol. Ecol. *67*, 6–20.

Smith, C.J., Nedwell, D.B., Dong, L.F., and Osborn, A.M. (2006). Evaluation of quantitative polymerase chain reaction-based approaches for determining gene copy and gene transcript numbers in environmental samples. Environ. Microbiol. *8*, 804–815.

Smyth, G.K. (2004). Linear models and empirical bayes methods for assessing differential expression in microarray experiments. Stat. Appl. Genet. Mol. Biol. 3, Article3.

Sohail, M., and Southern, E.M. (2000). Hybridization of antisense reagents to RNA. Curr. Opin. Mol. Ther. 2, 264–271.

Solanas, M., Moral, R., and Escrich, E. (2001). Unsuitability of using ribosomal RNA as loading control for Northern blot analyses related to the imbalance between messenger and ribosomal RNA content in rat mammary tumors. Anal. Biochem. *288*, 99–102.

Solinas, A., Brown, L.J., McKeen, C., Mellor, J.M., Nicol, J., Thelwell, N., and Brown, T. (2001). Duplex Scorpion primers in SNP analysis and FRET applications. Nucleic Acids Res. *29*, e96.

Srivastava, G.P., Guo, J., Shi, H., and Xu, D. (2008). PRIMEGENS-v2: genome-wide primer design for analyzing DNA methylation patterns of CpG islands. Bioinformatics *24*, 1837–1842.

Stahlberg, A., Hakansson, J., Xian, X., Semb, H., and Kubista, M. (2004a). Properties of the reverse transcription reaction in mRNA quantification. Clin. Chem. *50*, 509–515.

Stahlberg, A., Kubista, M., and Pfaffl, M. (2004b). Comparison of reverse transcriptases in gene expression analysis. Clin. Chem. *50*, 1678–1680.

Stangegaard, M., Dufva, I.H., and Dufva, M. (2006). Reverse transcription using random pentadecamer primers increases yield and quality of resulting cDNA. Biotechniques *40*, 649–657.

Stanley, K.K., and Szewczuk, E. (2005). Multiplexed tandem PCR: gene profiling from small amounts of RNA using SYBR Green detection. Nucleic Acids Res. *33*, e180.

Stenberg, J., Nilsson, M., and Landegren, U. (2005). ProbeMaker: an extensible framework for design of sets of oligonucleotide probes. BMC Bioinformatics *6*, 229.

Stephens, C. (2001). Bacterial cell cycle: seeing the big picture with microarrays. Curr. Biol. *11*, R222–225.

Stoeckert, C.J., Jr., Causton, H.C., and Ball, C.A. (2002). Microarray databases: standards and ontologies. Nat. Genet. 32 Suppl, 469–473.

Storhoff, J.J., Marla, S.S., Bao, P., Hagenow, S., Mehta, H., Lucas, A., Garimella, V., Patno, T., Buckingham, W., Cork, W., *et al.* (2004). Gold nanoparticle-based detection of genomic DNA targets on microarrays using a novel optical detection system. Biosensors Bioelectronics *19*, 875–883.

Streit, S., Michalski, C., Erkan, M., Friess, H., and Kleeff, J. (2009). Confirmation of DNA microarray-derived differentially expressed genes in pancreatic cancer using quantitative RT-PCR. Pancreatology *9*, 577–582.

Syvanen, A.C., Bengtstrom, M., Tenhunen, J., and Soderlund, H. (1988). Quantification of polymerase chain reaction products by affinity-based hybrid collection. Nucleic Acids Res. *16*, 11327–11338.

Takahashi, H., Norman, S.A., Mather, E.L., and Patterson, B.K. (2008). Evaluation of the NanoChip 400 system for detection of influenza A and B, respiratory syncytial, and parainfluenza viruses. J. Clin. Microbiol. *46*, 1724–1727.

Talaat, A.M., Hunter, P., and Johnston, S.A. (2000). Genome-directed primers for selective labeling of bacterial transcripts for DNA microarray analysis. Nat. Biotechnol. *18*, 679–682.

Talla, E., Tekaia, F., Brino, L., and Dujon, B. (2003). A novel design of whole-genome microarray probes for *Saccharomyces cerevisiae* which minimizes cross-hybridization. BMC Genomics *4*, 38.

Tan, P.K., Downey, T.J., Spitznagel, E.L., Jr., Xu, P., Fu, D., Dimitrov, D.S., Lempicki, R.A., Raaka, B.M., and Cam, M.C. (2003). Evaluation of gene expression measurements from commercial microarray platforms. Nucleic Acids Res. *31*, 5676–5684.

Taniguchi, M., Miura, K., Iwao, H., and Yamanaka, S. (2001). Quantitative assessment of DNA microarrays – comparison with Northern blot analyses. Genomics *71*, 34–39.

Tao, H., Gonzalez, R., Martinez, A., Rodriguez, M., Ingram, L.O., Preston, J.F., and Shanmugam, K.T. (2001). Engineering a homo-ethanol pathway in *Escherichia coli*: increased glycolytic flux and levels of expression of glycolytic genes during xylose fermentation. J. Bacteriol. *183*, 2979–2988.

Taton, T.A., Mirkin, C.A., and Letsinger, R.L. (2000). Scanometric DNA array detection with nanoparticle probes. Science *289*, 1757–1760.

Taylor, S. (2010). A practical guide to publishing RT-qPCR data that conform to the MIQE guidelines. Am. Biotechnol. Lab. *28*, 14–19.

Taylor, S., Wakem, M., Dijkman, G., Alsarraj, M., and Nguyen, M. (2010). A practical approach to RT-qPCR-publishing data that conform to the MIQE guidelines. Methods *50*, S1–5.

Thelwell, N., Millington, S., Solinas, A., Booth, J., and Brown, T. (2000). Mode of action and application of Scorpion primers to mutation detection. Nucleic Acids Res. *28*, 3752–3761.

Tichopad, A., Dilger, M., Schwarz, G., and Pfaffl, M.W. (2003). Standardized determination of real-time PCR efficiency from a single reaction set-up. Nucleic Acids Res. *31*, e122.

Tichopad, A., Kitchen, R., Riedmaier, I., Becker, C., Stahlberg, A., and Kubista, M. (2009). Design and optimization of reverse-transcription quantitative PCR experiments. Clin. Chem. 55, 1816–1823.

Tolstrup, N., Nielsen, P.S., Kolberg, J.G., Frankel, A.M., Vissing, H., and Kauppinen, S. (2003). OligoDesign: optimal design of LNA (locked nucleic acid) oligonucleotide capture probes for gene expression profiling. Nucleic Acids Res. *31*, 3758–3762.

Tong, W., Lucas, A.B., Shippy, R., Fan, X., Fang, H., Hong, H., Orr, M.S., Chu, T.M., Guo, X., Collins, P.J., *et al.* (2006). Evaluation of external RNA controls for the assessment of microarray performance. Nat. Biotechnol. *24*, 1132–1139.

Townsend, J.P., and Hartl, D.L. (2002). Bayesian analysis of gene expression levels: statistical quantification of relative mRNA level across multiple strains or treatments. Genome Biol. 3, RESEARCH0071.

Townsend, J.P., Cavalieri, D., and Hartl, D.L. (2003). Population genetic variation in genome-wide gene expression. Mol. Biol. Evol. *20*, 955–963.

Tracz, D.M., Backhouse, P.G., Olson, A.B., McCrea, J.K., Walsh, J.A., Ng, L.-K., and Gilmour, M.W. (2007). Rapid detection of *Vibrio* species using liquid microsphere arrays and real-time PCR targeting the *ftsZ* locus. J. Med. Microbiol. *56*, 56–65.

Tricarico, C., Pinzani, P., Bianchi, S., Paglierani, M., Distante, V., Pazzagli, M., Bustin, S.A., and Orlando, C. (2002). Quantitative real-time reverse transcription polymerase chain reaction: normalization to rRNA or single housekeeping genes is inappropriate for human tissue biopsies. Anal. Biochem. *309*, 293–300.

Troesch, A., Nguyen, H., Miyada, C.G., Desvarenne, S., Gingeras, T.R., Kaplan, P.M., Cros, P., and Mabilat, C. (1999). *Mycobacterium* species identification and rifampin resistance testing with high-density DNA probe arrays. J. Clin. Microbiol. 37, 49–55.

Tulpan, D., Andronescu, M., Chang, S.B., Shortreed, M.R., Condon, A., Hoos, H.H., and Smith, L.M. (2005). Thermodynamically based DNA strand design. Nucleic Acids Res. *33*, 4951–4964.

Tusher, V.G., Tibshirani, R., and Chu, G. (2001). Significance analysis of microarrays applied to the ionizing radiation response. Proc. Natl. Acad. Sci. U.S.A. *98*, 5116–5121.

Tyagi, S., and Kramer, F.R. (1996). Molecular beacons: probes that fluoresce upon hybridization. Nat. Biotechnol. *14*, 303–308.

Untergasser, A., Nijveen, H., Rao, X., Bisseling, T., Geurts, R., and Leunissen, J.A. (2007). Primer3Plus, an enhanced web interface to Primer3. Nucleic Acids Res. *35*, W71–74.

Urakawa, H., Matsumoto, J., Hoshino, T., and Tsuneda, S. (2007). Direct profiling of rRNA in saline wastewater treatment samples using an oligonucleotide microarray. Microb. Environ. *22*, 116–122.

Utaida, S., Dunman, P.M., Macapagal, D., Murphy, E., Projan, S.J., Singh, V.K., Jayaswal, R.K., and Wilkinson, B.J. (2003). Genome-wide transcriptional profiling of the response of *Staphylococcus aureus* to cell-wall-active antibiotics reveals a cell-wall-stress stimulon. Microbiology *149*, 2719–2732.

Uttamchandani, M., Neo, J.L., Ong, B.N., and Moochhala, S. (2009). Applications of microarrays in pathogen detection and biodefence. Trends. Biotechnol. *27*, 53–61.

Valasek, M.A., and Repa, J.J. (2005). The power of real-time PCR. Adv. Physiol. Educ. *29*, 151–159.

Van Hoeyveld, E., Houtmeyers, F., Massonet, C., Moens, L., Van Ranst, M., Blanckaert, N., and Bossuyt, X. (2004). Detection of single nucleotide polymorphisms in the mannose-binding lectin gene using minor groove binder-DNA probes. J. Immunol. Methods *287*, 227–230.

Vandesompele, J., De Preter, K., Pattyn, F., Poppe, B., Van Roy, N., De Paepe, A., and Speleman, F. (2002a). Accurate normalization of real-time quantitative RT-PCR data by geometric averaging of multiple internal control genes. Genome Biol. 3, RESEARCH0034.

Vandesompele, J., De Paepe, A., and Speleman, F. (2002b). Elimination of primer-dimer artifacts and genomic coamplification using a two-step SYBR Green I real-time RT-PCR. Anal. Biochem. *303*, 95–98.

VanGuilder, H.D., Vrana, K.E., and Freeman, W.M. (2008). Twenty-five years of quantitative PCR for gene expression analysis. Biotechniques *44*, 619–626.

Walker, M.S., and Hughes, T.A. (2008). Messenger RNA expression profiling using DNA microarray technology: diagnostic tool, scientific analysis or un-interpretable data? Int. J. Mol. Med. *21*, 13–17.

Wang, A.M., Doyle, M.V., and Mark, D.F. (1989). Quantitation of mRNA by the polymerase chain reaction. Proc. Natl. Acad. Sci. U.S.A. *86*, 9717–9721.

Wang, D., Coscoy, L., Zylberberg, M., Avila, P.C., Boushey, H.A., Ganem, D., and DeRisi, J.L. (2002). Microarray-based detection and genotyping of viral pathogens. Proc. Natl. Acad. Sci. U.S.A. 99, 15687–15692.

Wang, R.F., Beggs, M.L., Robertson, L.H., and Cerniglia, C.E. (2002). Design and evaluation of oligonucleotide-microarray method for the detection of human intestinal bacteria in fecal samples. FEMS Microbiol. Lett. *213*, 175–182.

Wang, T., and Brown, M.J. (1999). mRNA quantification by real time TaqMan polymerase chain reaction: validation and comparison with RNase protection. Anal. Biochem. *269*, 198–201.

Wang, X., and Seed, B. (2003). Selection of oligonucleotide probes for protein coding sequences. Bioinformatics *19*, 796–802.

Wang, Y., Barbacioru, C., Hyland, F., Xiao, W., Hunkapiller, K.L., Blake, J., Chan, F., Gonzalez, C., Zhang, L., and Samaha, R.R. (2006). Large scale real-time PCR validation on gene expression measurements from two commercial long-oligonucleotide microarrays. BMC Genomics 7, 59.

Weckx, S., De Rijk, P., Glassee, W., Van Broeckhoven, C., and Del-Favero, J. (2007). SNPbox: web-based high-throughput primer design with an eye for repetitive sequences. Meth. Mol. Biol. *402*, 179–200.

Wei, Y., Lee, J.M., Richmond, C., Blattner, F.R., Rafalski, J.A., and LaRossa, R.A. (2001). High-density microarray-mediated gene expression profiling of *Escherichia coli*. J. Bacteriol. *183*, 545–556.

Wendisch, V.F., Zimmer, D.P., Khodursky, A., Peter, B., Cozzarelli, N., and Kustu, S. (2001). Isolation of *Escherichia coli* mRNA and comparison of expression using mRNA and total RNA on DNA microarrays. Anal. Biochem. *290*, 205–213.

Whitcombe, D., Theaker, J., Guy, S.P., Brown, T., and Little, S. (1999). Detection of PCR products using self-probing amplicons and fluorescence. Nat. Biotechnol. *17*, 804–807.

Wilcox, J.L., Dunbar, H.E., Wolfinger, R.D., and Moran, N.A. (2003). Consequences of reductive evolution for gene expression in an obligate endosymbiont. Mol. Microbiol. *48*, 1491–1500.

Wilkening, S., and Bader, A. (2004). Quantitative real-time polymerase chain reaction: methodical analysis and mathematical model. J. Biomol. Tech. *15*, 107–111.

Wilson, A.C., Dunbar, H.E., Davis, G.K., Hunter, W.B., Stern, D.L., and Moran, N.A. (2006). A dual-genome microarray for the pea aphid, *Acyrthosiphon pisum*, and its obligate bacterial symbiont, *Buchnera aphidicola*. BMC Genomics 7, 50.

Wilson, I.G. (1997). Inhibition and facilitation of nucleic acid amplification. Appl. Environ. Microbiol. *63*, 3741–3751.

Wilson, K.H., Wilson, W.J., Radosevich, J.L., DeSantis, T.Z., Viswanathan, V.S., Kuczmarski, T.A., and Andersen, G.L. (2002). High-density microarray of small-subunit ribosomal DNA probes. Appl. Environ. Microbiol. *68*, 2535–2541.

Wilson, M., DeRisi, J., Kristensen, H.H., Imboden, P., Rane, S., Brown, P.O., and Schoolnik, G.K. (1999). Exploring drug-induced alterations in gene expression in *Mycobacterium tuberculosis* by microarray hybridization. Proc. Natl. Acad. Sci. U.S.A. *96*, 12833–12838.

Wilson, W.J., Strout, C.L., DeSantis, T.Z., Stilwell, J.L., Carrano, A.V., and Andersen, G.L. (2002). Sequence-specific identification of 18 pathogenic microorganisms using microarray technology. Mol. Cell Probes *16*, 119–127.

Wingren, C., and Borrebaeck, C.A. (2006). Antibody microarrays: current status and key technological advances. OMICS *10*, 411–427.

Winzeler, E.A. (2006). Applied systems biology and malaria. Nat. Rev. Microbiol. *4*, 145–151.

Wittwer, C.T., Herrmann, M.G., Moss, A.A., and Rasmussen, R.P. (1997). Continuous fluorescence monitoring of rapid cycle DNA amplification. Biotechniques 22, 130–*131*, 134–138.

Wojciechowski, J., Danley, D., Cooper, J., Yazvenko, N., and Taitt, C. (2010). Multiplexed electrochemical detection of *Yersinia pestis* and staphylococcal enterotoxin B using an antibody microarray. Sensors *10*, 3351–3362.

Wong, M.L., and Medrano, J.F. (2005). Real-time PCR for mRNA quantitation. Biotechniques *39*, 75–85.

Wu, C., Carta, R., and Zhang, L. (2005). Sequence dependence of cross-hybridization on short oligo microarrays. Nucleic Acids Res. *33*, e84.

Wu, X., and Munroe, D.J. (2006). EasyExonPrimer: automated primer design for exon sequences. Appl. Bioinformatics 5, 119–120.

Wu, Z., Irizarry, R., Gentleman, R., Murillo, F., and Spencer, F. (2004). A model based background adjustment for oligonucleotide expression arrays. J. Am. Stat. Assoc. *99*, 909–917.

Wu, Z., Siadaty, M.S., Riddick, G., Frierson, H.F., Jr., Lee, J.K., Golden, W., Knuutila, S., Hampton, G.M., El-Rifai, W., and Theodorescu, D. (2006). A novel method for gene expression mapping of metastatic competence in human bladder cancer. Neoplasia *8*, 181–189.

Wurmbach, E., Yuen, T., and Sealfon, S.C. (2003). Focused microarray analysis. Methods *31*, 306–316.

Yamada, T., Soma, H., and Morishita, S. (2006). PrimerStation: a highly specific multiplex genomic PCR primer design server for the human genome. Nucleic Acids Res. *34*, W665–669.

Yang, D.K., Kweon, C.H., Kim, B.H., Lim, S.I., Kim, S.H., Kwon, J.H., and Han, H.R. (2004). TaqMan reverse transcription polymerase chain reaction for the detection of Japanese encephalitis virus. J. Vet. Sci. *5*, 345–351.

Yang, Y.H., and Speed, T. (2002). Design issues for cDNA microarray experiments. Nat. Rev. Genet. 3, 579–588.

Yang, Y.H., Dudoit, S., Luu, P., Lin, D.M., Peng, V., Ngai, J., and Speed, T.P. (2002). Normalization for cDNA microarray data: a robust composite method addressing single and multiple slide systematic variation. Nucleic Acids Res. *30*, e15.

Yauk, C.L., Berndt, M.L., Williams, A., and Douglas, G.R. (2004). Comprehensive comparison of six microarray technologies. Nucleic Acids Res. *32*, e124.

Ye, R.W., Tao, W., Bedzyk, L., Young, T., Chen, M., and Li, L. (2000). Global gene expression profiles of *Bacillus subtilis* grown under anaerobic conditions. J. Bacteriol. *182*, 4458–4465.

Ye, R.W., Wang, T., Bedzyk, L., and Croker, K.M. (2001). Applications of DNA microarrays in microbial systems. J. Microbiol. Methods 47, 257–272.

Yokouchi, H., Fukuoka, Y., Mukoyama, D., Calugay, R., Takeyama, H., and Matsunaga, T. (2006). Whole-metagenome amplification of a microbial community associated with scleractinian coral by multiple displacement amplification using phi29 polymerase. Environ. Microbiol. *8*, 1155–1163.

You, F.M., Huo, N., Gu, Y.Q., Luo, M.C., Ma, Y., Hane, D., Lazo, G.R., Dvorak, J., and Anderson, O.D. (2008). BatchPrimer3: a high throughput web application for PCR and sequencing primer design. BMC Bioinformatics 9, 253.

You, Y., Fu, C., Zeng, X., Fang, D., Yan, X., Sun, B., Xiao, D., and Zhang, J. (2008). A novel DNA microarray for rapid diagnosis of enteropathogenic bacteria in stool specimens of patients with diarrhea. J. Microbiol. Methods 75, 566–571.

Yuan, J.S., Wang, D., and Stewart, C.N., Jr. (2008). Statistical methods for efficiency adjusted real-time PCR quantification. Biotechnol. J. *3*, 112–123.

Yuen, T., Wurmbach, E., Pfeffer, R.L., Ebersole, B.J., and Sealfon, S.C. (2002). Accuracy and calibration of commercial oligonucleotide and custom cDNA microarrays. Nucleic Acids Res. *30*, e48.

Zakharkin, S.O., Kim, K., Mehta, T., Chen, L., Barnes, S., Scheirer, K.E., Parrish, R.S., Allison, D.B., and Page, G.P. (2005). Sources of variation in Affymetrix microarray experiments. BMC Bioinformatics *6*, 214.

Zeeberg, B.R., Feng, W., Wang, G., Wang, M.D., Fojo, A.T., Sunshine, M., Narasimhan, S., Kane, D.W., Reinhold, W.C., Lababidi, S., *et al.* (2003). GoMiner: a resource for biological interpretation of genomic and proteomic data. Genome Biol. *4*, R28.

Zervakis, M., Blazadonakis, M.E., Tsiliki, G., Danilatou, V., Tsiknakis, M., and Kafetzopoulos, D. (2009). Outcome prediction based on microarray analysis: a critical perspective on methods. BMC Bioinformatics *10*, 53.

Zhang, J., and Byrne, C.D. (1999). Differential priming of RNA templates during cDNA synthesis markedly affects both accuracy and reproducibility of quantitative competitive reverse-transcriptase PCR. Biochem. J. *337*, 231–241.

Zhang, J., Finney, R.P., Clifford, R.J., Derr, L.K., and Buetow, K.H. (2005). Detecting false expression signals in high-density oligonucleotide arrays by an in silico approach. Genomics *85*, 297–308.

Zhang, L., Zhou, W., Velculescu, V.E., Kern, S.E., Hruban, R.H., Hamilton, S.R., Vogelstein, B., and Kinzler, K.W. (1997). Gene expression profiles in normal and cancer cells. Science *276*, 1268–1272.

Zhang, L., Miles, M.F., and Aldape, K.D. (2003). A model of molecular interactions on short oligonucleotide microarrays. Nat. Biotechnol. *21*, 818–821.

Zhang, L., Srinivasan, U., Marrs, C.F., Ghosh, D., Gilsdorf, J.R., and Foxman, B. (2004). Library on a slide for bacterial comparative genomics. BMC Microbiol. *4*, 12.

Zhang, L., Hurek, T., and Reinhold-Hurek, B. (2007). A *nifH*-based oligonucleotide microarray for functional diagnostics of nitrogen-fixing microorganisms. Microb. Ecol. *53*, 456–470.

Zhang, L., Foxman, B., Drake, D., Srinivasan, U., Henderson, J., Olson, B., Marrs, C., Warren, J., and Marazita, M. (2009). Comparative whole-genome analysis of *Streptococcus mutans* isolates within and among individuals of different caries status. Oral Microbiol. Immun. *24*, 197–203.

Zhao, J., Bai, Y., Zhang, Q., Wan, Y., Li, D., and Yan, X. (2005). Detection of hepatitis B virus DNA by real-time PCR using TaqMan-MGB probe technology. World J. Gastroentero. *11*, 508–510.

Zheng, J., Svensson, J.T., Madishetty, K., Close, T.J., Jiang, T., and Lonardi, S. (2006). OligoSpawn: a software tool for the design of overgo probes from large unigene datasets. BMC Bioinformatics *7*, 7.

Zhou, D., Han, Y., Song, Y., Tong, Z., Wang, J., Guo, Z., Pei, D., Pang, X., Zhai, J., Li, M., *et al.* (2004). DNA microarray analysis of genome dynamics in *Yersinia pestis*: insights into bacterial genome microevolution and niche adaptation. J. Bacteriol. *186*, 5138–5146.

Zhou, J., and Thompson, D.K. (2002). Challenges in applying microarrays to environmental studies. Curr. Opin. Biotechnol. *13*, 204–207.

Zhou, L., Lei, X.H., Bochner, B.R., and Wanner, B.L. (2003). Phenotype microarray analysis of *Escherichia coli* K-12 mutants with deletions of all two-component systems. J. Bacteriol. *185*, 4956–4972.

Zhu, B., Ping, G., Shinohara, Y., Zhang, Y., and Baba, Y. (2005). Comparison of gene expression measurements from cDNA and 60-mer oligonucleotide microarrays. Genomics *85*, 657–665.

Zimmer, D.P., Soupene, E., Lee, H.L., Wendisch, V.F., Khodursky, A.B., Peter, B.J., Bender, R.A., and Kustu, S. (2000). Nitrogen regulatory protein C-controlled genes of *Escherichia coli*: scavenging as a defense against nitrogen limitation. Proc. Natl. Acad. Sci. U.S.A. *97*, 14674–14679.

Zinn, K., DiMaio, D., and Maniatis, T. (1983). Identification of two distinct regulatory regions adjacent to the human beta-interferon gene. Cell *34*, 865–879.

Zuker, M. (2003). Mfold web server for nucleic acid folding and hybridization prediction. Nucleic Acids Res. *31*, 3406–3415.

Future Trends in Quantitative Real-time Polymerase Chain Reaction Technology and their Implication in Applied Microbiology

11

Vijay J. Gadkar and Martin Filion

Abstract

Real time-quantitative PCR (RT-qPCR) technology has revolutionized the detection landscape in every area of molecular biology. The fundamental basis of this technology has remained unchanged since its inception, however various modifications have enhanced the overall performance of this highly versatile technology. These improvements have ranged from changes in the individual components of the enzymatic reaction cocktail (polymerizing enzymes, reaction buffers, probes, etc.) to the detection system itself (instrumentation, software, etc.). The RT-qPCR technology currently available to researchers is more sensitive, faster and affordable than when this technology was first introduced. In this chapter, we summarize the developments of the last few years in RT-qPCR technology and nucleic acid amplification.

Introduction

Since its introduction in the late 1990s, real time-quantitative PCR (RT-qPCR) technology has revolutionized the detection landscape in every aspect of molecular biology. The fundamental basis of this technology has remained unchanged since its inception, which is fluorescence based detection of its target, amplified via classical multistep PCR. A steady modification(s) of individual components which makeup the RT-qPCR platform has however enhanced the overall performance of this highly versatile technology. These improvements have ranged from changes in the individual components of the enzymatic reaction cocktail (polymerizing enzymes, reaction buffers, probes, etc.) to the detection system itself (instrumentation, software, etc.). As a result, the RT-qPCR technology currently available to researchers is more sensitive, faster and most importantly, affordable for laboratories on moderate/modest budgets, than when this technology was first introduced.

In this final chapter, we seek to briefly summarize some of the developments which have happen within the last few years in the area of RT-qPCR technology and nucleic acid amplification (e.g. classical PCR). In our opinion, these developments might have a direct relevance for researchers working in the area of environmental transcriptomics. Some of described examples in this chapter have been validated on RT-qPCR platforms, making it relatively easy to adapt in their current state, while others have only been implemented on a conventional end-point detection platform.

Isothermal detection methods

The current RT-qPCR technology is based on the classical three-step thermal cycling process which is, template denaturation, followed by primer/probe annealing and finally, extension/detection of the fluorescence signal, to amplify and detect the target transcripts all under real-time conditions (Gelfand, 1992). A very commonly observed phenomenon in this multistep thermocycling amplification system is the generation of spurious fluorescence signal due to mispriming of primer/probes. To overcome such limitations, detection platforms have been proposed which amplify the target exponentially like PCR, but under isothermal conditions, i.e. at a fixed, user-defined temperature (reviewed by Gill and Ghaemi, 2008).

The helicase-dependent (HDA) amplification system is one such novel 'non-PCR' system for amplifying target DNA (Vincent *et al.*, 2004) and RNA (Goldmeyer *et al.*, 2007), under isothermal conditions. This revolutionary amplification system makes use of a novel enzymatic cocktail which does not require the DNA to be cycled between different temperatures, like that done for reactions based on Taq DNA polymerase amplification or any of its variants. In lieu of a standard denaturation step, the HDA system uses the helicase enzyme to unwind the double-stranded DNA and with the aid of other polymerizing enzymes, an exponential amplification is achieved (Vincent *et al.*, 2004). All these steps are performed at a fixed, user defined temperature. Though extremely novel when first introduced, the HDA system suffers from one major limitation- its inability to amplify DNA targets greater than 200 bp (Vincent *et al.*, 2004). As a result, in its present state, it is seriously unable to challenge and act as a viable alternative to the highly versatile PCR, or any amplification system based on it.

Despite this technical limitation, HDA-based amplification is increasingly being viewed as an attractive alternative in any amplification system, where length of the amplicon is less than 200 bp. One excellent example of such short amplicon size detection system is RT-qPCR. It is a well known fact that, for an efficient amplification, the primer/probe parameters of the RT-qPCR detection system is usually within the 50–150 bp size range. It is therefore not impossible to develop a RT-qHDA detection system, where the classical Taq DNA polymerase amplification cocktail is substituted by an enzymatic system, which amplifies its target at a fixed (isothermal) temperature.

Recently, Tong *et al.* (2008) showed that such enzymatic substitution is indeed feasible. In their work, primer/probes designed for TaqMan RT-qPCR fluorescence system were used to develop a novel RT-qHDA system to detect various biothreat microorganisms. Since pre-existing TaqMan primer/probes were used, the RT-qHDA system was optimized to work with the 60–65°C temperature range and detection sensitivity, similar to what one would expect from a classical RT-qPCR system. Though very preliminary, it is tempting to conclude that the RT-qHDA system, at least in theory, is 'backward compatible' with a pre-existing RT-qPCR system. In other words, the user does not have to separately invest in a new detection platform (i.e. instrumentation) or custom synthesize the flurogenic probes, both of which could be prohibitively expensive. The only modification one would have to make is to substitute the Taq DNA polymerase based reaction components, with a HDA amplification cocktail, which is now commercially available in various formats from Biohelix Corp, USA. Interestingly enough, isothermal amplification systems themselves are also constantly being refined to enhance fidelity (amplicon length) and robustness of the amplification process. Protocols which make use of novel enzymatic cocktails which are more efficient in unwinding and amplifying the target double helix are constantly being reported (Kiesling *et al.*, 2007; Tan *et al.*, 2008; Schaerli *et al.*, 2010; Joneja and Huang, 2011).

In our opinion, adapting such 'non-PCR' amplification technologies could in the near future lead to detection platforms (Mahalanabis *et al.*, 2010) which are more robust and would not suffer from the inherent drawbacks, for example spurious amplification, cycling parameter standardization, typically associated with the classical three-stage PCR system.

Development of novel normalization controls

The first step towards analysing microbial gene expression requires a quantitative extraction of RNA. This step has proven to be highly problematic for environmental matrices, due to compounded inefficiencies in individual steps which include, but not limited to, incomplete cell lysis, RNA degradation by ubiquitous RNases, co-extraction of inhibitors and their interaction with the enzymes used (Lorenz and Wackernagel, 1987; Moran *et al.*, 1993, Alm and Stahl, 2000). One straightforward approach to quantify such losses and apply the necessary correction is to include an internal amplification control (IAC), so as to make the final gene expression meaningful and reproducible. An IAC is essentially a non-target DNA/RNA sequence co-amplified, preferably in the same reaction tube, under the same reaction conditions (Hoorfar *et al.*, 2003, 2004; Nolte, 2004; van Doorn *et al.*, 2009). While attempts to develop IAC's have met with some success for experimental systems which are highly controlled (Oikonomou *et al.*, 2008; Scipioni *et al.*, 2008), developing such controls have proven to be highly problematic for certain experimental set-ups, for example complex environmental matrices. The main difficulty in these cases has been in our inability to identify an inert IAC which is able to (a) withstand the harsh nucleic acid extraction procedures usually employed for environmental matrices, and if such a sequence is indeed developed/identified (b) designing a primer/probe combination which would not cross-react with other non-target (nucleic acids) components of the matrices. While few potential IAC based solutions have been proposed, for example the Biotrove OpenArray platform (van Doorn *et al.*, 2009), high costs and proprietary issues of some IAC sequences (Nordstrom *et al.*, 2007) have served as a deterrent for researchers who are seriously interested in rigorously implementing this external normalization strategy.

Recently, Gonzalez-Escalona *et al.* (2009) proposed the implementation of a highly versatile RNA based IAC, called the myIC system. The myIC is essentially a 200 bp artificial sequence, with no known homology to any nucleic acid accession in the GENBANK (NCBI) database. As a result, primer/probes designed for this sequence are not expected to cross-react with any non-target nucleic acids, typically expected to be present in high amounts in RNA/DNA extracted from complex environmental matrices. Recently, its applicability was successfully demonstrated in quantifying bacterial gene transcripts from natural soil samples spiked with a defined amount of the bacterium *Pseudomonas* sp. LBUM300 (DeCoste *et al.*, 2011). This success should spur researchers to implement the myIC system to quantify microbial gene expression in other environmental matrices, but were unable due to lack of a suitable IAC. The added advantage of its non-proprietary usage and its applicability in both DNA/RNA formats (Gonzalez-Escalona *et al.*, 2009; Deer *et al.*, 2010), makes myIC a very attractive IAC to implement in various RT-qPCR based detection systems.

LNA/Zip nucleic acids

One critical factor for any PCR based detection platform, including RT-qPCR, is the invariant ability of the primer/probe to discern its target from a heterogeneous mixture and bind it with extreme specificity. In RT-qPCR, the design of these primers/probes is governed by strict guidelines (hybridization/sequence parameters) which, in turn, are based on the fluorescence chemistry used for that particular assay. For example, the hybridization parameters for TaqMan assays are different from the Molecular beacon based flurogenic assays. Empirically, these primers/probes are designed with the help of specific software's (e.g. PrimerExpress™ for TaqMan detection system) which are proprietary and usually supplied with the detection platform. Since these software use specific built-in algorithms, with a narrow set of highly defined binding parameters, it is not very uncommon to find these design software unable to identify optimal primer/probe combinations for a particular user input sequence.

While these design limitations have not exactly impeded our ability to vigorously implement the RT-qPCR technique, there has always been an ongoing endeavour to increase the sensitivity of an assay by enhancing the 'specificity and recognition' capacity of a primer/probe. The most conventional approach to achieve this goal has been to manipulate various components of the amplification reaction, for example use of reaction buffer which are highly optimized, increased level of divalent ions (Mg^{2+}) and use of specific additives which are known to prevent primer mispriming. Early research has however showed that it is also possible to enhance the specificity and recognition aspects of an oligonucleotide, by another novel approach- chemical modification of the oligonucleotide backbone itself. As a result of this modification, the oligonucleotide is able to bind to its target with extreme specificity and enhanced accuracy.

Of the several modifications developed till date (Hyrup and Neilson, 1996; Schulz and Gryaznov,

1996; VanAerschot *et al.*, 1996; Hendrix *et al.*, 1997; Nielson and Haaima, 1997), the locked nucleic acids (LNA) are probably seen as the most promising in terms of design simplicity and superior functionality, i.e. its unique ability to recognize and bind to its target with an extreme precision (Obika *et al.*, 1998; Singh *et al.*, 1998). The high specificity of LNAs have made it possible to apply it in single-nucleotide polymorphism (SNP) genotyping and detection of difficult templates which, otherwise is not feasible with conventional primer/probes (Simeonov and Nikiforov, 2002; Latorra *et al.*, 2003; Johnson *et al.*, 2004; Ugozzoli *et al.*, 2004). The recent introduction of zip nucleic acids (ZNA), which have similar functional properties of LNAs but, based on a different modification chemistry (Voirin *et al.*, 2007), is also an exciting development. The ZNA oligonucleotides have a spermine residues conjugated to the basic oligonucleotide backbone. This modification results in reduced electrostatic repulsion brought about by the polyanionic nature of nucleic acids. As a result, an enhanced sensitivity is routinely observed in detecting its target during both RT-qPCR and reverse transcription assay (Moreau *et al.*, 2009).

In our opinion, these 'enhanced specificity' oligonucleotides like LNAs, could be further exploited to design RT-qPCR primer/probes used to detect microbial gene expression from environmental matrices, especially complex types. As analysing gene expression from environmental samples entails discerning the target from a heterogeneous background, an enhanced specificity primer/probe would result in a very high signal-to-noise ratio and hence a more robust, accurate and reproducible detection of the target transcripts. The commercial availability of both the LNA and ZIP oligonucleotides should make it easier for researchers to develop specific applications to accurately detect its target from environmental matrices.

Use of polymerases engineered to resist inhibitors

As gene expression analysis using RT-qPCR entails the synthesis of cDNA by the reverse transcriptase (RT) enzyme, followed by an exponential amplification of the template by DNA polymerase, any chemical moiety, which is able to alter the fidelity (negatively) of these enzymes, has a direct bearing on the end results (Suslov and Steindler, 2005). The net effect of such interferences, at least in RT-qPCR, is the invariable quenching of the fluorescent signal, leading to sub-optimal or irreproducible gene expression data. Consequently, solutions are constantly being sought which could in some way, alleviate the effects of these interfering moieties on the normal functioning of these enzymes. Some of the commonly used inhibitor mitigation strategies include, use of RNA exaction protocols specifically designed to eliminate co-extracted inhibitors, and addition of specific additives in the RT-qPCR reaction which are known to bind the inhibitor moieties (e.g. BSA, T4 gene 32 protein). One attractive solution which could be an 'add-on' to these aforementioned solutions is the use of enzymes (reverse transcriptase and DNA polymerase enzymes) which are inherently 'resistant' to any of the signal quenching inhibitors.

The first step in RT-qPCR based gene expression analysis is the faithful conversion of RNA to cDNA by reverse transcription. Therefore, the first step would be the development of an RT enzyme which is able to resist a wide spectrum or a specific class of environmental inhibitors. While most RTs are highly sensitive to inhibitors, the recent report of a mutant Moloney murine leukaemia virus (MMLV)-RT enzyme, resistant to few common known inhibitors (Arezi *et al.*, 2010), is an exciting development. In the second step of RT-qPCR, the DNA polymerases are the most important component used for the exponential amplification of the target. While certain variants of DNA polymerases are capable of tolerating chemical inhibitors up to certain extent (Al-Soud and Rådström, 1998), the development of a DNA polymerase specifically engineered to resist inhibitors, has not been attempted till recently. Kermekchiev *et al.* (2009) using a site directed mutagenesis strategy, whereby a 'designer' DNA polymerase capable of resisting two recalcitrant inhibitors namely humic/fulvic acid (commonly found in soil) and haemoglobin (common in meat based matrices) was developed. Recently, Baar *et al.* (2011) showed through molecular breeding, development of a novel DNA polymerase called 2D9, which has sequence elements from four different DNA polymerases. This 'chimeric' polymerase exhibits a striking resistance to a broad spectrum of complex inhibitors including humic acid, bone dust, coprolite, peat extract, clay-rich soil, cave sediment and tar.

Taking both these development into consideration, it could be possible in the near future to develop RT-qPCR protocols which are specifically tailored for environmental microbiology. This modified approach would allow the end user to perform a more reproducible detection and quantification of gene expression from environmental matrices. Commercial availability, unit cost and proprietary issues are some of the factors which one would have to consider prior to implementing this approach.

Summary/conclusion

The application of the RT-qPCR technology has contributed immensely to our understanding of gene expression in various biological systems; however in certain areas of research, for example applied microbiology, application of this technique has not progressed as one would have liked to expect. This application gap could at best be attributed to the extreme difficulties in extracting nucleic acids from environmental samples and the high sensitivity of the RT-qPCR system towards chemical components inherently co-extracted from environmental samples. Development of a more robust RT-qPCR platform is one possible solution to overcome this problem. We believe that a cross-adaption of some new developments in amplification and enzymatic technology would alleviate some of the drawbacks inherent to the RT-qPCR technology so that it's potentially is fully realized in areas like applied microbiology.

References

Alm, E.W., and Stahl, D.A. (2000). Critical factors influencing the recovery and integrity of rRNA extracted from environmental samples: use of an optimized protocol to measure depth-related biomass distribution in freshwater sediments. J. Microbiol. Methods *40*, 153–162.

Al-Soud, W.A., and Rådström, P. (1998). Capacity of nine thermostable DNA polymerases to mediate DNA amplification in the presence of PCR-inhibiting samples. Appl. Environ. Microbiol. *64*, 3748–3753.

Arezi, B., McCarthy, M., and Hogrefe, H. (2010). Mutant of Moloney murine leukemia virus reverse transcriptase exhibits higher resistance to common RT-qPCR inhibitors. Anal. Biochem. *400*, 301–303.

Baar, C., d'Abbadie, M., Vaisman, A., Arana, M.E., Hofreiter, M., Woodgate, R., Kunkel, T.A., and Holliger, P. (2011). Molecular breeding of polymerases for resistance to environmental inhibitors. Nucleic Acids Res. *39*, e51.

DeCoste, N.J., Gadkar, V.J., and Filion, M. (2011). Relative and absolute quantitative real-time PCR-based quantifications of *hcnC* and *phlD* gene transcripts in natural soil spiked with *Pseudomonas* sp. strain LBUM300. Appl. Environ. Microbiol. *77*, 41–47.

Deer, D.M., Lampel, K.A., and González-Escalona, N. (2010). A versatile internal control for use as DNA in real-time PCR and as RNA in real-time reverse transcription PCR assays. Lett. Appl. Microbiol. *50*, 366–372.

van Doorn, R., Klerks, M.M., van Gent-Pelzer, M.P., Speksnijder, A.G., Kowalchuk, G.A., and Schoen, C.D. (2009). Accurate quantification of microorganisms in PCR-inhibiting environmental DNA extracts by a novel internal amplification control approach using Biotrove OpenArrays. Appl. Environ. Microbiol. *75*, 7253–7260.

Gelfand, D.H. (1992). Taq DNA polymerase, PCR technology. In Principles and Applications for DNA Amplification, Erlich, H.A. ed. (Freeman and Company, New York), pp. 17–22.

Gill, P., and Ghaemi, A. (2008). Nucleic acid isothermal amplification technologies – a review. Nucleosides Nucleotides Nucleic Acids. *27*, 224–243.

Goldmeyer, J., Kong, H., and Tang, W. (2007). Development of a novel one-tube isothermal reverse transcription thermophilic helicase-dependent amplification platform for rapid RNA detection. J. Mol. Diagn. *9*, 639–644.

González-Escalona, N., Hammack, T.S., Russel, M., Jacobson, A.P., De Jesús, A.J., Brown, E.W., and Lampel, K.A. (2009). Detection of live *Salmonella* sp. cells in produce by a TaqMan-Based quantitative reverse transcriptase real-time PCR targeting *invA* mRNA. Appl. Environ. Microbiol. *75*, 3714–3720.

Hendrix, C., Rosemeyer, H., De Bouvere, B., Van Aerschot, A., Seela, F., and Herdewijn, P. (1997). 1,5′-Anhydrohexitol oligonucleotides: hybridization and strand displacement with oligoribonucleotides, interaction with RNase H and HIV reverse transcriptase. Eur. J. Chem. 3, 1513–1520.

Hoorfar, J., Cook, N., Malorny, B., Wagner, M., De Medici, D., Abdulmawjood, A., and Fach, P. (2003). Making internal amplification control mandatory for diagnostic PCR. J. Clin. Microbiol. *41*, 5835.

Hoorfar, J., Cook, N., Malorny, B., Wagner, M., De Medici, D., Abdulmawjood, A., and Fach, P. (2004). Diagnostic PCR: making internal amplification control mandatory. Lett. Appl. Microbiol. *38*, 79–80.

Hyrup, B., and Nielson, P.E. (1996). Peptide nucleic acids (PNA): synthesis, properties and nucleic acids, a new promising antisense construct. Bioorg. Med. Chem. *4*, 5–23.

Johnson, M.P., Haupt, L.M., and Griffiths, L.R. (2004). Locked nucleic acids (LNA) single nucleotide polymorphism (SNP) genotype analysis and validation using real-time PCR. Nucleic Acids Res. *32*, e55.

Joneja, A., and Huang, X. (2011). Linear nicking endonuclease-mediated strand displacement DNA amplification, Anal. Biochem. *414*, 58–69.

Kermekchiev, M.B., Kirilova, L.I., Vail, E.E., and Barnes, W.M. (2009). Mutants of Taq DNA polymerase resistant to PCR inhibitors allow DNA amplification from whole blood and crude soil samples. Nucleic Acids Res. *37*, e40

Kiesling, T., Cox, K., Davidson, E.A., Dretchen, K., Grater, G., Hibbard, S., Lasken, R.S., Leshin, J., Skowronski, E., and Danielsen, M. (2007). Sequence specific detection of DNA using nicking endonuclease signal amplification (NESA). Nucleic Acids Res. *35*, e117.

Latorra, D., Campbell, K., Wolter, A., and Hurley, J.M. (2003). Enhanced allele-specific PCR discrimination in SNP genotyping using 3′ locked nucleic acid (LNA) primers. Hum. Mut. *22*, 79–85.

Lorenz, M.G., and Wackernagel, W. (1987). Adsorption of DNA to sand and variable degradation rates of adsorbed DNA. Appl. Environ. Microbiol. *53*, 2948–2952.

Mahalanabis M., Do, J., Al-Muayad H., Zhang, J.Y., and Klapperich, C.M. (2010). An integrated disposable device for DNA extraction and helicase dependent amplification. Biomed. Microdev. *12*, 353–359.

Moran, M.A., Torsvik, V.L., Torsvik, T., and Hodson, R.E. (1993). Direct extraction and purification of rRNA for ecological studies. Appl. Environ. Microbiol. *59*, 915–918.

Moreau, V., Voirin, E., Paris, C., Kotera, M., Nothisen, M., Rémy, J.S., Behr, J.P., Erbacher, P., and Lenne-Samuel, N. (2009). Zip Nucleic Acids: new high affinity oligonucleotides as potent primers for PCR and reverse transcription. Nucleic Acids Res. *37*, e130.

Nielson, P.E., and Haaima, G. (1997). Peptide nucleic acid (PNA): A DNA mimic with a pseudopeptide backbone. Chem. Soc. Rev. *26*, 73–78.

Nolte, F.S. (2004). Novel internal controls for real-time PCR assays. Clin. Chem. *50*, 801–802.

Nordstrom, J.L, Vickery, M.C, Blackstone, G.M., Murray, S.L., and DePaola, A. (2007). Development of a multiplex real-time PCR assay with an internal amplification control for the detection of total and pathogenic *Vibrio parahaemolyticus* bacteria in oysters. Appl. Environ. Microbiol. *73*, 5840–5847.

Obika, S., Nanbu, D., Hari, Y., Andoh, J-i., Morio, K-i., Doi, T., and Imanishi, T. (1998). Stability and structural features of the duplexes containing nulcoeside analogs with a fixed N-type conformation. 2′-O, 4′-C methylene ribonucleosides. Tetrahedron Lett. *39*, 5401–5404.

Oikonomou, I., Halatsi, K., and Kyriacou, A. (2008). Selective PCR: a novel internal amplification control strategy for enhanced sensitivity in *Salmonella* diagnosis. Lett. Appl. Microbiol. *46*, 456–461.

Schaerli Y., Stein, V., Spiering, M.M., Benkovic, S.J., Abell, C., and Hollfelder, F. (2010). Isothermal DNA amplification using the T4 replisome: circular nicking endonuclease-dependent amplification and primase-based whole-genome amplification. Nucleic Acids Res. *38*, e201.

Schulz, R.G., and Gryaznov, S.M. (1996). Oligo-2′-fluoro-2′-deoxynucleotide N3′->P5′ phosphoramidates: synthesis and properties. Nucleic Acids Res. *24*, 2966–2973.

Scipioni, A., Bourgot, I., Mauroy, A., Ziant, D., Saegerman, C., Daube, G., and Thiry, E. (2008). Detection and quantification of human and bovine noro viruses by a TaqMan RT-PCR assay with a control for inhibition. Mol. Cell. Probes *22*, 215–222.

Simeonov, A., and Nikiforov, T.T. (2002). Single nucleotide polymorphism genotyping using short, fluorescently labeled locked nucleic acid (LNA) probes and fluorescence polarization detection. Nucleic Acids Res. *30*, e91.

Singh, S.K., Nielsen, P., Koshkin, A.A., and Wengel, J. (1998). LNA (locked nucleic acids): synthesis and high affinity nucleic acid recognition. Chem. Commun. *4*, 455–456.

Suslov, O., and Steindler, D.A. (2005). PCR inhibition by reverse transcriptase leads to an overestimation of amplification efficiency. Nucleic Acids Res. *33*, e181.

Tan, E., Erwin, B., Dames, S., Ferguson, T., Buechel, M., Irvine, B., Voelkerding, K., and Niemz, A. (2008). Specific versus nonspecific isothermal DNA amplification through thermophilic polymerase and nicking enzyme activities. Biochem. *47*, 9987–9999.

Tong, Y., Tang, W., Kim, H.J., Pan, X., Ranalli, T., and Kong, H. (2008). Development of isothermal TaqMan assays for detection of biothreat organisms. Biotechniques *45*, 543–557.

Ugozzoli, L.A., Latorra, D., Pucket, R., Arar, K., and Hamby, K. (2004). Real-time genotyping with oligonucleotide probes containing locked nucleic acids. Anal. Biochem. *324*, 143–152.

Van Aerschot, A., Verheggen, I., Hendrix, C., and Herdewijn, P. (1995). 1,5-Anhydrohexitol nucleic acids, a new promising antisense construct. Angew Chem. Int. Ed. *34*, 1338–1339.

Vincent, M., Xu, Y., and Kong, H. (2004). Helicase-dependent isothermal DNA amplification. EMBO Rep. *5*, 795–800.

Voirin, E., Behr, J.P., and Kotera, M. (2007). Versatile synthesis of oligodeoxyribonucleotide–oligospermine conjugates. Nature Prot. *2*, 1360–1367.

Index

O

P

Q

R

S

T

V

W

Y

Z

www.ingramcontent.com/pod-product-compliance
Lightning Source LLC
LaVergne TN
LVHW062209200926
848484LV00006B/1357

* 9 7 8 1 9 0 8 2 3 0 0 1 0 *